AF356552

Spectroscopic, Thermodynamic and Molecular Docking Studies on Molecular Mechanisms of Drug Binding to Proteins

Spectroscopic, Thermodynamic and Molecular Docking Studies on Molecular Mechanisms of Drug Binding to Proteins

Editors

Tanveer A. Wani
Seema Zargar
Afzal Hussain

MDPI • Basel • Beijing • Wuhan • Barcelona • Belgrade • Manchester • Tokyo • Cluj • Tianjin

Editors

Tanveer A. Wani
Pharmaceutical Chemistry
College of Pharmacy King
Saud University
Riyadh
Saudi Arabia

Seema Zargar
Biochemistry
College of Science King Saud
University
Riyadh
Saudi Arabia

Afzal Hussain
Pharmacognosy
College of Pharmacy King
Saud University
Riyadh
Saudi Arabia

Editorial Office
MDPI
St. Alban-Anlage 66
4052 Basel, Switzerland

This is a reprint of articles from the Special Issue published online in the open access journal *Molecules* (ISSN 1420-3049) (available at: www.mdpi.com/journal/molecules/special_issues/drug_proteins).

For citation purposes, cite each article independently as indicated on the article page online and as indicated below:

LastName, A.A.; LastName, B.B.; LastName, C.C. Article Title. *Journal Name* **Year**, *Volume Number*, Page Range.

ISBN 978-3-0365-6226-1 (Hbk)
ISBN 978-3-0365-6225-4 (PDF)

Contents

About the Editors

Tanveer A. Wani

Tanveer A. Wani is a Professor at the Department of Pharmaceutical Chemistry, College of Pharmacy, King Saud University (KSU). He was associated with the pharmaceutical industry from 2004 to 2009 (Ranbaxy Laboratories Limited (RLL) (now Sun Pharma), India. He completed his master's and Ph.D. in a collaboration Programme between RLL and Hamdard University New Delhi. He joined the RLL in the Department of Clinical Pharmacology and pharmacokinetics as a Research Scientist and Quality Control and Project Manager. He worked in the intellectual property management group as a research executive for six months after his Masters'Degree. He was deputed to Fortis Clinical Research, India, to monitor quality control issues during contract studies conducted for RRL. He joined KSU in December 2009 as a professor, and has been a full professor since 2021. During his tenure at KSU, he published more than 106 research papers in top-indexed ISI journals along with some book chapters. He has an h-index of 24 (Scopus), and collaborated on several projects both within and outside the university. He is also an editor for several highly ranked ISI journals and has reviewed more than several hundred research articles as an act of service to the scientific community.

Seema Zargar

Seema Zargar is a Professor at the Department of Biochemistry, College of Science, King Saud University (KSU). She joined KSU in 2011, and prior to that she worked as an Associate Research Scientist at King Faisal Specialist Hospital and Research Center in Riyadh, Saudi Arabia. She completed her Ph.D. from Hamdard University, New Delhi, India in 2009. She joined King Saud University Riyadh in 2011 and is currently working as a professor. She is an esteemed academician with a track record of 72 publications, two book chapters, and a few conference proceedings. Her main research interests include molecular biology, toxicology, and biophysics. She is an academic editor for several reputed journals and has voluntarily reviewed a number of research articles as an act of service to the scientific community. In addition. She is an active member of the Indian Education Forum, which works on the improvement of education in various forms.

Afzal Hussain

Afzal Hussain is an Associate Professor at the Department of Pharmacognosy, College of Pharmacy, King Saud University, Riyadh, Saudi Arabia. He completed his Ph.D. in Chemistry at Jamia Millia Islamia, New Delhi, India in 2010. He has worked as a postdoctoral fellow at Sultan Qaboos University, Muscat, Oman, and Aligarh Muslim University, Aligarh, India. Having been admitted as a Member of the Royal Society of Chemistry (MRSC) in 2018, he has 12 years of experience in teaching and research. His areas of research include separation sciences, bioanalysis, drug analysis, cosmeceutical analysis, nutraceutical analysis, and synthesis of novel nanomaterials as drug delivery carriers. Dr. Hussain as a co-investigator recently completed a research project based on the Novel Metal-Based Anticancer Drugs. Dr. Hussain has published more than 100 papers in international journals and conference proceedings and is an active reviewer for many prominent journals.

Preface to "Spectroscopic, Thermodynamic and Molecular Docking Studies on Molecular Mechanisms of Drug Binding to Proteins"

Protein–ligand interactions essentially play a significant role in various biological functions, such as signal transduction, cell regulation, and immune response. Both the exogenous and endogenous molecules interact with proteins, and understanding of these interactions is therefore central to understanding biology at the molecular level. Hence, protein–ligand recognition and binding is essential in drug discovery development. The structural information of proteins and protein–ligand complexes by various methodologies, such as molecular modelling, molecular dynamics, and cheminformatic approaches, are very helpful for a competent analysis of such complexes and to provide greater insights in the molecular mechanisms of these interactions. Current progress in experimental methodologies for identifying and characterizing ligand binding sites on protein targets has provided biological insights that are significant for drug discovery. In addition, an advancement in new hardware, software and chemoinformatic technologies, the in silico approaches in drug design nowadays is one of the central scientific fields in the pharmaceutical industry.

A Special Issue in *Molecules* was guest edited by us and collected 15 research articles which focused on various aspects of protein ligand interactions using wide range of experimental and computational tools. We are grateful to Jessica Wang, Section Managing Editor of MDPI who provided a dedicated and wonderful professional support for organizing this work. I hope that this reprint makes for interesting reading and will be useful to fellow scientists and graduate students interested in protein–ligand interactions.

Tanveer A. Wani, Seema Zargar, and Afzal Hussain
Editors

Editorial

Spectroscopic, Thermodynamic and Molecular Docking Studies on Molecular Mechanisms of Drug Binding to Proteins

Tanveer A. Wani [1,*], Seema Zargar [2] and Afzal Hussain [3]

[1] Department of Pharmaceutical Chemistry, College of Pharmacy, King Saud University, P.O. Box 2457, Riyadh 11451, Saudi Arabia
[2] Department of Biochemistry, College of Science, King Saud University, P.O. Box 22452, Riyadh 11451, Saudi Arabia
[3] Department of Pharmacognosy, College of Pharmacy, King Saud University, P.O. Box 2457, Riyadh 11451, Saudi Arabia
* Correspondence: twani@ksu.edu.sa

Citation: Wani, T.A.; Zargar, S.; Hussain, A. Spectroscopic, Thermodynamic and Molecular Docking Studies on Molecular Mechanisms of Drug Binding to Proteins. *Molecules* **2022**, *27*, 8405. https://doi.org/10.3390/ molecules27238405

Received: 17 November 2022
Accepted: 29 November 2022
Published: 1 December 2022

Publisher's Note: MDPI stays neutral with regard to jurisdictional claims in published maps and institutional affiliations.

1. Introduction

Molecular recognition, which is the process of biological macromolecules interacting with each other or various small molecules with a high specificity and affinity to form a specific complex, constitutes the basis of all processes in living organisms. Proteins, an important class of biological macromolecules, realize their functions through binding to themselves or other molecules. Protein–ligand interactions play an important role in most biological processes, such as signal transduction, cell regulation, and immune response. Therefore, the study of protein–ligand interactions continues to be very important in life science fields. Since the recognition of their importance at the beginning of the 20th century, investigations into binding parameters have received significant attention, and understanding of these interactions is therefore central to understanding biology at the molecular level. Moreover, knowledge of the mechanisms responsible for protein–ligand recognition and binding also facilitates the discovery, design, and development of drugs.

Quantifying the binding of chemical entities to a protein is an important early screening step during drug discovery and is of fundamental interest for estimating safety margins during drug development. Current progress in experimental and computational methods for identifying and characterizing ligand binding sites on protein targets has provided biological insights that are significant for drug discovery. In addition, because the aim of rational drug design is to make use of knowledge of the structural data and protein–ligand binding mechanisms to optimize the process of finding new drugs, an in-depth understanding of the nature of molecular recognition/interactions is also of great importance. For a deeper understanding of the molecular recognition between a protein and its ligand, physicochemical mechanisms underlying the protein–ligand interaction, the binding kinetics, the basic thermodynamic concepts and relationships relevant to protein–ligand binding, and the driving forces/factors of binding and enthalpy–entropy compensation are investigated using experimental and theoretical methods and using various techniques such as isothermal titration calorimetry (ITC), surface plasmon resonance (SPR), and fluorescence, circular dichroism, and UV absorption spectroscopy.

2. Contributions

Fourteen research papers covering various aspects of drug development that help in understanding an interaction of various ligands to proteins were published in this Special Issue. These studies investigated the molecular mechanisms and structural changes in target proteins involved in the interaction with various viruses and bacteria. In addition, drug–protein interactions, drug–drug interaction mechanisms, anti-glycating and antioxidant activity, binding affinity, network-pharmacology-driven investigations, and targeted anticancer treatments were also investigated.

Proteins are essential biological macromolecules and have enormous importance in all physiochemical process. The native or folded forms of proteins are necessary to perform their biological functions. Pharmacologically profiling drugs offers an understanding of the interactions of vital therapeutic drugs or their derivatives with either plasma or target tissue proteins. In medicinal chemistry, studies pertaining to plasma proteins and drug binding are attracting researchers across the globe because these studies provide a platform to study drugs' behaviors and actions, thereby delineating their transport and distribution characteristics in the circulatory system. The simultaneous administration of two or more drugs might lead to competition between the two drugs to bind to a similar site. The conformation of the protein binding pockets can be altered on interaction with ligands, thereby influencing the binding of other ligands to the same binding pockets. These studies will help in understanding the various binding mechanisms involved in the interaction between various ligands and proteins.

Respiratory syncytial virus (RSV) is a lower respiratory tract pathogen that causes pneumonia and bronchiolitis in children, especially those younger than five years of age. There is no licensed vaccine or any therapeutics available against RSV. The only available preventive measure is the injection of a monoclonal antibodies cocktail (palivizumab) specific to the fusion glycoprotein, which may decrease the severity of disease and rate of hospitalization in infants. In the their study, Hamza et al. [1] examined the structural changes in ectodomain G protein (edG) in a wide pH range. The absorbance results revealed that protein maintains its tertiary structure at physiological and highly acidic and alkaline pH. Fluorescence quenching, molecular docking, and MD simulation studies suggested strong binding between the edG and heparan sulfate. The study suggested that heparan-sulfate-mimicking compounds can be used to target the effective host–pathogen interaction.

Oxidative stress is an imbalance between ROS production and antioxidant defense, which causes cell damage at high levels. The ROS and their pathophysiological effects depend on the concentration, type, and specific production site. When ROS are at a high level, they react with DNA, proteins, cell membranes, and other molecules, causing cellular damage and producing other more reactive radicals. Zargar and Wani [2] investigated protection given by quercetin against CCl4-induced neurotoxicity in rats. In addition to various biochemical parameters, the authors also used VirtualToxLab to study the interaction between quercetin and various proteins to understand the molecular basis of the protective potential of quercetin against carbon tetrachloride toxicity on rat brains. The authors also hypothesized a possible mechanism for quercetin protection against CCl4 toxicity in the rat brain.

Human serum albumin (HSA) is a major transporter and the most abundant protein in the plasma. It exhibits ligand-binding capabilities, acting as a warehouse and transporter of many endo- and exogenous compounds. Shamsi et al. [3] investigated Huperzine A (HpzA), a natural sesquiterpene alkaloid (*Huperzia serrata*) used in various neurological conditions, including Alzheimer's disease (AD), binding to HSA using advanced computational approaches such as molecular docking and molecular dynamic (MD) simulation, followed by fluorescence-based binding assays. The binding of HpzA with HSA showed an appreciable binding affinity and many intermolecular interactions. MD trajectory analyses (i.e., RMSD, RMSF, R_g, SASA, and hydrogen bonding) suggested that the HSA–HpzA docked complex was quite stable with minimal conformational alterations. Fluorescence-based binding ascertained the actual binding affinity between HpzA and HSA, suggesting that HpzA binds to HSA with a significant affinity. The results provide a basis for setting up an experimental platform and information regarding the mechanism of HpzA interactions with has.

Simultaneous administration of two or more drugs might lead to a competition between the two drugs to bind to a similar site present on the serum albumin. The interaction between erlotinib (ERL) and bovine serum albumin (BSA) was studied in the presence of quercetin (QUR), a flavonoid with antioxidant properties, by Wani et al. [4]. Quercetin may interfere with the binding of the other concomitantly used drugs, impacting their pharmacokinetics and displacing them from their albumin binding sites, leading to higher free drug fractions of ERL in the system in the presence of QUR. Such interference may lead to a toxic or a subtherapeutic response of the drug in the presence of QUR. The presence of quercetin in the BSA-ERL system reduces the binding constant of the BSA-ERL system to almost half of what was observed in its absence. Thus, co-administration of QUR and ERL might influence the pharmacokinetics of ERL and needs to be investigated by in vivo studies.

Globally, hepatitis E accounts for an estimated mortality rate of ~3.3% in the infected population. HEV has a genome of ~7.2 kb plus-stranded RNA with a $5'$-methylguanine (m^7G) cap accorded by guanylyltransferase (GTase) and methyltransferase (MTase). RNA capping is essential for the viruses' evasion of the host immune system and to produce other viral proteins by protecting the viral mRNA from nucleases. In the case of HEV, the $5'$ m^7G cap has been demonstrated to play an additional role, increasing infectivity in non-human primates and cultured hepatoma cells. Despite being an important enzyme, few studies have examined the functional and structural aspects of the HEV MTase. In the present study, Hooda et al. [5] cloned, expressed, and purified MTase spanning 33–353 amino acids of HEV genotype 1. The activity of the purified enzyme and the conformational changes were established through biochemical and biophysical studies. This study suggested an indispensable role of Mg^{2+} in MTase activity and stability. This work also established the optimal experimental conditions helpful for the screening of inhibitor libraries against HEV MTase to identify potential inhibitors.

The inhibition of glycation has also been shown to be beneficial in the treatment of diabetic complications. The medicinal properties of garlic (*Allium sativum*) are well documented and were further investigated by Khan et al. [6] as a natural plant phytochemical that can prevent glycation with fewer adverse side effects. Biophysical, biochemical, and molecular docking investigations were conducted to assess the antiglycating, antioxidant, and protein structural protection activities of garlic. Garlic extract exhibited significant levels of protection in protein structural stabilization against glycation. This research significantly supports the numerous therapeutic properties of garlic and addresses the question of the interdependence of various biological activities and their antioxidant capacity. The authors also suggested to include garlic into any existing preventative or treatment approach for glycation-induced health complications in diabetic patients.

Quetiapine (QTP) is a second-generation (short-acting atypical) antipsychotic drug of dibenzothiazepine (class) used to treat schizophrenia, acute bipolar disorder, and major depression in adolescents and adults. In their study, Zargar et al. [7] investigated the binding of QTP to human serum albumin (HSA). The QTP–HSA binding interactions showed moderate binding affinity of QTP toward HSA and involved hydrogen bonding and hydrophobic interactions. The also showed the complex formation between QTP and HSA and a static quenching mechanism. The QTP binding region at subdomain IB of HSA was investigated. This study is expected to help understand the drug's mechanisms and pharmacokinetics for further clinical research and novel drug delivery systems.

Donor–acceptor complexation plays an important role, especially in the field of biochemical energy transfer processes. The formation of brilliantly colored CT complexes that absorb visible light is frequently linked to charge transfer interactions between electron acceptors and donors. In biological systems, mechanisms requiring molecular complexation and structural recognition include drug design, enzyme catalysis, and ion exchanges via lipophilic membranes. Alamri et al. [8] investigated a typical antipsychotic drug, Haloperidol's (HPL), solid charge transfer (CT) products with 7,7,8,8-tetracyanoquinodimethane (TCNQ) and picric acid (PA). The findings of the study suggest that [(HPL)(TCNQ)] coupled with serotonin and dopamine more efficiently than HPL alone. In addition, [(HPL)(TCNQ)]–dopamine had a higher binding energy value than HPL–dopamine. The molecular dynamic simulation at 100 ns demonstrated that the [(HPL)(TCNQ)]–dopamine complex had a more stable interaction with the dopamine receptor than the HPL–dopamine complex.

Khan et al. [9] investigated the binding of caffeic acid and coumaric acid with α-amylase and analyzed the effect of these compounds on the formation of advanced glycation end-products (AGEs). Caffeic and coumaric acid bind with α-amylase and also inhibit the formation of AGEs to a certain extent. Caffeic acid possesses more inhibitory activity, which could be due to its planarity and hydrogen bonding potential. Van der Waals and hydrogen bonding are the major forces in polyphenols–protein interactions. Molecular docking, along with fluorescence quenching and synchronous fluorescence, displayed the ability of phenolics to form stable complexes with amylase. Moreover, these phenolics decrease AGE formation by inhibiting fructosamine. Furthermore, the oxidation of proteins boosted the effect of glycation; caffeic and coumaric acid, on the other hand, attenuate it by protecting thiol and carbonyl groups. The authors suggest more research on similar structures along with in vivo studies to design inhibitors for diabetic complications.

Belal et al. [10] implemented the repurposing of various FDA-approved ophthalmic medications for targeting MMP-2 and MMP-9 to establish a treatment outline for keratoconus (KC), a primary cause of corneal ectasia. Although the exact etiology and pathogenic mechanism are unknown, environmental and genetic variables are considered to play a role in the disease's progression. The authors selected a group of thirty-two FDA-approved drugs and subjected them to virtual screening through docking against MMP-2 and MMP-9 proteins to identify the most promising inhibitors as a proposed computational mechanism to treat KC. The results indicate that some drugs may have potential activities against these proteins, opening the field to further biological studies. The docking results showed the ability of atenolol and ampicillin to accommodate well into the active sites of MMP-2 and MMP-9, respectively. Molecular dynamic simulations and MM-GBSA calculations point to the stability of the binding of both drugs to the respective enzyme, thus adding to the potential of both compounds in KC management. These encouraging findings pave the way for additional clinical investigations to confirm such theoretical findings.

Zhang et al. [11] employed HPLC-MS and network pharmacology to identify the active components and key signaling pathways of DGS. Transgenic zebrafish and HUVECs cell assays were used to evaluate the effectiveness of DGS. A total of 37 potentially active compounds were identified that interacted with 112 potential targets of CAD. Furthermore, PI3K-Akt, MAPK, relaxin, VEGF, and other signal pathways were determined to be the most promising DGS-mediated pathways. The findings of the study suggest that DGS may exert proangiogenic and vasodilatory effects through the activation of the VEGF/VEGFR2/Akt/Erk/eNOS signaling pathway. Molecular docking and molecular dynamics suggest that salvianolic acid C may be a key component in exerting angiogenic and vasodilatory effects.

NIMA-related kinase7 (NEK7) plays a multifunctional role in cell division and NLRP3 inflammasone activation. Any mutation or atypical expression of NEK7 leads to the development of cellular oncogenesis and may provoke a fatal inflammatory response, causing the tumorigenesis of multiple organs. These findings lend testimony to the involvement of NEK7 in the progression and development of numerous deadly diseases.

NEK7 is a promising therapeutic target for preventing and treating NEK7-related diseases. A few medications have recently been developed to target the NEK7-mediated inflammasome pathway, but the mechanism and treatment outcomes are not specific and consistent. Aziz et al. [12] focused on the virtual screening of 1200 benzene sulphonamide derivatives retrieved from the PubChem database by selecting and docking validation of the crystal structure of NEK7 protein (PDB ID: 2WQN). The compounds library was subjected to virtual screening using Auto Dock Vina. The binding energies of screened compounds were compared to standard dabrafenib. This study utilized deep learning models for the prediction of binding affinity, pIC$_{50}$, and ADMET properties. Among the studied compounds, 762 showed good binding affinity and demonstrated a promising ADMET profile. Molecular dynamics simulations determined the protein–ligand complex was stable.

The increased incidence of tuberculosis cases and drug-resistant strains prompted researchers to identify novel drug targets and natural therapeutics with lesser toxicity. Therefore, three medicinal plants, *Achyranthes aspera* (*A. aspera*), *Calotropis gigantea* (*C. gigantea*), and *Calotropis procera* (*C. procera*), were investigated by Beg et al. [13] for their potential therapeutic intervention in Tuberculosis. The plants' extracts were tested against different mycobacterial strains. *A. aspera* aerial and *C. gigantea* flower ash were found to be active against the *M. tuberculosis* $H_{37}Rv$ ATCC 27294 strains with an MIC value of 64 mg/L. A multitarget assessment study was used to identify the possible mycobacterial target proteins. Ten proteins, viz., *BpoC, RipA, MazF4, RipD, TB15.3, VapC15, VapC20, VapC21, TB31.7*, and *MazF9*, were found in the intersection of two categories, viz., available PDB dataset proteins and proteins classified in virulence, detoxification, and adaptation. In silico characterization identified *TB15.3* (Rv1636) in the intersection of the PPI network, which are the universal stress proteins. The MD simulation was used to determine the stability and accuracy of the complex. The results showed that the complex of β-amyrin and *Rv1636* was a stable complex, and the protein did not undergo unfolding during the simulation run. This study established a significant bridge in the field of mycobacterial biology in the targeting of Rv1636, a universal stress protein of mycobacteria, through natural phytoconstituents.

Author Contributions: Conceptualization, T.A.W. and S.Z.; validation, T.A.W., S.Z., A.H.; formal analysis, T.A.W., S.Z., A.H.; resources, data curation, T.A.W., S.Z., A.H.; writing—original draft preparation, T.A.W.; writing—review and editing, T.A.W., S.Z.; supervision, T.A.W. All authors have read and agreed to the published version of the manuscript.

Funding: This research received no external funding.

Conflicts of Interest: The authors declare no conflict of interest.

References

1. Hamza, A.; Samad, A.; Imam, A.; Faizan, I.; Ahmed, A.; Almajhdi, F.N.; Hussain, T.; Islam, A.; Parveen, S. Structural Characterization of Ectodomain G Protein of Respiratory Syncytial Virus and Its Interaction with Heparan Sulfate: Multi-Spectroscopic and In Silico Studies Elucidating Host-Pathogen Interactions. *Molecules* **2021**, *26*, 7398. [CrossRef] [PubMed]
2. Zargar, S.; Wani, T.A. Protective Role of Quercetin in Carbon Tetrachloride Induced Toxicity in Rat Brain: Biochemical, Spectrophotometric Assays and Computational Approach. *Molecules* **2021**, *26*, 7526. [CrossRef] [PubMed]
3. Shamsi, A.; Shahwan, M.; Khan, M.S.; Alhumaydhi, F.A.; Alsagaby, S.A.; Al Abdulmonem, W.; Abdullaev, B.; Yadav, D.K. Mechanistic Insight into Binding of Huperzine A with Human Serum Albumin: Computational and Spectroscopic Approaches. *Molecules* **2022**, *27*, 797. [CrossRef] [PubMed]
4. Wani, T.A.; Alanazi, M.M.; Alsaif, N.A.; Bakheit, A.H.; Zargar, S.; Alsalami, O.M.; Khan, A.A. Interaction Characterization of a Tyrosine Kinase Inhibitor Erlotinib with a Model Transport Protein in the Presence of Quercetin: A Drug–Protein and Drug–Drug Interaction Investigation Using Multi-Spectroscopic and Computational Approaches. *Molecules* **2022**, *27*, 1265. [CrossRef] [PubMed]
5. Hooda, P.; Ishtikhar, M.; Saraswat, S.; Bhatia, P.; Mishra, D.; Trivedi, A.; Kulandaisamy, R.; Aggarwal, S.; Munde, M.; Ali, N.; et al. Biochemical and Biophysical Characterisation of the Hepatitis E Virus Guanine-7-Methyltransferase. *Molecules* **2022**, *27*, 1505. [CrossRef] [PubMed]
6. Khan, M.W.A.; Otaibi, A.A.; Alsukaibi, A.K.D.; Alshammari, E.M.; Al-Zahrani, S.A.; Sherwani, S.; Khan, W.A.; Saha, R.; Verma, S.R.; Ahmed, N. Biophysical, Biochemical, and Molecular Docking Investigations of Anti-Glycating, Antioxidant, and Protein Structural Stability Potential of Garlic. *Molecules* **2022**, *27*, 1868. [CrossRef] [PubMed]
7. Zargar, S.; Wani, T.A.; Alsaif, N.A.; Khayyat, A.I.A. A Comprehensive Investigation of Interactions between Antipsychotic Drug Quetiapine and Human Serum Albumin Using Multi-Spectroscopic, Biochemical, and Molecular Modeling Approaches. *Molecules* **2022**, *27*, 2589. [CrossRef] [PubMed]
8. Alamri, A.S.; Alhomrani, M.; Alsanie, W.F.; Alyami, H.; Shakya, S.; Habeeballah, H.; Alamri, A.; Alzahrani, O.; Alzahrani, A.S.; Alkhatabi, H.A.; et al. Enhancement of Haloperidol Binding Affinity to Dopamine Receptor via Forming a Charge-Transfer Complex with Picric Acid and 7,7,8,8-Tetracyanoquinodimethane for Improvement of the Antipsychotic Efficacy. *Molecules* **2022**, *27*, 3295. [CrossRef] [PubMed]
9. Khan, M.S.; Alokail, M.S.; Alenad, A.M.H.; Altwaijry, N.; Alafaleq, N.O.; Alamri, A.M.; Zawba, M.A. Binding Studies of Caffeic and p-Coumaric Acid with α-Amylase: Multispectroscopic and Computational Approaches Deciphering the Effect on Advanced Glycation End Products (AGEs). *Molecules* **2022**, *27*, 3992. [CrossRef] [PubMed]
10. Belal, A.; Elanany, M.A.; Santali, E.Y.; Al-Karmalawy, A.A.; Aboelez, M.O.; Amin, A.H.; Abdellattif, M.H.; Mehany, A.B.M.; Elkady, H. Screening a Panel of Topical Ophthalmic Medications against MMP-2 and MMP-9 to Investigate Their Potential in Keratoconus Management. *Molecules* **2022**, *27*, 3584. [CrossRef] [PubMed]
11. Zhang, Y.-G.; Liu, X.-X.; Zong, J.-C.; Zhang, Y.-T.-J.; Dong, R.; Wang, N.; Ma, Z.-H.; Li, L.; Wang, S.-L.; Mu, Y.-L.; et al. Investigation Driven by Network Pharmacology on Potential Components and Mechanism of DGS, a Natural Vasoprotective Combination, for the Phytotherapy of Coronary Artery Disease. *Molecules* **2022**, *27*, 4075. [CrossRef]

12. Aziz, M.; Ejaz, S.A.; Zargar, S.; Akhtar, N.; Aborode, A.T.; Wani, T.A.; Batiha, G.E.-S.; Siddique, F.; Alqarni, M.; Akintola, A.A. Deep Learning and Structure-Based Virtual Screening for Drug Discovery against NEK7: A Novel Target for the Treatment of Cancer. *Molecules* **2022**, *27*, 4098. [CrossRef]
13. Beg, M.A.; Shivangi; Afzal, O.; Akhtar, M.S.; Altamimi, A.S.A.; Hussain, A.; Imam, M.A.; Ahmad, M.N.; Chopra, S.; Athar, F. Potential Efficacy of β-Amyrin Targeting Mycobacterial Universal Stress Protein by In Vitro and In Silico Approach. *Molecules* **2022**, *27*, 4581. [CrossRef] [PubMed]

Article

Potential Efficacy of β-Amyrin Targeting Mycobacterial Universal Stress Protein by In Vitro and In Silico Approach

Md Amjad Beg [1], Shivangi [2], Obaid Afzal [3,*], Md Sayeed Akhtar [4], Abdulmalik S. A. Altamimi [3], Afzal Hussain [5], Md Ali Imam [1], Mohammad Naiyaz Ahmad [6,7], Sidharth Chopra [6,7] and Fareeda Athar [1,*]

[1] Centre for Interdisciplinary Research in Basic Science, Jamia Millia Islamia, Jamia Nagar, New Delhi 110025, Uttar Pradesh, India; md164429@st.jmi.ac.in (M.A.B.); ali.imamuit@gmail.com (M.A.I.)
[2] CSIR-Institute of Genomics and Integrative Biology, Mall Road, Delhi 110007, Uttar Pradesh, India; mathuria93shivangi@gmail.com
[3] Department of Pharmaceutical Chemistry, College of Pharmacy, Prince Sattam Bin Abdulaziz University, Al-Kharj 11942, Saudi Arabia; as.altamimi@psau.edu.sa
[4] Department of Clinical Pharmacy, College of Pharmacy, King Khalid University, Abha 61421, Saudi Arabia; mdhusain@kku.edu.sa
[5] Department of Pharmaceutics, College of Pharmacy, King Saud University, Riyadh 11451, Saudi Arabia; afzal.pharma@gmail.com
[6] Division of Molecular Microbiology and Immunology, CSIR-Central Drug Research Institute, Sector 10, Janakipuram Extension, Sitapur Road, Lucknow 226031, Uttar Pradesh, India; naiyaz.ahmad@gmail.com (M.N.A.); skchopra007@gmail.com (S.C.)
[7] Academy of Scientific and Innovative Research (AcSIR), Ghaziabad 201002, Uttar Pradesh, India
* Correspondence: o.akram@psau.edu.sa (O.A.); fathar@jmi.ac.in (F.A.); Tel.: +96-61-1588-6094 (O.A.); +91-11-2698-4492 (F.A.)

Citation: Beg, M.A.; Shivangi; Afzal, O.; Akhtar, M.S.; Altamimi, A.S.A.; Hussain, A.; Imam, M.A.; Ahmad, M.N.; Chopra, S.; Athar, F. Potential Efficacy of β-Amyrin Targeting Mycobacterial Universal Stress Protein by In Vitro and In Silico Approach. *Molecules* **2022**, *27*, 4581. https://doi.org/10.3390/molecules27144581

Academic Editor: Anna Maria Almerico

Received: 29 May 2022
Accepted: 14 July 2022
Published: 18 July 2022

Publisher's Note: MDPI stays neutral with regard to jurisdictional claims in published maps and institutional affiliations.

Abstract: The emergence of drug resistance and the limited number of approved antitubercular drugs prompted identification and development of new antitubercular compounds to cure *Tuberculosis* (TB). In this work, an attempt was made to identify potential natural compounds that target mycobacterial proteins. Three plant extracts (*A. aspera*, *C. gigantea* and *C. procera*) were investigated. The ethyl acetate fraction of the aerial part of *A. aspera* and the flower ash of *C. gigantea* were found to be effective against *M. tuberculosis* $H_{37}Rv$. Furthermore, the GC-MS analysis of the plant fractions confirmed the presence of active compounds in the extracts. The *Mycobacterium* target proteins, i.e., available PDB dataset proteins and proteins classified in virulence, detoxification, and adaptation, were investigated. A total of ten target proteins were shortlisted for further study, identified as follows: *BpoC, RipA, MazF4, RipD, TB15.3, VapC15, VapC20, VapC21, TB31.7,* and *MazF9*. Molecular docking studies showed that β-amyrin interacted with most of these proteins and its highest binding affinity was observed with *Mycobacterium* Rv1636 (*TB15.3*) protein. The stability of the protein-ligand complex was assessed by molecular dynamic simulation, which confirmed that β-amyrin most firmly interacted with *Rv1636* protein. *Rv1636* is a universal stress protein, which regulates *Mycobacterium* growth in different stress conditions and, thus, targeting *Rv1636* makes *M. tuberculosis* vulnerable to host-derived stress conditions.

Keywords: *A. aspera*; β-amyrin; *C. gigantea*; *C. procera*; GC-MS; Minimum inhibitory concentration (MIC); Molecular docking; MD simulations; *Tuberculosis* (TB)

1. Introduction

Tuberculosis (TB) is primarily caused by *Mycobacterium tuberculosis* (*M. tuberculosis*) and considered to be an airborne disease that spreads through sneezing (air borne fine droplets), direct contact, and sharing personal daily use items [1]. Many attempts have been made to control and cure TB and its related critical consequences. However, several factors complicate the treatment strategy, such as the emergence of multidrug resistance against established drugs due to regular mutations, patients having poor access to drugs, long-term therapy, poor patient adherence, and severe dose-dependent side effects [2]. These factors have resulted in progressive growth of latent and active TB cases annually, as reported by the World Health Organization in 2021 (WHO, 2021). The WHO estimated that there were approximately 9.9 million TB cases in 2020, wherein there were about 1.3 million with HIV-negative and 0.214 million with HIV-positive cases [3–5]. Notably, the disease affects patients of all ages, the elderly, and the immunocompromised [6,7]. Circulating TB strains are now resistant to a variety of therapeutic combinations and because the discovery of novel drugs takes time, an alternative approach to provide adjuvants that can boost antibiotics potency is to be considered. It was reported that bacterial susceptibility to antibiotics increases with the co-administration of some natural products [8–10].

In this study, we investigated three medicinal plants, *Achyranthes aspera* (*A. aspera*), *Calotropis gigantea* (*C. gigantea*) and *Calotropis procera* (*C. procera*), for their potential therapeutic intervention in TB. *A. aspera* is a widely known weed in many southeast Asian countries. The plant has been used as a folk medicine in Australia, Kenya, and India, since ancient times. In India, it has been used to treat dropsy, hydrophobia, snake bites, ophthalmic, and cutaneous diseases [11–13]. As a part of the Amaranthaceous family, it is also used to treat asthma, kidney stones, skin diseases, and epilepsy [14]. Previous studies have also reported it as being antidepressant, antioxidant, anxiolytic, anticonvulsant, antihyperglycemic, antiallergenic, anti-obese, hypolipidemic, and hepatoprotective [15–19].

Calotropis is a common wasteland weed, commonly known as milkweed or swallowwort [20]. It is a member of the Asclepiadaceae family (Milkweeds), which has almost 2000 species globally and has a common place in the tropics, and subtropics, but is rare in cold climates [21]. Traditionally, *Calotropis* has been used to cure common ailments, such as cold, asthma, vomiting, and diarrhea. The dried whole plant is a tonic, expectorant, depurative, and anthelmintic, according to Ayurveda. Asthma, bronchitis, and dyspepsia are treated with the powdered root. Paralysis, arthralgia, swellings, and intermittent fevers can be treated with the leaves and its flowers, due to its bitter, digestive, astringent, stomachic, anthelmintic, and tonic properties. Moreover, it is a well-known homoeopathic remedy [22].

C. procera is an Asclepiadaceae shrub native to Egypt. It has purgative, antibacterial, anthelmintic, anticoagulant, antipyretic, anti-inflammatory, analgesic, and neuromuscular blocking properties [23–27]. The plant's extract has physiological effects on cardiac, soft, and skeletal muscular tissues. In traditional medicine, the genus *Calotropis* is used to treat leprosy, ulcers, tumors, liver, and piles problems. It also has potential anticancer effects [28]. The key phytoconstituents are practically found in every *Calotropis* species. However, the relative distribution in individual plants can vary depending upon environmental conditions [29–32].

The increased incidence of TB cases and drug resistant strains prompted us to identify novel drug targets and natural therapeutics with lesser toxicity. Considering the broad spectrum of biological properties of *A. aspera*, *C. gigantea* and *C. procera* medicinal plants, we investigated the effect of their phytoconstituents on mycobacterial proteins and their survival.

2. Results

2.1. Identification and Procurement of the Plant Materials

The aerial and root parts of *A. aspera*, and flowers of *C. gigantea* and *C. procera* were collected from the burial ground in Shahjahanpur, Uttar Pradesh. They were identified by their flower and inflorescence. *C. gigantea* and *C. procera* contain white and purple flowers, respectively, as the major phenotypic differentiation [33,34].

2.2. Preparation of Plant Extracts

The plant was dried in shade followed by cutting into small pieces and grinding into fine powder. The dried powdered material of aerial and root parts were 1500 g and 300 g, respectively. The dried powdered material of the flower parts of *C. procera*, *C. gigantea* and *C. gigantea ash* were 200 g each. The flower ash was obtained by burning the flowers in petroleum ether. The powder of *A. aspera* was soaked in methanol (10 times *w/v*) and the powders of *C. procera* and *C. gigantea* were soaked in methanol (5 times *w/v*), at room temperature, for 10 days and 5 days, respectively, depending upon the weight of the powders. The extracted content was first filtered by Whatman filter paper (150 mm) and then evaporated using a Heidolph rotary evaporator. The crude methanol extract was successively fractionated in various solvents, i.e., hexane, ethyl acetate, ethanol and water, in order of their increasing polarity [35,36]. Total yield of the extractable components (EC) from *A. aspera* (aerial and root parts), *C. procera* (flower), *C. gigantea* (flower) and *C. gigantea* (flower ash) in various solvents are shown in Table 1.

Table 1. Total yield of the extractable components (EC) from *A. aspera* (aerial and root parts) *C. procera* (flower), *C. gigantea* (flower) and *C. gigantea* (flower ash).

S. No.	Extracted Fractions of *A. aspera* Aerial Part (1.5 kg)	EC (in g)	Extracted Fractions *A. aspera* of Roots (300 g)	EC (in g)
1	Methanol	58.85	Methanol	39.93
2	Aqueous	54.60	Aqueous	35.86
3	Hexane	7.14	Hexane	1.82
4	Ethyl acetate	3.18	Ethyl acetate	0.93
5	Ethanol	1.73	Ethanol	0.46

S. No.	Extracted fractions of *C. procera* Flower (200 g)	EC (in g)	Extracted Fractions *C. gigantea* Flower (200 g)	EC (in g)
1	Methanol	13.26	Methanol	11.33
2	Aqueous	10.83	Aqueous	9.66
3	Hexane	1.02	Hexane	3.82
4	Ethyl acetate	3.56	Ethyl acetate	3.33
5	Ethanol	1.20	Ethanol	1.26

S. No.	Extracted fractions *C. gigantea* Flower (200 g)	EC (in g)
1	Methanol	12.56
2	Aqueous	9.89
3	Ethyl acetate	3.69

2.3. Phytochemical Screening of Plant Extracts

Phytochemical tests of each extract fraction were performed as mentioned in the method section for alkaloids, tannins, saponins, terpenoids, and organic acids. The phytochemical screening of *A. aspera*, *C. procera* and *C. gigantea* plant extracts in various fractions showed the presence of alkaloids, tannins, saponins, terpenoids, and organic acid, as reported in literature [37,38] (Table 2).

Table 2. Phytochemical screening of alkaloids, tannins, saponins, and terpenoids present in the various plant fractions.

S. No.	Phytochemicals	Methanol	Aqueous	Ethanol	EtOAc	Hexane
			Aerial Part Extracts of *Achyranthes aspera*			
1	Alkaloids	+	−	+	−	−
2	Tannins	+	−	−	+	−
3	Saponins	−	−	+	+	−
4	Terpenoids	+	−	+	−	−
			Roots Extracts of *Achyranthes aspera*			
1	Alkaloids	+	+	−	+	−
2	Tannins	+	−	+	+	−
3	Saponins	+	+	+	−	−
4	Terpenoids	−	+	−	+	−

Table 2. *Cont.*

S. No.	Phytochemicals	Methanol	Aqueous	Ethanol	EtOAc	Hexane
			Flower Extracts of *Calotropis procera*			
1	Alkaloids	+	+	+	−	−
2	Tannins	+	+	+	+	+
3	Saponins	+	−	−	−	+
4	Terpenoids	−	+	−	+	+
			Flower Extracts of *Calotropis gigantea*			
1	Alkaloids	−	+	+	+	−
2	Tannins	+	−	−	−	−
3	Saponins	+	+	−	−	+
4	Terpenoids	+	−	−	+	−

2.4. Detection of Total Flavonoid Content (TFC) of Plants

Flavonoids are secondary metabolites having multiple physiological activities in plant development, pigmentation, and UV protection, and in defense and signaling pathways between plants and microbes [39,40]. The total flavonoid contents of *A. aspera* (aerial and root), *C. procera*, *C. gigantea* and *C. gigantea ash* parts were analyzed in water, methanol, ethyl acetate, ethanol, and hexane fractions, as shown in Figure 1. The highest TFC was found in the root-based ethyl acetate fraction (103.76 ± 14.8 mg RE g^{-1}) and the aerial ethanol fraction (97.61 ± 10.65 mg RE g^{-1}). Similarly, the highest TFC was found in the ethyl acetate fraction of *C. gigantea* ash (184.28 ± 11.64 mg RE g^{-1}), followed by the ethyl acetate fractions of *C. gigantea* (112.48 ± 4.28 mg RE g^{-1}) and *C. procera* (118.89 ± 0.44 mg RE g^{-1}).

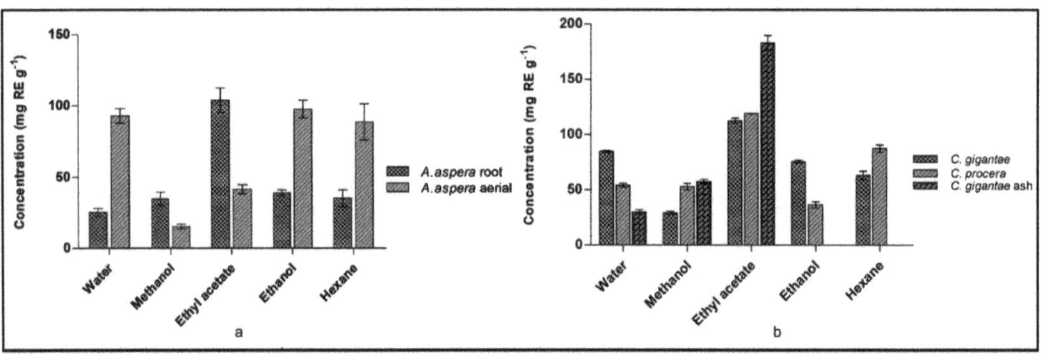

Figure 1. Total flavonoid content determination (**a**) *A. aspera* (**b**) *C. gigantea* and *C. procera*.

2.5. Detection of Total Polyphenolic Content (TPC) of Plants

Polyphenolic compounds are key constituents, known to protect plants from reactive oxygen species. This feature enables the extracts to act as reducing agents and as free radical scavengers [39,40]. The total phenolic content (TPC) of *A. aspera* (aerial and root), *C. procera*, *C. gigantea* and *C. gigantea* ash parts were analyzed in water, methanol, ethyl acetate, ethanol, and hexane fractions, shown in Figure 2. The highest TPC was found in the ethyl acetate fraction of *A. aspera* i.e., 15.1 ± 0.36 mg GAE g^{-1} and 18.27 ± 0.56 mg GAE g^{-1} in the aerial and root parts, respectively. *C. gigantea*, *C. gigantea* ash and *C. procera* showed the highest TPC measures (in mg GAE g^{-1}) in the ethanol fraction (36.47 ± 0.93), the ethyl acetate fraction (37.45 ± 0.94), and the ethyl acetate fraction (25.85 ± 0.22), respectively.

2.6. Minimum Inhibitory Concentrations (MICs)

Extract codes 3 and 23 of the ethyl acetate fractions of *A. aspera* aerial and *C. gigantea* flower ash were found to be active against the *M. tuberculosis* H$_{37}$Rv ATCC 27294 strains, respectively, having an MIC value of 64 mg/L. The rest of the extracts were inactive against tuberculosis and non-tuberculosis strains and showed the MIC value > 64 mg/L. None of the compounds screened were active against non-tuberculous *Mycobacterium*. The known anti-tuberculosis drugs for virulent strains, such as isoniazid, rifampicin, streptomycin, ethambutol showed significant MIC against *M. tuberculosis* $H_{37}Rv$, and levofloxacin (drug used for avirulent strains) showed significant MIC values against all the tested avirulent strains [41,42] (Table S1).

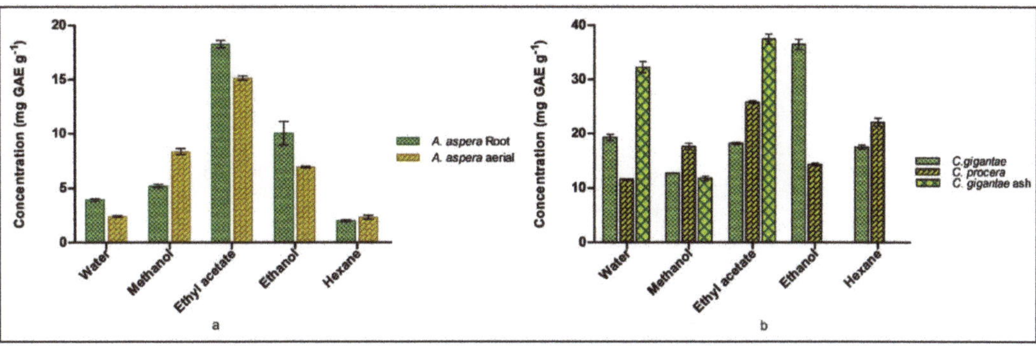

Figure 2. Total Flavonoid content determination (**a**) *A. aspera* (**b**) *C. gigantea* and *C. procera*.

2.7. Gas Chromatography-Mass Spectrometry (GC-MS) Studies

To get the chemical profile of the phytoconstituents, GC-MS analysis of the ethyl acetate plant extracts was performed, due to these having the occurrence of highest TPC and TFC contents. The chromatogram obtained from GC-MS analysis of the ethyl acetate fractions of *A. aspera* and *C. gigantea* plants showed 87 and 68 peaks, respectively. The height of the individual peak resembled the comparative concentration of the compound in the extract. The GC-MS analysis chromatogram is shown in Figure 3. The studied chromatogram peak identified the phytochemical constituents from the ethyl acetate fractions of the aerial part of *A. aspera* (Table S3) and the flower ash of *C. gigantea* (Table S4).

2.8. Assessment of Multitarget Signature Mtb Proteins: An In-Silico Approach

Mycobrowser was analyzed for the availability of genes or proteins responsible for mycobacterial virulence and these were categorized into virulence, detoxification, and adaptation categories [43]. The Venn-diagram used to categorize V.D.A category and known *Mtb* PDB structure proteins is provided in the supporting material (Figure S1). The PDB structures were assessed by employing different in silico methods (Table 3). The crystal structures of the selected *M. tuberculosis* $H_{37}Rv$ proteins (PDB: 7LD8, 4Q4N, 5XE2, 4LJ1, 1TQ8, 4CHG, 5WZ4, 2JAX, 5SV2, 6L2A) [44] are provided in Figure S2.

Table 3. Assessment of multitarget signature *Mtb* proteins based on virulence, detoxification, and adaptation category.

S. No.	Rv No.	Name	PDB ID	Information	Ref.
1	0554	*BpoC*	7LD8	Possible peroxidase BpoC (Non-essential gene for in vitro growth of $H_{37}Rv$)	[45]
2	1477	*RipA*	4Q4N	Peptidoglycan hydrolase; (essential gene for in vitro growth of $H_{37}Rv$)	[46]
3	1495	*MazF4*	5XE2	Possible toxin MazF4 (non-essential gene for in vitro growth of $H_{37}Rv$)	[47]
4	1566c	*RipD*	4LJ1	Possible *Inv* protein (non-essential gene for in vitro growth of $H_{37}Rv$)	[48]

Table 3. *Cont.*

S. No.	Rv No.	Name	PDB ID	Information	Ref.
5	1636	*TB15.3*	1TQ8	Iron-regulated universal stress protein family protein TB15.3 (non-essential gene for in vitro growth of $H_{37}Rv$)	[49]
6	2010	*VapC15*	4CHG	Toxin VapC15 (Non-essential gene for in vitro growth of $H_{37}Rv$)	[50]
7	2549c	*VapC20*	5WZ4	Possible toxin VapC20 (non-essential gene for in vitro growth of $H_{37}Rv$)	[51]
8	2623	*TB31.7*	2JAX	Universal stress protein family protein TB31.7 (non-essential gene for in vitro growth of $H_{37}Rv$)	[52]
9	2757c	*VapC21*	5SV2	Possible toxin VapC21 (non-essential gene for in vitro growth of $H_{37}Rv$)	[53]
10	2801c	*MazF9*	6L2A	Toxin MazF9 (non-essential gene for in vitro growth of $H_{37}Rv$)	[54]

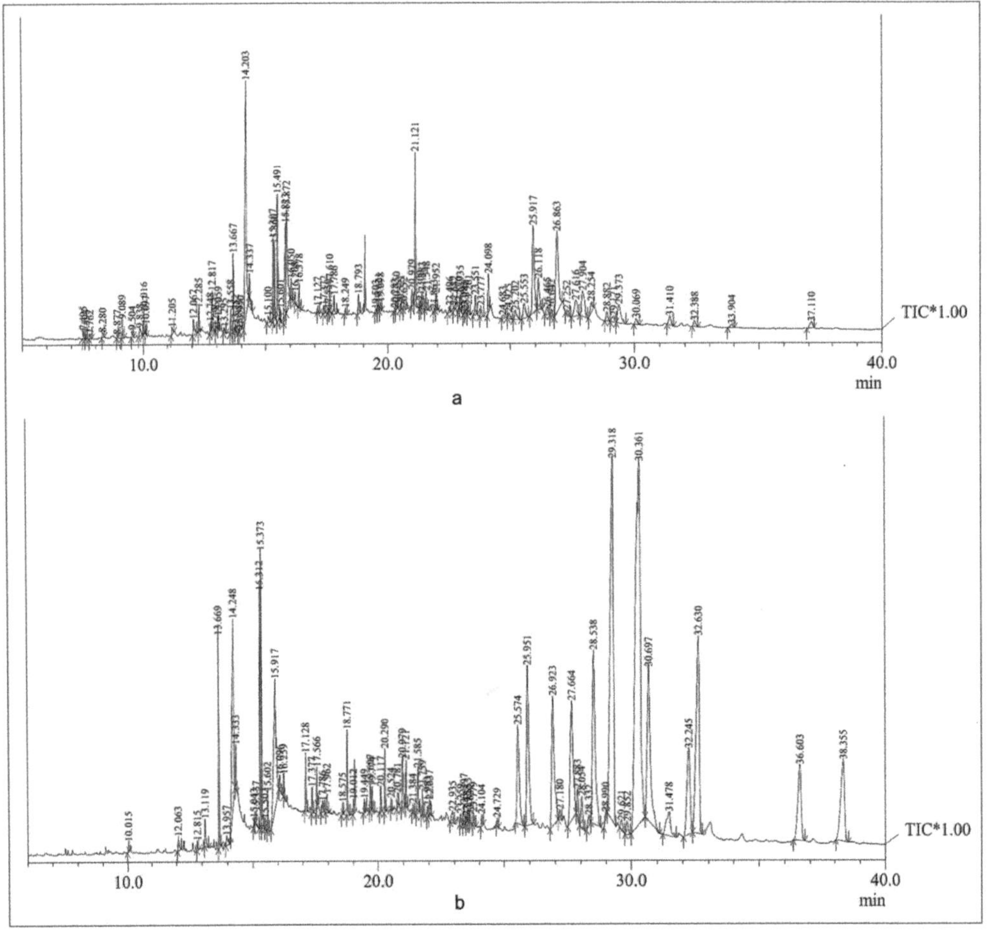

Figure 3. GC-MS chromatogram of ethyl acetate fractions of (**a**) *A. aspera* (aerial) and (**b**) *C. gigantea* (flower ash). TIC* indicates Total Ion Chromatogram.

2.8.1. Physiochemical Parameters

Mycobacterial virulent proteins are the major factors that aid in pathogenesis. The amino acid sequence of virulent proteins was obtained from the Mycobrowser database [55]. The Protparam tool was used to calculate the physicochemical parameters for these proteins, as shown in Table S4 [56]. The physiochemical parameter of the proteins calculated as the isoelectric points (pIs) of *Rv1477*, *Rv1495*, *Rv1566* and *Rv2801c* were greater than 7 (pI > 7), which means these proteins had more basic amino acids. The instability index value was also calculated and four unstable proteins (Instability index > 40) were found, viz. *Rv1477*, *Rv1566c*, *Rv2010*, and *Rv2623*. The calculated aliphatic index of some of these proteins were very high, which means they were thermostable. The GRAVY (grand average hydropathy) value was also determined, which indicated that *Rv0554*, *Rv1636*, *Rv2010*, and *Rv2623* were polar, and rest of the six proteins were non-polar (Figure 4a).

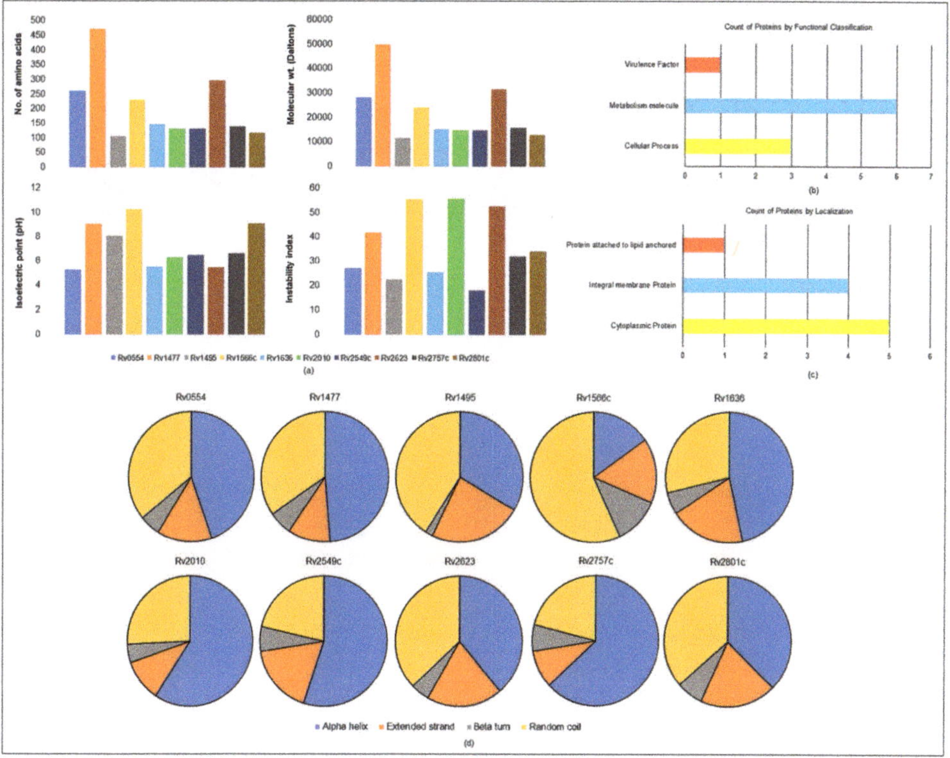

Figure 4. The selected virulence, detoxification, adaptation (VDA) category of *M. tuberculosis* H₃₇Rv proteins. (**a**) Physiochemical parameters of VDA proteins (**b**) Functional classification of VDA proteins (**c**) Subcellular localization of VDA proteins and (**d**) Secondary structure analysis of VDA proteins.

2.8.2. Functional Classification

The protein targets were classified functionally under three categories, viz. virulence factors, proteins involved in metabolism, and those involved in cellular processes, and were regarded as an effective target for emerging TB treatments [43]. VICMpred webserver predicted six proteins involved in metabolism, three in cellular processes, and one protein as a virulence factor (*Rv0554*) [57–60]. The comprehensive analysis of these functional proteins is shown in (Figure 4b).

2.8.3. Subcellular Localization

For forecasting a protein's function, it is crucial to know where it is located. We used the TBpred server in this investigation to predict localization, which is based on the support vector machine (SVM)-based subcellular localization prediction of the mycobacterial protein [61,62]. The protein's location in the cytoplasm, integral membrane, lipid anchoring, and secretory pathways is predicted. The cytoplasmic proteins (Rv1636, Rv2010, Rv2549c, Rv2623, Rv2757c), integral membrane proteins (Rv0554, Rv1477, Rv1566, Rv2801c), and lipid anchored proteins (Rv1495) are displayed in (Figure 4c). The prediction accuracy was 86.62%.

2.8.4. Secondary Structure Prediction

The two-dimensional (2D) structure prediction was performed by the SOPMA web-server for all the ten proteins. The SOPMA server is a straightforward and accurate method for predicting the 2D structure of a protein. The predicted 2D structures with default parameters were set to analyze the structural patterns of the proteins. The structural studies concluded that these proteins have more alpha helix regions compared to extended strands in beta-sheet. The details of these predicted 2D structures are given in Table S6 and the schematic representation of the 2D representation is in Figure 4d.

2.8.5. Phylogenetic Analysis

Phylogenetic analysis of all ten selected genes was conducted using the Mega11 program. The homologs in different mycobacterial species were analyzed and the pathlengths were specified for each interaction, and are shown in Figure 5.

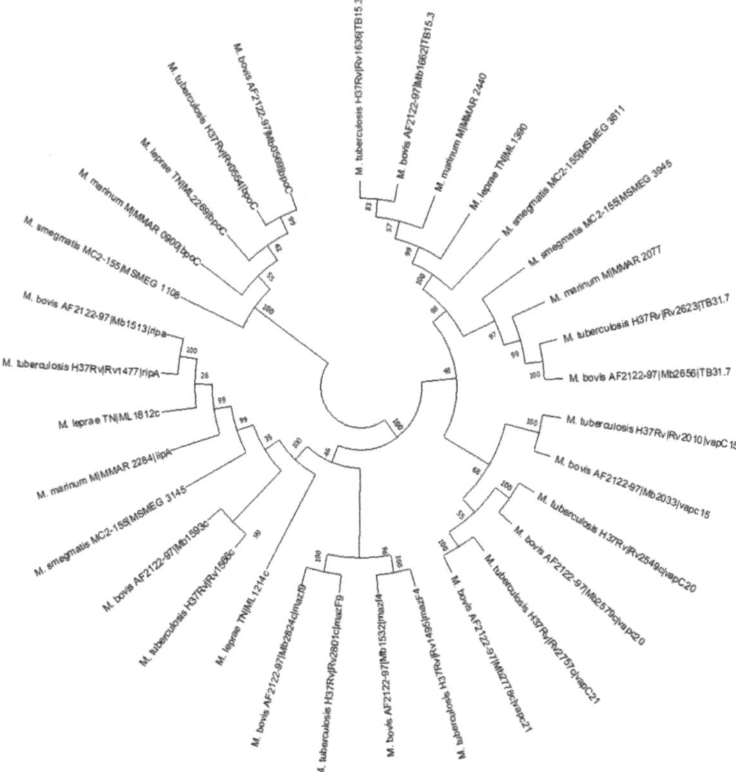

Figure 5. Phylogenetic tree depicting the relationships between virulence categories proteins in *M. tuberculosis* H$_{37}$Rv.

2.8.6. PPI Network Analysis

The protein-protein interaction (PPI) network analysis was done using the STRING v11.0 database. The input proteins showed a confidence score of 0.6–0.9. In the first category of "betweenness, closeness, radiality and degree", there were 3 genes found in the intersection (Rv1636, Rv2623 and Rv1566). In the second category of "betweenness, closeness, stress and degree" only one gene (Rv1636) was found in the intersection, as shown in Figure 6.

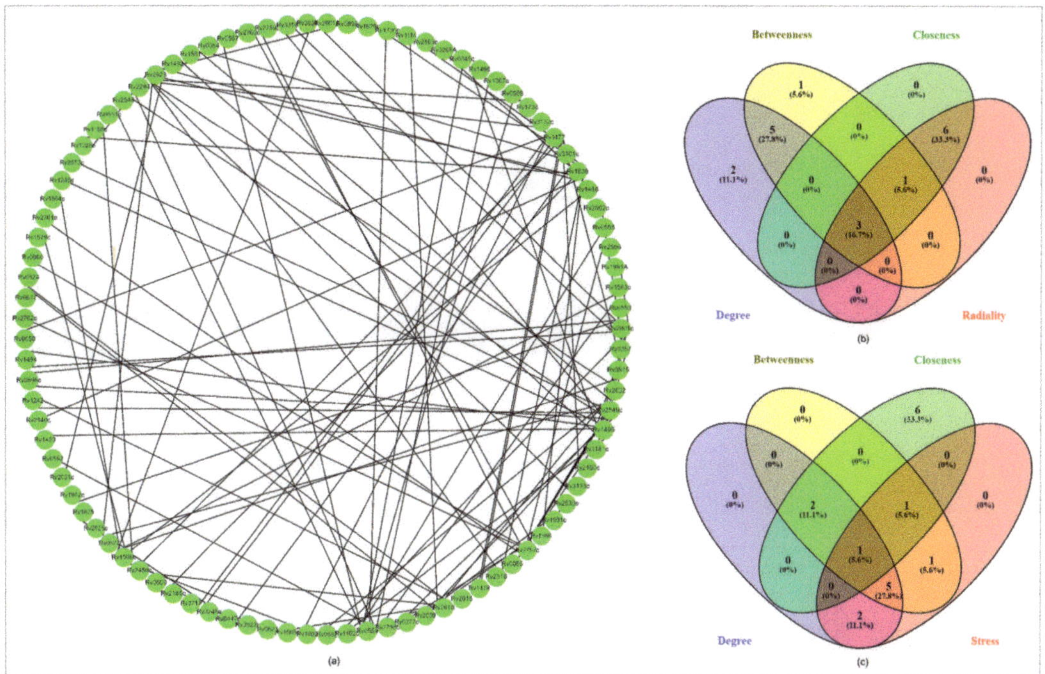

Figure 6. Network analysis: (**a**) PPI network of the selected ten virulent genes which were involved in significant pathways comprised a total number of 93 nodes and 101 edges. (**b**) Venn plot showing the intersection of degree-based radiality category found 3 genes. (**c**) Venn plot showing the intersection degree-based stress category found 3 genes.

2.8.7. Structural Classification of the Selected VDA Proteins

Rv0554 is an integral part of the menaquinone regulatory operon in Mycobacterium. Menaquinone is a crucial factor in the mycobacterial electron transport chain [45]. *Rv1477* is a mycobacterial resuscitation promoting factor interacting protein (*RipA*) and participates in the cleavage of peptide cross linkages between peptidoglycans similar to other cleavage enzymes, such as *OwlT* and *Spr* in *B. subtilis* and *E. coli*, respectively [46]. *Rv1495* is designated as *mazEF*, which is a subcomponent of the type II toxin–antitoxin system in mycobacteria [47]. *Rv1566c* (*RipD*) is a predicted peptidoglycan specific peptidase of the NlpC/p60 family. The unusual peptide linkages of *Mycobacterium* i.e., L-D and D-D and other iso-peptide linkages makes the peptide resilient to cleavage by mycobacterial peptidases [48]. *Rv1636* is a universal stress protein (USP) in *M. tuberculosis*. In *M. tuberculosis*, it is classified into class I USP, based on the presence of only a single conserved domain of USP, which is of similar size to *UspA*. CAMP is bound to *Rv1636*, which regulates the signaling associated with the cAMP molecule [49]. *Rv2010* of *M. tuberculosis* codes for *VapC*-15 toxin. These proteins belong to the PIN domain family proteins and contain ribonuclease activity [50].

Rv2549c is designated as *VapC20* of *M. tuberculosis* and along *VapB*, it forms *VapBC* toxin antitoxin complex [51]. *Rv2623* is a universal stress protein of *M. tuberculosis*, that helps mycobacteria in different stress conditions and is also important for mycobacterial growth and persistence [52]. *Rv2757c* of *M. tuberculosis* codes for *VapC*-21 toxin. These proteins also belong to the PIN domain family proteins and contain ribonuclease activity [53]. *Rv2801c* encodes for the *MazF*-mt1 protein of *M. tuberculosis*. *MazF* is a known component of the *MazEF* toxin–antitoxin system in many prokaryotic cells [54].

2.8.8. Validation of the Selected Proteins' Structures

The PROCHECK server was used to validate the models, and a Ramachandran plot of the projected models revealed that the modeled structure of mycobacterial virulence proteins had ~90% residues in the most favored region, indicating the modeled structure was good [63,64] (Table 4).

Table 4. Ramachandran plot statistics of selected virulent category proteins.

S. No.	Protein	Most Favoured Region	Additional Allowed Region	Generously Allowed Region	Disallowed Region
1	Rv0554	92.1%	7.5%	0.0%	0.4%
2	Rv1477	92.5%	6.9%	0.0%	0.6%
3	Rv1495	95.3%	4.7%	0.0%	0.0%
4	Rv1566c	98.0%	2.0%	0.0%	0.0%
5	Rv1636	91.0%	9.0%	0.0%	0.0%
6	Rv2010	94.2%	5.0%	0.0%	0.7%
7	Rv2549c	95.8%	4.2%	0.0%	0.0%
8	Rv2623	74.5%	22.7%	1.8%	0.9%
9	Rv2757c	96.7%	3.3%	0.0%	0.0%
10	Rv2801c	94.8%	5.2%	0.0%	0.0%

2.8.9. Molecular Docking

The molecular docking analysis was performed by using InstaDock v1.0 software, New Delhi, India (https://hassanlab.org, accessed on 1 April 2022), ChemBioDraw Ultra 14.0, in-house python script and Discovery Studio as mentioned in the experimental section [65–69]. It included the fundamental orientations between the receptor and the ligands (plant extract phytoconstituents). The 10 shortlisted proteins were used for molecular docking in order to explain how these proteins interact with the phytoconstituents identified in the ethyl acetate aerial part of *A. aspera* and flower ash extract of *C. gigantea*. The high negative docking score (binding free energy, kcal/mol) indicated that their binding was steady. In Table S7, the comparative analysis of docking results on the multitarget proteins is provided. The docking scores are shown in Figure 7. The 2D structure of the phytoconstituents is provided in Figure S3. The selected top hit phytoconstituents had binding free energies ranging from -7.5 to -10.6 kcal/mol, and the correspomding proteins were chosen for further studies. β-amyrin (PubChem CID: 225689) was found to have higher binding free energies against *Rv1636*, *Rv1566*, *Rv2549c*, and *Rv1495* proteins. The 3D crystal structures of VDA proteins, showing different conformational changes, were identified and structural analysis was carried out using a PyMOL visualizer (Figure 8).

2.8.10. Determination of the Selected Phytoconstituents' Drug Abilities

The drug-likeness properties of the selected phytoconstituents were assessed using the Swiss ADME, pkCSM and PASS webservers [70–72]. The top 10 hit compounds' physicochemical or drug likeness properties demonstrated that most of the phytoconstituents followed Lipinski's rule. All the selected compounds' physiochemical parameters are shown in Table 5. The ADMET properties of the identified phytoconstituents were further investigated in order to rule out any potentially harmful patterns in the molecular structures (Table 6).

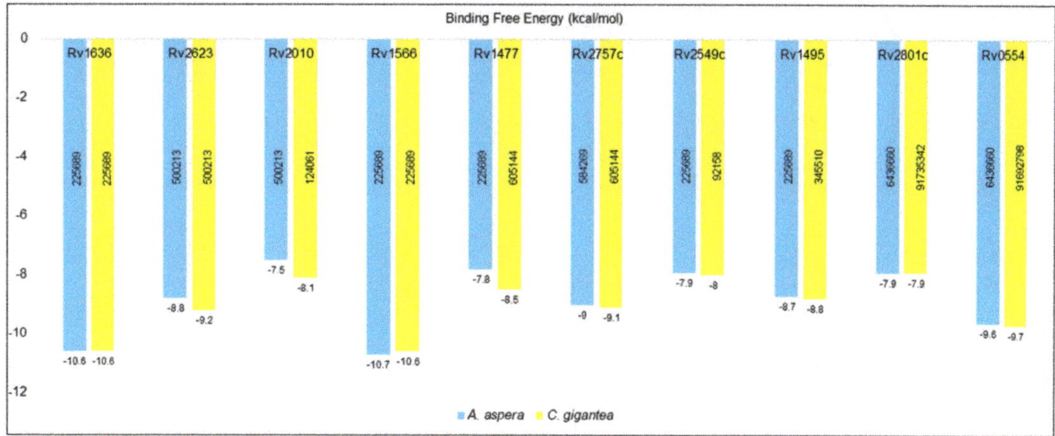

Figure 7. Comparative docking score analysis of the VDA proteins against top hit phytoconstituents.

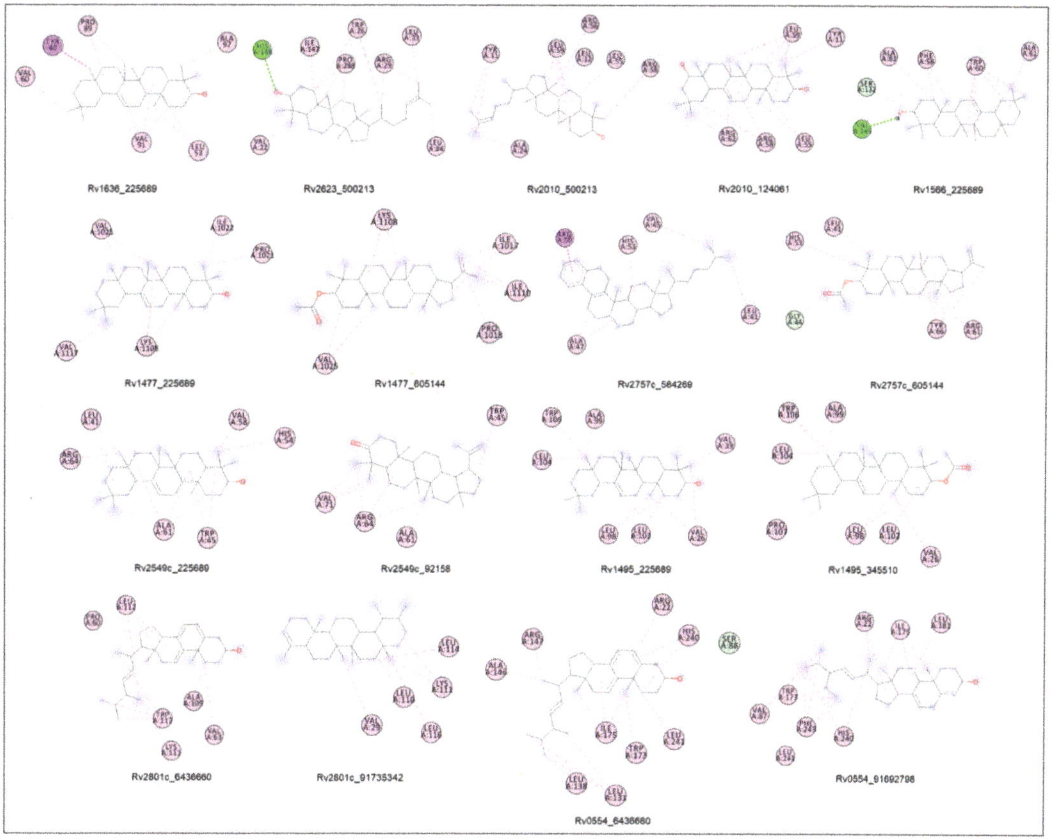

Figure 8. Docking interaction analysis: 2D structural representation of Protein-ligand complexes having pi-Alkyl bonds (purple) and Hydrogen bond (green) along with interacted residues.

Table 5. The physicochemical properties of the selected top hit compounds.

PubChem ID	Name	#M.W.	#Rot. Bond	#HBA	#HBD	LogP
225689	Beta-Amyrin	426	0	1	1	4.74
500213	Handianol	426.72	4	1	1	5.17
584269	-	472.79	5	0	0	5.88
6436660	Dehydroergosterol	394.63	4	1	1	4.68
124061	Olean-12-ene-3, 22-diol	442.72	0	2	2	4.65
605144	-	468.75	3	2	0	4.95
92158	Lupenone	424.70	1	1	0	4.54
345510	Beta-Amyrenyl acetate	468.75	2	2	0	5.19
91537342	24-Norursa-3, 12-diene	394.68	0	0	0	4.76
91692798	Stigmasta-4,7,22-triene-3. alpha.-ol	410.67	5	1	1	4.70

#M.W. (Molecular Weight (Da); #Rot. bond (Rotatable bond); #HBA (hydrogen bond acceptor); #HBD (hydrogen bond donor).

Table 6. ADMET properties of the top hit selected compounds.

PubChem ID	Absorption		Distribution (BBB/CNS Permeation)	Metabolism (CYP2D6 Inhibitor)	Excretion OCT2 Substrate	Toxicity A/H/S
	GI abs.	W.S.				
225689	93.733	−6.531	No	No	No	No
500213	95.248	−5.762	No	No	No	No
584269	97.43	−4.664	No	No	No	No
6436660	94.999	−7.112	No	No	No	No
124061	92.522	−6.351	No	No	No	No
605144	98.182	−5.878	No	No	No	No
92158	98.467	−5.828	No	No	No	No
345510	97.342	−6.649	No	No	No	No
91537342	95.778	−6.925	No	No	No	No
91692798	95.604	−6.696	No	No	No	No

GI abs. (Gastrointestinal absorption percentage); W.S. (Water Solubility (log mol/L); BBB/CNS permeation (blood brain barrier/central nervous system); Toxicity A/H/S (toxicity AMES/Hepatotoxicity/Skin sensitization).

A prediction of structure activity relations (SARs) was performed, by a machine learning program using the PASS online webserver, to investigate the biological activities of the selected phytoconstituents. β-amyrin was shown to have multiple biological activities, such as insulin promoter, caspase-3 stimulant, transcription factor NF kappa B stimulant, mucomembranous protector, hepatoprotection, apoptosis agonist, antineoplastic, oxidoreductase inhibitor, membrane integrity antagonist, and chemoprevention, with a Pa score ranging from 0.903 to 0.977. The biological activity predictions of all the ten phytoconstituents are provided in Table S8.

Moreover, β-amyrin, in combination with lupeol, was reported to have antibacterial activity, including antimycobacterial activity. The mixture of both compounds showed modest antibacterial activity against most of the bacteria, with MIC of 62.5 µg/mL, for *Staphylococcus aureus*, *Pseudomonas aeruginosa*, *Mycobacterium fortuitum* and *Mycobacterium smegmatis*. β-amyrin is abundantly found in plants with varied pharmacological activities. The compounds revealed after GC-MS analysis were used for molecular docking analysis with 10 shortlisted proteins. The docked complexes showed that most of the proteins had significant binding affinity with β-amyrin. Thus, β-amyrin was selected for further MD simulation study.

2.9. MD Simulation

The structural conformational stability of a protein-ligand was investigated using molecular dynamics (MD) simulation. The structural insight of Rv1636 and Rv1636 with β-amyrin in water at 298 K was analyzed. RMSD, RMSF, Rg, SASA, and funnel energy of landscape (FEL) were tracked throughout 50 ns [73–76].

2.9.1. Average Potential Energy of System

The average potential energy of Rv1636 and Rv1636_β-amyrin complex were monitored to ensure that the system was equilibrated. A constant fluctuation for each system at constant temperature (298 K) and pH (7.0), indicated a steady and accurate MD simulation. The average potential energy of *Rv1636* was -0.061205×10^7 kJ/mol and for the Rv1636_β-amyrin complex it was -1.48416×10^7 kJ/mol.

2.9.2. Root Mean Square Distance (RMSD)

The interaction of a ligand with a protein can result in considerable conformational change in the structure. The RMSD was measured as a function of time with respect to the initial conformation and is illustrated in Figure 9a. The RMSD average value for *Rv1636* was 0.71379 nm and for the Rv1636_β-amyrin complex it was 0.61481 nm. The RMSD plot evidently implied that the Rv1636 and Rv1636_β-amyrin complexes were stable during the simulation time frame till 45 ns, but after 45 ns there was some hindrance showing fluctuation in the Rv1636_β-amyrin complex, as compared to Rv1636.

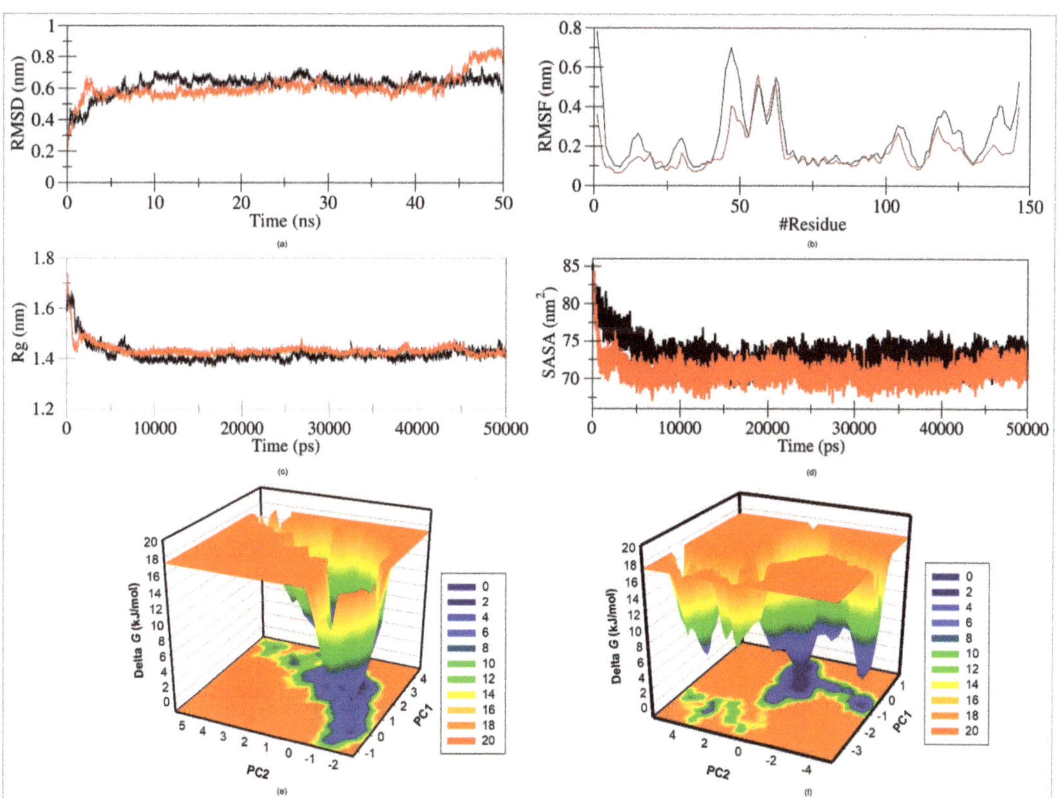

Figure 9. Structural dynamics of universal stress protein (Rv1636) upon β-amyrin binding as a function of time. (**a**) RMSD plot of Rv1636 in complex with β-amyrin. (**b**) RMSF plot of Rv1636 in complex with β-amyrin. (**c**) Structural compactness and folding Rv1636 upon β-amyrin. (**d**) SASA plot of Rv1636 as a function of time before and after β-amyrin binding. (**e**) The time evolution of projections of trajectories on both EVs. The free energy landscapes of free Rv1636 protein (**f**) The free energy landscapes of free Rv1636 with β-amyrin.

2.9.3. Root Mean Square Fluctuation (RMSF)

The RMSF of the Rv1636 and Rv1636_β-amyrin complexes were displayed as a function of residue number to obtain the average fluctuation of all residues throughout the simulation time (Figure 9b). The RMSF revealed that residual fluctuations existed in various regions. These residual fluctuations were observed to be reduced when Rv1636 and β-amyrin was bound, rather than when the Rv1636 protein was alone.

2.9.4. Radius of Gyration

Rg is the RMS distance between a group of atoms and their collective center of mass, and it is linked to a protein's tertiary structure stability. It is one of the most extensively used criteria for determining the compactness of a protein structure. The values of Rg of Rv1636 and the Rv1636_β-amyrin complex were 1.42461 nm and 1.43199 nm, respectively. A minor increase in the Rg values of the Rv1636_β-amyrin complex was directly in agreement with the RMSF values. The Rg plot suggested that Rv1636 protein was stably folded with the R1636_β-amyrin complex, as shown in Figure 9c.

2.9.5. Solvent Accessible Surface Area (SASA)

Solvent accessible surface area (SASA) analysis was performed to investigate protein folding behavior and stability. While studying the plot, no significant changes in SASA values were observed during the simulation time, suggesting a stable complex. The SASA of native protein was found to be 72.4474 nm^2 and the SASA of the protein-β-amyrin complex was 70.9425 nm^2. A slight decrease in the average SASA signified enhanced packing of the protein-ligand complex, which could also be correlated with the Rg (Figure 9d).

2.9.6. Free Energy Landscape (FEL)

The free energy landscape mechanism was used to define the protein folding into native conformation and protein denaturing landscapes. The two PCs were used to design the FELs and energy minima of Rv1636 and the Rv1636_β-amyrin complex. C-alpha atoms were focused to project the conformational sampling of Rv1636 and the Rv1636_β-amyrin complex, as illustrated in Figure 9e,f. The FEL plot showed that the binding of β-amyrin to Rv1636 slightly distorted the complex size and position of the phase's minimum. A deep blue plot in Rv1636 free landscape signified the conformation with less energy towards native conformation. The plot showed that Rv1636 showed single global minima confines with single basin. On the other hand, the Rv1636_β-amyrin complex also showed single global minima but with more than 1 basin. Finally, this FEL study concluded that the binding of β-amyrin to Rv1636 did not cause protein unfolding throughout the simulation run.

3. Discussion

The medications used for TB were beneficial in the past but, at present, the insufficiency of these medications has raised the need for the identification and discovery of novel therapeutics. *A. aspera* is a widely known medicinal plant with various anti-bacterial activities. The plant is famous for its contents of alkaloids, saponins, carbohydrates, glycosides, flavonoids, tannins, and triterpenoids [11,12]. *C. gigantea* is also a traditional medicinal herb, used to cure common ailments, such as fevers, cough, cold, asthma etc. [20]. The root bark is an expectorant, febrifuge, anthelmintic, depurative, and laxative. Asthma, bronchitis, and dyspepsia are treated with the powdered root. Paralysis, arthralgia, swellings, and intermittent fevers can all be treated with the leaves. The flowers have a range of properties, being bitter, digestive, astringent, stomachic, anthelmintic, and tonic [22] and make a well-known homoeopathic remedy [22]. *C. procera* is a shrub with purgative, anthelmintic, anticoagulant, palliative (for respiratory and blood pressure disorders), antipyretic, analgesic and neuromuscular blocking properties [24]. The family members of this plant are high in cardiac glycosides [25,26].

The aerial and root parts of *A. aspera* and the flowers of *C. gigantea* and *C. procera* are widely grown, and are commonly used herbal medications [33]. The plants are distinguished through the different appearances of their flowers [34]. The plant parts were used to prepare the extracts in hexane, ethyl acetate, ethanol, methanol and water. The phytochemical analysis showed the presence of flavonoids, alkaloids, phenol, anthraquinone, terpenoids, tannins, steroids, saponins, carbohydrates, glycosides in the extracts [35]. Many flavonoids are key components of medicinal plants and are employed in the regulation of inflammation and cancer prevention, due to their ubiquity in the human diet [40]. The highest total flavonoid content in *A. aspera* was found in the root ethyl acetate fraction (103.76 ± 14.8 mg RE g^{-1}) and the aerial ethanol fraction (97.61 ± 10.65 mg RE g^{-1}). The highest TFC was found in the ethyl acetate fraction of *C. gigantea* ash (184.28 ± 11.64 mg RE g^{-1}) (Figure 1). The total polyphenolic content of the plant aerial and root parts was analyzed in water, methanol, ethyl acetate, ethanol and hexane fractions. Polyphenols are important components of plants as they defend plants from reactive oxygen species [39]. The highest TPC was found in the ethyl acetate fraction i.e., 15.1 ± 0.36 mgGAE g^{-1} and 18.27 ± 0.56 mgGAE g^{-1} in the aerial and root parts of *A. aspera*, respectively (Figure 2). The flavonoid and phenolic content of the extracts was higher in the ash of *C. gigantea*, indicating heating the extracts caused solubilization of these organic compounds. *M. tuberculosis* H_{37}Rv ATCC 27294, *M. fortuitum* ATCC 6841, *M. abscessus* ATCC 19977 and *M. chelonae* ATCC 35752 were used to determine the minimum inhibitory concentrations of the extracts against the mycobacterial strains. *A. aspera*'s aerial ethyl acetate fraction (3) and *C. gigantea*'s flower ash ethyl acetate fraction (23) was found to be active against the *M. tuberculosis* H_{37}Rv ATCC 27294 strain with an MIC value of 64 mg/L. The non-tuberculous bacteria showed the maximum growth against all extracted compounds (Table S1). Anti-tuberculosis drugs, such as isoniazid, rifampicin, streptomycin, and ethambutol, with MIC values of 0.03 mg/L, 0.3 mg/L, 1 mg/L, and 1 mg/L, respectively, were used as positive control [41,42].

As the ethyl acetate fraction of *A. aspera* and *C. gigantea* plants were the most effective fractions in containing the flavonoid, phenolic content and bacteriostatic effects against *M. tuberculosis* H_{37}Rv, it was further analyzed by GC-MS for the detection of various phytoconstituents (Figure 3). The presence of the compounds, as confirmed in the GC-MS analysis, are listed in Tables S3 and S4.

The mycobacterial proteins were categorized into virulence, detoxification, adaptation (VDA) categories, and the proteins having known structural information. The VDA category is a vast family of proteins that participates in maintaining mycobacterial metabolism and, therefore, targeting these proteins would be an asset in targeting mycobacterial cells [43]. After the evaluation of the complete mycobacterial database, 238 proteins were found to belong to the VDA category and 135 proteins had available structural information in the RSCB-PDB databank. In Table 3, a counteractive study on these proteins led to the determination of 10 proteins, which were *Rv1477*, *Rv1495*, *Rv1566*, *Rv2801c*, *Rv2010*, *Rv2623*, *Rv0554*, *Rv1636*, *Rv2549c*, and *Rv2757c* [44]. *Rv0554* is an integral part of the menaquinone regulatory operon in Mycobacterium. Menaquinone is a crucial factor in the mycobacterial electron transport chain. There are many genes that were identified, which participate in the menaquinone biosynthesis pathway and the operon is present between *Rv0534c* and *Rv0558*. A gene known as *yfbB* in *E. coli* is known as *menH* and *Rv0554* in *M. tuberculosis* and was predicted to encode an enzyme with a similar function to *menH*. Structurally, it was categorized in the alpha beta hydrolase fold of *E. coli* and is the only protein which is located nearest to the *M. tuberculosis* menaquinone synthesis pathway [45]. *Rv1477* is a mycobacterial resuscitation promoting factor interacting protein (*RipA*) that participates in the cleavage of peptide cross linkages between peptidoglycans similar to other cleavage enzymes, such as *OwlT* and *Spr* in *B. subtilis* and *E. coli*, respectively. *Rv1477* was found to colocalize at bacterial septa with resuscitation promoting factor B (*RpfB*). This protein is important for mycobacterial growth as the depletion strains of this protein in *M. smegmatis* showed abnormal phenotype and decreased growth pattern and, therefore, this protein is a wonderful candidate for drug target strategies [46]. *Rv1495* is designated as *mazEF*, which is a subcomponent of type II toxin antitoxin system in mycobacteria. Out of 7 homologs of *mazF* which are identified in *M. tuberculosis*, four homologs comprise endoribonuclease activity. *MazF* acts as a toxin and it recognizes and cleaves to the intracellular RNA sequence in a ribosome-independent manner. As it is a sequence specific ribonuclease, it inhibits translation with a lesser degree than other non-specific toxins. Previous studies have also reported extracellular death factor functioning in mediated quorum sensing [47]. *Rv1566c* (*RipD*) is a predicted peptidoglycan specific peptidase of NlpC/p60 family. The unusual peptide linkages of *Mycobacterium* i.e., L-D and D-D and other iso-peptide linkages makes the peptide resilient to cleavage by mycobacterial peptidases. However, during the cell division process, the mycobacterial cells produce specific peptidases that weaken these linkages and help in generation of daughter cells. *Rv1566c* is such a specific peptidase. The known specific peptidases in mycobacterium are *RipA* and *RipB*, which cleave the peptide stem between D-glutamic acid and meso-Dap residue. *Rv1566c* containing peptidoglycan specific peptidase domain has 52% and 51% similarity with the *RipB* and *RipA* proteins, respectively. The *Rv1566c* is the first example of a peptidase domain, which binds to the peptidoglycan in a non-catalytic manner, and this feature is specific to mycobacteria only [48]. *Rv1636* is a universal stress protein (USP) in *M. tuberculosis*. In *M. tuberculosis* it is classified into class I USP, based on the presence of only a single conserved domain of USP which is of similar size of *UspA*. Rv1636 amino acid sequence contains the GXXG-9X-G-S/T conserved domain. The exact role of *Rv1636* in stress condition is yet to be detected but it was predicted that this protein might exclusively be expressed in hypoxia and other stress conditions. One interesting feature of *Rv1636* is its cAMP binding activity. The significant fraction of intracellular cAMP bound to *Rv1636* regulates the signaling associated with the cAMP molecule [49]. *Rv2010* of *M. tuberculosis* codes for *VapC*-15 toxin. These proteins belong to the PIN domain family proteins and contain ribonuclease activity. *VapC* toxin is deleterious to cells, but its effect gets neutralized by *VapB* antitoxin. This protein works in a similar manner to T4 RNase and Mja FEN-1 endonuclease. *VapBC* complex is a significant toxin–antitoxin system and an important participator in mycobacterial metabolism [50]. *Rv2549c* is designated as *VapC20* of *M. tuberculosis* and along *VapB* it forms the *VapBC* toxin–antitoxin complex. *VapC20* interacts with its cognate partner *VapB20* to form a stable complex. Both proteins in their individual states are present in dimer states, which form stable homo-tetramers or homo-octamers upon interaction. *Rv2623* is a universal stress protein of *M. tuberculosis* that helps mycobacteria in different stress conditions. This protein is also important for mycobacterial growth and persistence. *Rv2623* protein is a highly induced protein of mycobacteria in response to stress conditions, such as hypoxia and nitrosative stress, which the bacteria face in infected host cells. Apart from its role as a USP, this protein also has the ability to bind ATP. *Rv2623* also interacts with *Rv1747*, which is an ABC transporter protein and helps in exporting lipo-oligosaccharides to negatively regulate mycobacterial growth [52]. *Rv2757c* of *M. tuberculosis* codes for *VapC*-21 toxin. These proteins belong to the PIN domain family of proteins and have ribonuclease activity. *VapC* toxin is deleterious to cells, but its effects become neutralized by *VapB* antitoxin. *VapC21* is similar in function to the other known *VapC* proteins of *M. tuberculosis* [53]. *Rv2801c* encodes for *MazF*-mt1 protein of *M. tuberculosis*. *MazF* is a known component of *MazEF* toxin–antitoxin system in many prokaryotic cells. *MazEF* is part of the TA system that forms persister cells of *M. tuberculosis*. *M. tuberculosis* has ten such *MazEF* proteins from numbers 1 to 10 and all *MazF* proteins are RNases. *MazF*-mt1 specifically cleaves mRNA. *MazF* family members play important roles in antibiotic and immune tolerance mechanisms [54].

In silico characterization was performed to determine the secondary structure, polarity, instability index and localization of all the selected 10 proteins. *Rv1477*, *Rv1566c*, *Rv2010*, and *Rv2623* were found to be unstable proteins, based on their instability indices, which were based on protein sequence information. *Rv2010* and *Rv2623* were categorized into polar proteins and, therefore, these proteins might be more vulnerable towards surrounding nature (Figure 4a) [56]. Rv0554 was also found to be a non-essential gene for mycobacterial growth, but it is listed as an important virulence factor that codes for a peroxidase. Most of the proteins participated in mycobacterial metabolism (Figure 4b) [57]. Rv1495 was found to be a lipid anchored protein, and it is a probable toxin (MazF4). Rv1636, Rv2010, Rv2549c, Rv2623, Rv2757c are cytoplasmic proteins and, therefore, components of the secretory system of mycobacteria (Figure 4c) [58]. The secondary structure analysis showed that most of the regions of the proteins were comprised of an alpha helical pattern, which confirmed the stable structural state of the proteins (Figure 4d) [59].

The phylogenetic analysis was performed by Mega11 program and showed close proximity of Rv1636 of *M. tuberculosis* H_{37}Rv and *Mycobacterium bovis* Rv1662, where both code for USP TB15.3. *M. tuberculosis* H_{37}Rv Rv1636, *M. marinum* 2440, *Mycobacterium bovis* 1662, *M. leprae* 1390, *M. smegmatis* mc^{2}155 3811 showed significant identity percentages among their USP domain (Figure 5) [60]. The protein interaction was determined by STRING server and properties like betweenness, closeness, radiality, stress and degree were used as the parameters for the interactive analysis. The analysis configured three proteins Rv1636, Rv2623 and Rv1566 in betweenness, closeness, radiality, and degree category, whereas Rv1636 was the only protein which was highlighted in the stress parameter (Figure 6) [61,62].

Molecular docking studies were executed to determine the highly interactive compound for their binding capacity with the proteins. The structure of the compound is mentioned in (Figure S3). *β*-amyrin (PubChem CID: 225689) was found to have higher binding free energies against *Rv1636, Rv1566, Rv2549c,* and *Rv1495* proteins (Figure 7). *β*-amyrin strongly integrated with most of the proteins and the interaction involved the pi-alkyl bonds, and hydrogen bonds (Table S7, Figure 8). The ADMET properties confirmed that shortlisted and highly interactive compounds can be a putative drug candidate, as they passed all the qualifying parameters (Tables 5 and 6) [70–72]. Most of the proteins showed high and significant binding affinity with *β*-amyrin, and, thus, it was selected for further analysis [77].

Rv1636 protein was found to be the top candidate in all the examinations (interaction, docking, biological process etc.), and, therefore, this protein was further analyzed for its stability with *β*-amyrin by molecular dynamic simulation. The RMSD plot showed that the protein and its complex were stable in the initial period till 45 ns, but started to experience a little destabilizing after 45 ns (Figure 9a). This destabilization might be due to the change in the protein structure, as in the instability index *Rv1636* was found to be an unstable protein. The RMSF plot showed instability in most of the residues, whereas the complex showed lesser fluctuation as compared to the protein alone (Figure 9b) [73–75]. The SASA result suggested that the binding of *β*-amyrin to the protein stabilizes the complex (Figure 9d) and this stability was further confirmed by Rg plot (Figure 9c) and FEL, which also confirmed the compactness and folding of the protein in complex form with *β*-amyrin (Figure 9e,f) [76].

4. Materials and Methods

4.1. Plant Collection and Identification

The medicinal plants *Achyranthes aspera, Calotropis gigantea* and *Calotropis procera* were collected from a burial ground in Shahjahanpur, Uttar Pradesh, India. The plants were identified by their flowers and inflorescence and the authenticity of these medicinal plants was established from previous literatures [33,34].

4.2. Plant Extraction

A. aspera aerial and root parts, as well as *C. gigantea* and *C. procera* flowers, were carefully cleaned with running tap water and then with sterile autoclaved water. The material was shade-dried, indelicately pulverised with a motor and pestle, and then extracted. Using a Soxhlet extractor, a weighed amount (500 g) of the substance was extracted using solvents of different polarity, including water, methanol, hexane, ethyl acetate, and ethanol. Nearly 48 extraction cycles were completed, under reduced pressure and at a controlled temperature, using a rotatory evaporator. The extracts were then concentrated, dried, packaged, and kept in a refrigerator at 4 °C for use [35,36].

4.3. Secondary Metabolite Identification

To identify various phytoconstituents, all extracts were subjected to a preliminary phytochemical examination using conventional techniques. Many antioxidants, such as alkaloids, terpenoids, saponins, and other compounds with varied pharmacological effects, were found in the plants [37,38].

4.3.1. Alkaloids Presence: Mayer's Reagent Test

An amount of 5000 µL of extract solution was warmed in a water bath with 2% HCl, and some droplets of Mayer's reagent was added. The sample was examined for the existence of turbidness or yellow precipitation.

4.3.2. Tannins Presence: Ferric Chloride Test

An amount of 500 µL of plant extract was added to 1000 µL of distilled water and some droplets of ferric chloride were mixed in. The presence of a green black colour showed the presence of tannins.

4.3.3. Saponins Presence: Frothing Test

An amount of 1000 µL plant extract was added to 4000 µL of distilled water and shaken vigorously. The appearance of foam showed the presence of saponins, which persisted for at least 15 min.

4.3.4. Terpenoids Presence: Salkowski Test

Intp 5000 µL extract solution and 2000 µL of chloroform, 3000 µL sulphuric acid was carefully added. The formation of a layer with a greyish colour indicated the presence of terpenoids.

4.4. Total Flavonoid Content (TFC)

The TFC of the medicinal plant extracts was determined by the aluminum chloride calorimetric method [39]. A 100 µL solution of 2% aluminum chloride in methanol was added to 100 µL of extract samples. The solution was incubated for 30 min at room temperature (RT) and the optical density was measured at 415 nm. Before adding the aluminum chloride solution, a pre-plate reading was taken. The standard curve was built using five different Rutin concentrations. Extract TFC was measured in mg Rutin equivalents per gram of extract [40].

4.5. Total Polyphenolic Content (TPC)

The TPC of the medicinal plant extracts was estimated by slightly changing the Folin-Ciocalteu method used by Siddhuraju et al. [39]. A pre-plate reading was taken and after that 20 µL of each plant extracts were added to 110 µL of ten times diluted newly made Folin-Ciocalteu reagent. After that, 70 µL of sodium carbonate solution was added and incubated for 30 min at RT and the optical density (absorbance) was determined at 765 nm. Gallic acid (GA) was used as a standard to plot standard curves with five different concentrations. The medicinal plant extract TPC was quantified in milligram (mg) of GA equivalents per gram (g) of extract [40].

4.6. Minimal Inhibitory Concentration (MIC) Assay

The mycobacterial strains used were *M. tuberculosis* $H_{37}Rv$ ATCC 27294, *M. fortuitum* ATCC 6841, *M. abscessus* ATCC 19977 and *M. chelonae* ATCC 35752. The mycobacterial strains were cultured in Middlebrook 7H9 enriched (Difco, Becton, NJ, USA) 10% Oleic acid, Dextrose, 0.2% glycerol, BSA and 0.05% Tween-80 supplemented medium.

The antibacterial susceptibility testing was performed using a broth microdilution technique. Stock solutions of plant extracts and control substances at 10 mg/mL in DMSO were prepared and kept at $-20\,°C$. Bacterial cultures were put into appropriate media and their absorbance was measured at OD_{600}, before diluting the culture to reach a concentration of 10^5 CFU/mL. The plant extracts were evaluated in a two-fold serial dilution method from 64 to 0.5 mg/L, with 2.5 µL of every individual concentration added in each well of Elisa plate. Each well contained bacterial culture around 97.5 µL with the test drug and associated controls. Resazurin-based dye (Thermo Fisher, Waltham, MA, USA) was applied to visually identify active phytoconstituents. The lowest concentration of active substance that prevented observable development after an incubation period was established as the MIC of the active plant extracts. The MIC experiment was done in triplicate and independently on duplicate samples for each drug. The MIC 96 well (Elisa) microtiter plate was incubated for non-tuberculous mycobacteria for 24–48 h and slow growers for 7 days [41,42].

4.7. GC-MS Analysis of Plant Extracts

The ethyl acetate fraction of *A. aspera* aerial part and the *C. gigantea* flower part were analyzed using Shimadzu GCMS-QP2010 Ultra, furnished with a Flame Thermionic Detector (FTD detector), to identify the chemical composition of the fractions. Helium gas was employed as the carrier with a 0.7 mL min^{-1} flow rate. The injection temperature was 260 °C, and the preliminary column temperature of 100 °C was kept for two minutes before ramping to 250 °C at a rate of 10 °C min^{-1} and hold on for 19 min before increasing to 290 °C at a rate of 10 °C min^{-1}. A solvent delay of 3.5 min was used. Mass spectra were documented in the range of 40–650 m/z and compounds were recognized by using NIST11 Library.

4.8. Determination of the Target Proteins: Using In Silico Approaches

To study the mycobacterial target, the proteins of virulence, detoxification, adaptation (VDA) functional category and proteins having PDB structures from the mycobacterial database were selected and an array of in silico analysis was performed. The mycobacterial genome database Mycobrowser was analyzed for the availability of genes responsible for mycobacterial virulence and categorized into virulence, detoxification, adaptation category [43]. *M. tuberculosis* $H_{37}Rv$ genome had a total 4173 proteins, out of which 238 proteins belonged to virulence, detoxification, adaptation category. On the other hand, the RSCB-PDB databank contained PDB structures of 135 *M. tuberculosis* $H_{37}Rv$ proteins excluding repeated or mutated ones, which were categorized in the mentioned categories (Figure S1). The co-integrative analysis of both these categories showed that there were 10 VDA functional category proteins, having PDB structures that were also present, and these 10 proteins were, therefore, present in both categories. We employed various bioinformatics tools which might empower experimental work to identify the prospective targets of this bacterium and illuminated the efficacy of significant regulators of mycobacterial pathways (Table S2) [44].

4.8.1. Retrieval of the Protein Sequence

The sequence of *M. tuberculosis* $H_{37}Rv$ proteins was obtained from the Mycobrowser (Mycobacterial browser) online database, consisting of different types of pathogenic and non-pathogenic mycobacterial strains in a repository for genomic and proteomic comprehensive analysis [55].

4.8.2. Physiochemical Parameters

The proteins identified were investigated for their physiochemical properties. The ProtParam server was used for calculating the theoretical parameters, such as molecular weight, amino acids, pI, instability index etc. [56].

4.8.3. Functional Classification

The primary identifying mechanism for understanding bacterial pathogenesis in prokaryotes is the distinction of virulent and non-virulent proteins. The VICMpred online prediction server was used for functional classification of bacteria using a bi-layer cascade SVM approach, which applies the sequence information for the prediction of different virulent factors. The VICMpred webserver used amino acid sequence in a pattern-based approach that showed extremely important values of functional classification (median values > 1.0) [57].

4.8.4. Subcellular Localization

Protein localization is an important aspect in identifying new drug targets. Since no information regarding the subcellular localization of these protein sequences was available, the TBpred webserver was used to predict the localization of selected proteins. Multiple prediction approaches for analyzing the localization of mycobacterial proteins were used to predict the presence of protein, whether in membrane, cytoplasm, lipid-anchored and secreted categories, based on the scores [58].

4.8.5. Secondary (2D) Structure Prediction

The SOPMA webserver was used to predict the 2D structure of target proteins. This online server is simple and accurate, and predicts different characteristics in secondary structure, such as alpha-helix, beta turns, extended strands or random coil region [59].

4.8.6. Phylogenetic Analysis

Distant and close relatives of virulence proteins were searched in the Mycobrowser database. The boundaries of these genes were specified through the Pfam database and multiple sequence alignments were performed. Phylogenetic analysis was performed by the Mega11 server neighbor joining method [60].

4.8.7. Virulent Genes Regulating Network Analysis

A protein-protein interaction network of virulent proteins was established using STRING v11.5 database, accessed on 7 January 2022 (https://string-db.org/) and visualized by using STRING application, available in Cytoscape. A cutoff score of 0.6–0.9 was selected that showed interaction with high confidence. The studied genes were further analyzed by enrichment analysis by setting a significant statistical threshold less than 0.05. The resulting proteins were then classified by four intrinsic factors, such as "betweenness, closeness, radiality and degree" and betweenness, closeness, stress, and degree. The top virulent proteins were ranked using these factors by the CytoHubba application in the CytoScape [61,62].

4.8.8. Retrieval of the 3D Protein's Structure

The PDB structure of all the available proteins were retrieved from the RCSB Protein Databank (PDB ID: 7LD8, 4Q4N, 5XE2, 4LJ1, 1TQ8, 4CHG, 5WZ4, 2JAX, 5SV2, 6L2A) [45–54]. Out of these ten proteins, only *Rv1477* was essential for in vitro growth of *M. tuberculosis* $H_{37}Rv$ (PDB ID: 4Q4N). The 3D structure of these proteins was visualized by using PyMOL (Figure S2).

4.8.9. Validation of the Selected Protein's Structure

The model validation of a protein structure was performed by SAVES6.0 webserver, that estimated various characteristics, especially the stereo-chemical feature of a protein structure by residue geometry. SAVES6.0 server had PROCHECK, and analyzed the Ramachandran plot. It endorsed the protein structure on the premise of φ, ψ values of an individual deposit. The inclusive structure geometry established the validation score of the Ramachandran plot for a protein structure depending on number of amino acids present in favored, allowed, and disallowed regions [63,64].

4.8.10. Molecular Docking

Molecular docking studies were carried out in order to identify the top hit phytoconstituents present in the aerial part extract of *A. aspera* (ethyl acetate fraction) and flower part of *C. gigantea* (ethyl acetate fraction) [65]. The three-dimensional (3D) crystal structure of target proteins (PDB: 7LD8, 4Q4N, 5XE2, 4LJ1, 1TQ8, 4CHG, 5WZ4, 2JAX, 5SV2, 6L2A) was retrieved from RCSB PDB and refined before performing molecular docking. Binding sites/pockets were determined by using CASTp 3.0 server, which examined the geometric and topological properties of the protein structures, including surface pockets, interior cavities and cross channels, as they are fundamentally important for the proteins to carry out their functions. The GC-MS analyzed phytoconstituents (as ligand) were downloaded from PubChem database. The compound structures unavailable in PubChem database were drawn by ChemBioDraw Ultra 14.0 [66]. For the receptor preparation, the water molecules and co-crystallized ligands were removed from the PDB file and polar hydrogens were added. The receptor protein (target) was transformed from pdb format into pdbqt format using in-house protocol [67]. In molecular docking was performed by InstaDock. Results were evaluated from the log files via Python script [68]. The blind docking mechanism was used to explore the binding site(s) in the protein structures. PyMOL was used to visualize protein-ligand interactions. The receptor-ligand complex was prepared by Discovery Studio and 2D interaction of docked conformations was analyzed to understand the ligand binding amino acid residues [69]. The best-fitting conformation related to the binding affinity of the ligand-receptor complex was identified, while keeping the receptor as a rigid entity and ligand as flexible. The top 10 hits showing strong binding affinity to the binding sites were selected.

4.8.11. Determination of the Selected Phytoconstituent Drug-Ability

The physiochemical properties of the chosen compounds were calculated through online SwissADME software. Further, ADME properties and toxicity of the selected compounds were also calculated using the freely available online sever pkCSM. An online server, PASS, was also used for predicting the biological activity of these natural compounds [70–72].

4.9. MD Simulation

To conduct MD simulations, GROMACS 5.1.2 Bio-Simulation package was used. To clarify the molecular dynamic characteristics and different computations of proteins and ligands employed in this in silico study, the force field GROMOS96 43a2 was used [73]. The receptor-ligand docked complexes files were retrieved using the gmx grep module. To create ligand topology and force-field conditions, PRODRG server was used [74]. To solvate the protein, the water model SPC216 was employed. A 50 ns MD simulation in water at 298 K was used as a control. All protein and ligand atoms were equilibrized in a three-dimensional box with a range of almost 10.5 Å from all side. The protein was thoroughly equilibrated in water, and redundant molecules were removed. To eliminate all poor contacts, energy was minimized for each system with the steepest decline up to a forbearance of 1000 kJ mol^{-1}nm^{-1} and the overall charge was neutralized in the system by adding ionic concentrations of NaCl. To perform the simulation, the sizes (x, y, and z) of the simulation frame were established depending on the size and 3D positioning of the protein. All the systems were produced in a specified box, with the protein in the center and water and co-solvents padded around it. The energy minimization method was carried out using the steepest-descent algorithm and conjugate gradient. Two troupe methods, NVT and NPT, were used to equilibrize the system. Before beginning the MD run, environments, such as pH and temperature, were pre-defined. All this evidence was contained in the NVT, NPT, and MD criteria files. The binary trajectory file was generated after the production run for additional examination [75].

5. Conclusions

The effect of the phytoconstituents of *C. procera*, *C. gigantea* and *A. aspera* plant extracts on the *M. tuberculosis* $H_{37}Rv$ cell proteins was investigated in this study. The phytochemical analysis of all plant extracts showed the presence of a significant content of phenols and flavonoids, especially in the ethyl acetate fraction of *A. aspera* and the ash of *C. gigantea* fractions. The plants extracts were tested against different mycobacterial strains. *A. aspera* aerial and *C. gigantea* flower ash was found to be active against the *M. tuberculosis* $H_{37}Rv$ ATCC 27294 strains with an MIC value of 64 mg/L. A multitarget assessment study was used to identify the possible mycobacterial target proteins. Ten proteins, viz. *BpoC*, *RipA*, *MazF4*, *RipD*, *TB15.3*, *VapC15*, *VapC20*, *VapC21*, *TB31.7*, *MazF9*, were found in the intersection of two categories, viz. available PDB dataset proteins and proteins classified in virulence, detoxification, adaptation. In silico characterization identified *TB15.3* (Rv1636) in the intersection of PPI network, which are the universal stress proteins. The phylogenetic analysis showed Rv1636 is a conserved protein among different mycobacterial strains. The molecular docking study of β-amyrin revealed its highest binding affinity with *Rv1636*. Furthermore, MD simulation was used to determine the stability and accuracy of the complex and it showed that the complex of β-amyrin and *Rv1636* was a stable complex, and the protein did not undergo unfolding during the simulation run. On a final note, this study established a significant bridge in the field of mycobacterial biology, which focused on targeting Rv1636, a universal stress protein of mycobacteria, through natural phytoconstituents.

Supplementary Materials: The following supporting information can be downloaded at: https://www.mdpi.com/article/10.3390/molecules27144581/s1, Table S1: Minimum inhibitory concentration (MIC) of different plants extracts in various solvent fractions against *M. tuberculosis* strains; Figure S1: Venn-diagram to categorize V.D.A category and known *Mtb* PDB structure protein; Figure S2: Crystal structure of selected *M. tuberculosis* $H_{37}Rv$ proteins (PDB: 7LD8, 4Q4N, 5XE2, 4LJ1, 1TQ8, 4CHG, 5WZ4, 2JAX, 5SV2, 6L2A); Figure S3: In multitarget protein docking, the top hit selected phytoconstituents; Table S2: List of the bioinformatics tools and databases for learning substantial outcome of Hypothetical protein from *M. tuberculosis* $H_{37}Rv$; Table S3: Phytochemical constituents identified in the ethyl acetate aerial part extract of *A. aspera* using gas chromatography–mass spectrometry; Table S4: Phytochemical constituents identified in the ethyl acetate flower ash extract of *C. gigantea* using gas chromatography–mass spectrometry; Table S5: Physiochemical parameters of selected virulence, detoxification, adaptation category proteins; Table S6: Secondary structure analysis of selected virulent proteins of *Mtb*; Table S7: Selected hits and their binding free energies (kcal/mol) toward multiple target proteins; and Table S8: Selected compounds and their biological properties.

Author Contributions: M.A.B.: Conceptualization, Methodology, Software, Formal analysis, Data curation, Writing—original draft; S.: Methodology, Data curation, Writing—review & editing; O.A.: Methodology, Formal analysis, Data curation, Writing—review & editing. M.S.A.: Methodology, Software, Formal analysis, Project administration; A.S.A.A.: Software, Formal analysis, Data curation; A.H.: Formal analysis, Data curation, Writing—review & editing; M.A.I.: Methodology, Software, Formal analysis, Data curation; M.N.A.: Methodology, Formal analysis, Data curation, Writing—review & editing; S.C.: Methodology, Formal analysis, Data curation, Writing—review & editing; F.A.: Conceptualization, Supervision. All authors have read and agreed to the published version of the manuscript.

Funding: This work is supported by the Deanship of Scientific Research, King Khalid University, Saudi Arabia [Grant number RGP.1/242/43].

Institutional Review Board Statement: Not applicable.

Informed Consent Statement: Not applicable.

Data Availability Statement: The data that supports the findings of this study are contained within the article and supporting information.

Acknowledgments: The authors extend their appreciation to the Deanship of Scientific Research at King Khalid University for funding this work through Small Groups. (Project under grant number RGP.1/242/43).

Conflicts of Interest: The authors declare no conflict of interest.

Sample Availability: Not applicable.

References

1. Paik, S.; Kim, J.K.; Chung, C.; Jo, E.-K. Autophagy: A new strategy for host-directed therapy of tuberculosis. *Virulence* **2019**, *10*, 448–459. [CrossRef] [PubMed]
2. Sieniawska, E.; Sawicki, R.; Swatko-Ossor, M.; Napiorkowska, A.; Przekora, A.; Ginalska, G.; Augustynowicz-Kopec, E. The effect of combining natural terpenes and antituberculous agents against reference and clinical *Mycobacterium tuberculosis* strains. *Molecules* **2018**, *23*, 176. [CrossRef] [PubMed]
3. Brown, C.A.; Draper, P.; Hart, P.A. Mycobacteria and lysosomes: A paradox. *Nature* **1969**, *221*, 658–660. [CrossRef] [PubMed]
4. WHO. WHO Report of Tuberculosis 2021. Available online: https://www.who.int/publications/i/item/9789240037021 (accessed on 10 March 2022).
5. WHO. WHO Report of Tuberculosis 2020. Available online: https://www.who.int/publications/i/item/9789240013131 (accessed on 10 March 2022).
6. Dou, H.-Y.; Tseng, F.-C.; Lin, C.-W.; Chang, J.-R.; Sun, J.-R.; Tsai, W.-S.; Lee, S.-Y.; Su, I.-J.; Lu, J.-J. Molecular epidemiology and evolutionary genetics of *Mycobacterium tuberculosis* in Taipei. *BMC Infect. Dis.* **2008**, *8*, 170. [CrossRef] [PubMed]
7. Brosch, R.; Gordon, S.V.; Marmiesse, M.; Brodin, P.; Buchrieser, C.; Eiglmeier, K.; Garnier, T.; Gutierrez, C.; Hewinson, G.; Kremer, K.; et al. A new evolutionary scenario for the *Mycobacterium tuberculosis* complex. *Proc. Natl. Acad. Sci. USA* **2002**, *99*, 3684–3689. [CrossRef]
8. Gutierrez, M.C.; Brisse, S.; Brosch, R.; Fabre, M.; Omaïs, B.; Marmiesse, M.; Supply, P.; Vincent, V. Ancient origin and gene mosaicism of the progenitor of *Mycobacterium tuberculosis*. *PLoS Pathog.* **2005**, *1*, e5. [CrossRef]
9. Ge, F.; Zeng, F.; Liu, S.; Guo, N.; Ye, H.; Song, Y.; Fan, J.; Wu, X.; Wang, X.; Yu, L.; et al. In vitro synergistic interactions of oleanolic acid in combination with isoniazid, rifampicin or ethambutol against *Mycobacterium tuberculosis*. *J. Med. Microbiol.* **2010**, *59*, 567–572. [CrossRef]
10. Knezevic, P.; Aleksic, V.; Simin, N.; Svircev, E.; Petrovic, A.; Mimica-Dukic, N. Antimicrobial activity of *Eucalyptus camaldulensis* essential oils and their interactions with conventional antimicrobial agents against multi-drug resistant *Acinetobacter baumannii*. *J. Ethnopharmacol.* **2016**, *178*, 125–136. [CrossRef]
11. Hasan, S. Pharmacological and medicinal uses of *Achyranthes aspera*. *Int. J. Sci. Environ. Technol.* **2014**, *3*, 123–129.
12. Sharma, V.; Chaudhary, U. An overview on indigenous knowledge of *Achyranthes aspera*. *J. Crit. Rev.* **2015**, *2*, 7–19.
13. Maiden, J.H. *The Useful Native Plants of Australia: (Including Tasmania)*; Turner and Henderson: Sydney, Australia, 1889.
14. Sharma, J.; Gairola, S.; Gaur, R.D.; Painuli, R.M.; Siddiqi, T.O. Ethnomedicinal plants used for treating epilepsy by indigenous communities of sub-Himalayan region of Uttarakhand, India. *J. Ethnopharmacol.* **2013**, *150*, 353–370. [CrossRef] [PubMed]
15. Gawande, D.Y.; Druzhilovsky, D.; Gupta, R.C.; Poroikov, V.; Goel, R.K. Anticonvulsant activity and acute neurotoxic profile of *Achyranthes aspera* Linn. *J. Ethnopharmacol.* **2017**, *202*, 97–102. [CrossRef] [PubMed]
16. Barua, C.C.; Talukdar, A.; Begum, S.A.; Borah, P.; Lahkar, M. Anxiolytic activity of methanol leaf extract of *Achyranthes aspera* Linn in mice using experimental models of anxiety. *Indian J. Pharmacol.* **2012**, *44*, 63. [CrossRef] [PubMed]
17. Barua, C.C.; Talukdar, A.; Begum, S.A.; Lahon, L.C.; Sarma, D.K.; Pathak, D.C.; Borah, P. Antinociceptive activity of methanolic extract of leaves of *Achyranthes aspera* Linn. (Amaranthaceae) in animal models of nociception. *Indian J. Exp. Biol.* **2010**, *48*, 817–821.
18. Bhosale, U.; Yegnanarayan, R.; Prachi, P.; Zambare, M.; Somani, R.S. Study of CNS depressant and behavioral activity of an ethanol extract of *Achyranthes aspera* (Chirchita) in mouse model. *Ann. Neurosci.* **2011**, *18*, 44. [CrossRef]
19. Khan, N.; Akhtar, M.S.; Khan, B.A.; de Andrade Braga, V.; Reich, A. Antiobesity, hypolipidemic, antioxidant and hepatoprotective effects of *Achyranthes aspera* seed saponins in high cholesterol fed albino rats. *Arch. Med. Sci. AMS* **2015**, *11*, 1261–1271. [CrossRef]
20. Mainasara, M.M.; Aliero, B.L.; Aliero, A.A.; Yakubu, M. Phytochemical and antibacterial properties of root and leaf extracts of *Calotropis procera*. *Niger. J. Basic Appl. Sci.* **2012**, *20*, 1–6.
21. Rathore, M.; Meena, R.K. Potential of utilizing *Calotropis procera* flower biomass as a renewable source of energy. *J. Phytol.* **2010**, *2*, 78–83.
22. Dwivedi, B.; Singh, A.; Mishra, S.; Singh, R.; Pant, P.; Thakur, L.K.; Padhi, M.M. Evaluation of phytochemical constituents by gas chromatography-mass spectroscopy & HPTLC of *Calotropis procera*. *World J. Pharm. Res.* **2014**, *3*, 708–715.
23. Kumar, V.L.; Shivkar, Y.M. In vivo and in vitro effect of latex of *Calotropis procera* on gastrointestinal smooth muscles. *J. Ethnopharmacol.* **2004**, *93*, 377–379. [CrossRef]
24. Jalalpure, S.S.; Salahuddin, M.; Imtiyaz Shaikh, M.; Manvi, F.V. Anticonvulsant effects of *Calotropis procera* root in rats. *Pharm. Biol.* **2009**, *47*, 162–167. [CrossRef]

25. Mukherjee, P.K.; Sahoo, A.K.; Narayanan, N.; Kumar, N.S.; Ponnusankar, S. Lead finding from medicinal plants with hepatoprotective potentials. *Expert Opin. Drug Discov.* **2009**, *4*, 545–576. [CrossRef] [PubMed]

26. Patra, A.; Jha, S.; Murthy, P.N.; Vaibhav, A.D.; Chattopadhyay, P.; Panigrahi, G.; Roy, D. Anti-inflammatory and antipyretic activities of *Hygrophila spinosa* T. Anders leaves (Acanthaceae). *Trop. J. Pharm. Res.* **2009**, *8*, 133–137. [CrossRef]

27. Mahmoud, M.E.; Shiina, T.; Hirayama, H.; Iwami, M.; Miyazawa, S.; Nikami, H.; Takewaki, T.; Shimizu, Y. Extract from *Calotropis procera* latex activates murine macrophages. *J. Nat. Med.* **2009**, *63*, 297–303.

28. Barrett, B.; Kieffer, D. Medicinal plants, science, and health care. *J. Herbs Spices Med. Plants* **2001**, *8*, 1–36. [CrossRef]

29. Oluwaniyi, O.O.; Ibiyemi, S.A. Extractability of *Thevetia peruviana* glycosides with alcohol mixture. *Afr. J. Biotechnol.* **2007**, *6*, 2166–2170.

30. Larhsini, M.; Oumoulid, L.; Lazrek, H.B.; Wataleb, S.; Bousaid, M.; Bekkouche, K.; Jana, M. Antibacterial activity of some Moroccan medicinal plants. *Phytother. Res.* **2001**, *15*, 250–252. [CrossRef]

31. Ibrar, M. Ethobotanic study of the weeds of five crops in district Abbottabad, N-WPakistan. *Pak. J. Weed Sci. Res.* **2003**, *9*, 229–240.

32. Ramos, M.V.; Freitas, C.D.; Stanisçuaski, F.; Macedo, L.L.; Sales, M.P.; Sousa, D.P.; Carlini, C.R. Performance of distinct crop pests reared on diets enriched with latex proteins from *Calotropis procera*: Role of laticifer proteins in plant defense. *Plant Sci.* **2007**, *173*, 349–357. [CrossRef]

33. Priyamvada, P.M.; Sha, A.; Mohapatra, A.K. Evaluation of antidiabetic and antioxidant activities of *Achyranthes aspera* leaf extracts: An in vitro study. *J. Pharmacogn. Phytochem.* **2021**, *10*, 103–110.

34. Singh, N.; Jain, N.K.; Kannojia, P.; Garud, N.; Pathak, A.K.; Mehta, S.C. In vitro antioxidant activity of *Calotropis gigantea* hydroalcohlic leaves extract. *Der Pharm. Lett.* **2010**, *2*, 95–100.

35. Pattnaik, P.K.; Kar, D.; Chhatoi, H.; Shahbazi, S.; Ghosh, G.; Kuanar, A. Chemometric profile & antimicrobial activities of leaf extract of *Calotropis procera* and *Calotropis gigantea*. *Nat. Prod. Res.* **2017**, *31*, 1954–1957. [PubMed]

36. Moustafa, A.M.Y.; Ahmed, S.H.; Nabil, Z.I.; Hussein, A.A.; Omran, M.A. Extraction and phytochemical investigation of *Calotropis procera*: Effect of plant extracts on the activity of diverse muscles. *Pharm. Biol.* **2010**, *48*, 1080–1190. [CrossRef] [PubMed]

37. Abbas, M.N.; Rana, S.A.; Mahmood-Ul-Hassan, M.; Rana, N.; Iqbal, M. Phytochemical constituents of weeds: Baseline study in mixed crop zone agroecosystem. *Pak. J. Weed Sci. Res.* **2013**, *19*, 231–238.

38. Dhale, D.A.; Bhoi, S. Pharmacognostic Characterization and Phytochemical Screening of *Achyranthes aspera* Linn. *Curr. Agric. Res. J.* **2013**, *1*, 51. [CrossRef]

39. Siddhuraju, P.; Becker, K. Antioxidant properties of various solvent extracts of total phenolic constituents from three different agroclimatic origins of drumstick tree (*Moringa oleifera* Lam.) leaves. *J. Agric. Food Chem.* **2003**, *51*, 2144–2155. [CrossRef] [PubMed]

40. Dissanayake, D.P.A.; Sivaganesh, S.; Tissera, M.H.A.; Handunnetti, S.M.; Arawwawala, L.D.A.M. Comparison of antioxidant properties of *Cyathula prostrata* Linn and *Achyranthes aspera* Linn grown in Sri Lanka. *Res. Rev. Insights* **2018**, *2*, 1–3. [CrossRef]

41. Chopra, S.; Matsuyama, K.; Hutson, C.; Madrid, P. Identification of antimicrobial activity among FDA-approved drugs for combating *Mycobacterium abscessus* and *Mycobacterium chelonae*. *J. Antimicrob. Chemother.* **2011**, *66*, 1533–1536. [CrossRef]

42. Sahoo, S.K.; Rani, B.; Gaikwad, N.B.; Ahmad, M.N.; Kaul, G.; Shukla, M.; Nanduri, S.; Dasgupta, A.; Chopra, S.; Yaddanapudi, V.M. Synthesis and structure-activity relationship of new chalcone linked 5-phenyl-3-isoxazolecarboxylic acid methyl esters potentially active against drug resistant Mycobacterium tuberculosis. *Eur. J. Med. Chem.* **2021**, *222*, 113580. [CrossRef]

43. Mészáros, B.; Tóth, J.; Vértessy, B.G.; Dosztányi, Z.; Simon, I. Proteins with complex architecture as potential targets for drug design: A case study of *Mycobacterium tuberculosis*. *PLoS Comput. Biol.* **2011**, *7*, e1002118. [CrossRef]

44. Beg, M.A.; Hejazi, I.I.; Thakur, S.C.; Athar, F. Domain-wise differentiation of Mycobacterium tuberculosis H_{37} Rv hypothetical proteins: A roadmap to discover bacterial survival potentials. *Biotechnol. Appl. Biochem.* **2022**, *69*, 296–312. [CrossRef] [PubMed]

45. Johnston, J.M.; Jiang, M.; Guo, Z.; Baker, E.N. Structural and functional analysis of Rv0554 from Mycobacterium tuberculosis: Testing a putative role in menaquinone biosynthesis. *Acta Crystallogr. Sect. D Biol. Crystallogr.* **2010**, *66 Pt 8*, 909–917. [CrossRef] [PubMed]

46. Ruggiero, A.; Marasco, D.; Squeglia, F.; Soldini, S.; Pedone, E.; Pedone, C.; Berisio, R. Structure and functional regulation of RipA, a mycobacterial enzyme essential for daughter cell separation. *Structure* **2010**, *18*, 1184–1190. [CrossRef] [PubMed]

47. Ahn, D.-H.; Lee, K.-Y.; Lee, S.J.; Park, S.J.; Yoon, H.-J.; Kim, S.-J.; Lee, B.-J. Structural analyses of the MazEF4 toxin-antitoxin pair in *Mycobacterium tuberculosis* provide evidence for a unique extracellular death factor. *J. Biol. Chem.* **2017**, *292*, 18832–18847. [CrossRef] [PubMed]

48. Böth, D.; Steiner, E.M.; Izumi, A.; Schneider, G.; Schnell, R. RipD (Rv1566c) from Mycobacterium tuberculosis: Adaptation of an NlpC/p60 domain to a non-catalytic peptidoglycan-binding function. *Biochem. J.* **2014**, *457*, 33–41. [CrossRef] [PubMed]

49. Banerjee, A.; Adolph, R.S.; Gopalakrishnapai, J.; Kleinboelting, S.; Emmerich, C.; Steegborn, C.; Visweswariah, S.S. A universal stress protein (USP) in mycobacteria binds cAMP. *J. Biol. Chem.* **2015**, *290*, 12731–12743. [CrossRef]

50. Das, U.; Pogenberg, V.; Subhramanyam, U.K.; Wilmanns, M.; Gourinath, S.; Srinivasan, A. Crystal structure of the VapBC-15 complex from Mycobacterium tuberculosis reveals a two-metal ion dependent PIN-domain ribonuclease and a variable mode of toxin-antitoxin assembly. *J. Struct. Biol.* **2014**, *188*, 249–258. [CrossRef]

51. Deep, A.; Kaundal, S.; Agarwal, S.; Singh, R.; Thakur, K.G. Crystal structure of Mycobacterium tuberculosis VapC20 toxin and its interactions with cognate antitoxin, VapB20, suggest a model for toxin-antitoxin assembly. *FEBS J.* **2017**, *284*, 4066–4082. [CrossRef]

52. Glass, L.N.; Swapna, G.; Chavadi, S.S.; Tufariello, J.M.; Mi, K.; Drumm, J.E.; Lam, T.T.; Zhu, G.; Zhan, C.; Vilchéze, C.; et al. Mycobacterium tuberculosis universal stress protein Rv2623 interacts with the putative ATP binding cassette (ABC) transporter Rv1747 to regulate mycobacterial growth. *PLoS Pathog.* **2017**, *13*, e1006515. [CrossRef]

53. Jardim, P.; Santos, I.C.; Barbosa, J.A.; de Freitas, S.M.; Valadares, N.F. Crystal structure of VapC21 from *Mycobacterium tuberculosis* at 1.31 Å resolution. *Biochem. Biophys. Res. Commun.* **2016**, *478*, 1370–1375. [CrossRef]

54. Chen, R.; Zhou, J.; Sun, R.; Du, C.; Xie, W. Conserved Conformational Changes in the Regulation of *Mycobacterium tuberculosis* MazEF-mt1. *ACS Infect. Dis.* **2020**, *6*, 1783–1795. [CrossRef] [PubMed]

55. Kapopoulou, A.; Lew, J.M.; Cole, S.T. The MycoBrowser portal: A comprehensive and manually annotated resource for mycobacterial genomes. *Tuberculosis* **2011**, *91*, 8–13. [CrossRef] [PubMed]

56. Garg, V.K.; Avashthi, H.; Tiwari, A.; Jain, P.A.; Ramkete, P.W.; Kayastha, A.M.; Singh, V.K. MFPPI—Multi FASTA ProtParam Interface. *Bioinformation* **2016**, *12*, 74–77. [CrossRef] [PubMed]

57. Saha, S.; Raghava, G.P. VICMpred: An SVM-based method for the prediction of functional proteins of Gram-negative bacteria using amino acid patterns and composition. *Genom. Proteom. Bioinform.* **2006**, *4*, 42–47. [CrossRef]

58. Rashid, M.; Saha, S.; Raghava, G.P. Support Vector Machine-based method for predicting subcellular localization of mycobacterial proteins using evolutionary information and motifs. *BMC Bioinform.* **2007**, *8*, 337. [CrossRef] [PubMed]

59. Geourjon, C.; Deléage, G. SOPMA: Significant improvements in protein secondary structure prediction by consensus prediction from multiple alignments. *Comput. Appl. Biosci. CABIOS* **1995**, *11*, 681–684. [CrossRef] [PubMed]

60. Tamura, K.; Stecher, G.; Kumar, S. MEGA11: Molecular Evolutionary Genetics Analysis Version 11. *Mol. Biol. Evol.* **2021**, *38*, 3022–3027. [CrossRef] [PubMed]

61. Szklarczyk, D.; Gable, A.L.; Nastou, K.C.; Lyon, D.; Kirsch, R.; Pyysalo, S.; Doncheva, N.T.; Legeay, M.; Fang, T.; Bork, P.; et al. The STRING database in 2021: Customizable protein-protein networks, and functional characterization of user-uploaded gene/measurement sets. *Nucleic Acids Res.* **2021**, *49*, D605–D612. [CrossRef]

62. Sharma, K.; Singh, P.; Beg, M.A.; Dohare, R.; Athar, F.; Syed, M.A. Revealing new therapeutic opportunities in hypertension through network-driven integrative genetic analysis and drug target prediction approach. *Gene* **2021**, *801*, 145856. [CrossRef]

63. Laskowski, R.A.; Rullmannn, J.A.; MacArthur, M.W.; Kaptein, R.; Thornton, J.M. AQUA and PROCHECK-NMR: Programs for checking the quality of protein structures solved by NMR. *J. Biomol. NMR* **1996**, *8*, 477–486. [CrossRef]

64. Wlodawer, A. Stereochemistry and Validation of Macromolecular Structures. *Methods Mol. Biol.* **2017**, *1607*, 595–610. [PubMed]

65. Meng, X.-Y.; Zhang, H.-X.; Mezei, M.; Cui, M. Molecular docking: A powerful approach for structure-based drug discovery. *Curr. Comput.-Aided Drug Des.* **2011**, *7*, 146–157. [CrossRef] [PubMed]

66. Li, Z.; Wan, H.; Shi, Y.; Ouyang, P. Personal experience with four kinds of chemical structure drawing software: Review on ChemDraw, ChemWindow, ISIS/Draw, and ChemSketch. *J. Chem. Inf. Comput. Sci.* **2004**, *44*, 1886–1890. [CrossRef]

67. Forli, S.; Huey, R.; Pique, M.E.; Sanner, M.F.; Goodsell, D.S.; Olson, A.J. Computational protein-ligand docking and virtual drug screening with the AutoDock suite. *Nat. Protoc.* **2016**, *11*, 905–919. [CrossRef] [PubMed]

68. Mohammad, T.; Mathur, Y.; Hassan, M.I. InstaDock: A single-click graphical user interface for molecular docking-based virtual high-throughput screening. *Brief. Bioinform.* **2021**, *22*, bbaa279. [CrossRef]

69. Athar, F.; Ansari, S.; Beg, M.A. Molecular docking studies of *Calotropis gigantea* phytoconstituents against *Staphylococcus aureus* tyrosyl-tRNA synthetase protein. *J. Bacteriol. Mycol. Open Access.* **2020**, *8*, 78–91. [CrossRef]

70. Daina, A.; Michielin, O.; Zoete, V. SwissADME: A free web tool to evaluate pharmacokinetics, drug-likeness and medicinal chemistry friendliness of small molecules. *Sci. Rep.* **2017**, *7*, 42717. [CrossRef]

71. Pires, D.E.; Blundell, T.L.; Ascher, D.B. pkCSM: Predicting Small-Molecule Pharmacokinetic and Toxicity Properties Using Graph-Based Signatures. *J. Med. Chem.* **2015**, *58*, 4066–4072. [CrossRef]

72. Baswar, D.; Sharma, A.; Mishra, A. In silico Screening of Pyridoxine Carbamates for Anti-Alzheimer's Activities. *Cent. Nerv. Syst. Agents Med. Chem.* **2021**, *21*, 39–52. [CrossRef]

73. Hejazi, I.I.; Beg, M.A.; Imam, M.A.; Athar, F.; Islam, A. Glossary of phytoconstituents: Can these be repurposed against SARS CoV-2? A quick in silico screening of various phytoconstituents from plant Glycyrrhiza glabra with SARS CoV-2 main protease. *Food Chem. Toxicol.* **2021**, *150*, 112057. [CrossRef]

74. Schüttelkopf, A.W.; van Aalten, D.M. PRODRG: A tool for high-throughput crystallography of protein-ligand complexes. *Acta Crystallogr. Sect. D Biol. Crystallogr.* **2004**, *60 Pt 8*, 1355–1363. [CrossRef] [PubMed]

75. Khan, A.; Mohammad, T.; Shamsi, A.; Hussain, A.; Alajmi, M.F.; Husain, S.A.; Iqbal, M.A.; Hassan, M.I. Identification of plant-based hexokinase 2 inhibitors: Combined molecular docking and dynamics simulation studies. *J. Biomol. Struct. Dyn.* **2021**, 1–13. [CrossRef] [PubMed]

76. Marsh, J.A.; Teichmann, S.A. Relative solvent accessible surface area predicts protein conformational changes upon binding. *Structure* **2011**, *19*, 859–867. [CrossRef] [PubMed]

77. Ramadwa, T.E.; Awouafack, M.D.; Sonopo, M.S.; Eloff, J.N. Antibacterial and Antimycobacterial Activity of Crude Extracts, Fractions, and Isolated Compounds from Leaves of Sneezewood, *Ptaeroxylon obliquum* (Rutaceae). *Nat. Prod. Commun.* **2019**, *14*, 1–7. [CrossRef]

molecules

Article

Investigation Driven by Network Pharmacology on Potential Components and Mechanism of DGS, a Natural Vasoprotective Combination, for the Phytotherapy of Coronary Artery Disease

You-Gang Zhang [1,†], Xia-Xia Liu [1,2,†], Jian-Cheng Zong [3,†], Yang-Teng-Jiao Zhang [1], Rong Dong [1], Na Wang [1,2], Zhi-Hui Ma [1,4], Li Li [3], Shang-Long Wang [3], Yan-Ling Mu [1], Song-Song Wang [1], Zi-Min Liu [5,*] and Li-Wen Han [1,*]

[1] School of Pharmacy and Pharmaceutical Science, Shandong First Medical University and Shandong Academy of Medical Sciences, Jinan 250000, China; zmc19941114@gmail.com (Y.-G.Z.); liuxiaxia0415@126.com (X.-X.L.); zytj1136091171@163.com (Y.-T.-J.Z.); doloda56@163.com (R.D.); 17862954657@163.com (N.W.); 17854111663@163.com (Z.-H.M.); myling501@hotmail.com (Y.-L.M.); wangsongsong@sdfmu.edu.cn (S.-S.W.)
[2] School of Pharmaceutical Science, Shanxi Medical University, Taiyuan 030000, China
[3] Chenland Research Institute, Irvine, CA 92697, USA; jczong@chenland.com (J.-C.Z.); lli@chenland.com (L.L.); mwang@chenland.com (S.-L.W.)
[4] School of Pharmacy, Shandong University of Traditional Chinese Medicine, Jinan 250000, China
[5] Chenland Nutritionals Inc., Irvine, CA 92697, USA
*   Correspondence: dliu@chenland.com (Z.-M.L.); hanliwen@sdfmu.edu.cn (L.-W.H.)
†   These authors contributed equally to this work.

Citation: Zhang, Y.-G.; Liu, X.-X.; Zong, J.-C.; Zhang, Y.-T.-J.; Dong, R.; Wang, N.; Ma, Z.-H.; Li, L.; Wang, S.-L.; Mu, Y.-L.; et al. Investigation Driven by Network Pharmacology on Potential Components and Mechanism of DGS, a Natural Vasoprotective Combination, for the Phytotherapy of Coronary Artery Disease. *Molecules* 2022, 27, 4075. https://doi.org/10.3390/molecules27134075

Academic Editors: Tanveer A. Wani, Seema Zargar and Afzal Hussain

Received: 11 May 2022
Accepted: 21 June 2022
Published: 24 June 2022

Publisher's Note: MDPI stays neutral with regard to jurisdictional claims in published maps and institutional affiliations.

Abstract: Phytotherapy offers obvious advantages in the intervention of Coronary Artery Disease (CAD), but it is difficult to clarify the working mechanisms of the medicinal materials it uses. DGS is a natural vasoprotective combination that was screened out in our previous research, yet its potential components and mechanisms are unknown. Therefore, in this study, HPLC-MS and network pharmacology were employed to identify the active components and key signaling pathways of DGS. Transgenic zebrafish and HUVECs cell assays were used to evaluate the effectiveness of DGS. A total of 37 potentially active compounds were identified that interacted with 112 potential targets of CAD. Furthermore, PI3K-Akt, MAPK, relaxin, VEGF, and other signal pathways were determined to be the most promising DGS-mediated pathways. NO kit, ELISA, and Western blot results showed that DGS significantly promoted NO and VEGFA secretion via the upregulation of VEGFR2 expression and the phosphorylation of Akt, Erk1/2, and eNOS to cause angiogenesis and vasodilation. The result of dynamics molecular docking indicated that Salvianolic acid C may be a key active component of DGS in the treatment of CAD. In conclusion, this study has shed light on the network molecular mechanism of DGS for the intervention of CAD using a network pharmacology-driven strategy for the first time to aid in the intervention of CAD.

Keywords: coronary artery disease; phytotherapy; network pharmacology; angiogenesis; zebrafish; dynamics molecular docking

1. Introduction

Coronary Artery Disease (CAD) is a cardiovascular disease with significant morbidity and mortality, especially among the elderly [1], and imposes a heavy burden on health care systems worldwide [2–4]. CAD is caused by excessive lipid accumulation in the vessel wall because of long-term exposure to lifestyle risk factors such as high sugar and fat. This accumulation results in reduced endothelial function, which in turn leads to stenosis and blockage [5]. A study has shown that vascular regeneration in the infarcted areas of the heart and the construction of new vascular transport channels at the onset of CAD are essential for alleviating CAD symptoms [6]. In this regard, phytotherapy has been used in China for thousands of years to regulate human health, especially in the case of cardiovascular and cerebrovascular diseases [7].

As an important part of phytotherapy, Traditional Chinese Medicine (TCM) provides us with a vast resource to develop supplements for health care and disease treatment [8]. In previous research in our laboratory, DGS was screened as a novel natural vasoprotective combination. DGS was composed of three Chinese medicinal materials, namely, Salviae Miltiorrhizae Radix et Rhizoma (Danshen), Puerariae Lobatae Radix (Gegen), and Crataegi Folium (Shanzhaye). Modern pharmacological studies have shown that Danshen can increase blood flow [9], improve circulation [10], and preserve endothelial function [11]. Danshen is the most representative herbal medicine in the Chinese medicine formula repertoire for the treatment of ischemic diseases [12]. Gegen is usually used as a "medicinal pair" with Danshen. Shanzhaye is also commonly utilized for promoting blood circulation and resolving blood stasis. All three herbs are used as dietary supplements for the prevention and treatment of diseases [13–15]. Further studies have reported salvianolic acid B in Danshen exerts significant anti-myocardial ischemic effects and can alleviate oxidative stress damage caused by myocardial ischemia [16]. Hyperoside, one of the important flavonoids in Gegen, is known for its vasoprotective effects [17]. However, the potential components and mechanisms of this combination have not been thoroughly reported.

TCM exerts a multi-component, multi-target, and multi-pathway synergistic effect [18]. However, this complexity is a bottleneck for revealing its modern scientific significance. Network pharmacology is a new approach that analyzes the complexities of drugs and diseases and visualizes the drug treatment mechanisms through the construction of biological association networks, which are in line with TCM philosophy [19]. Therefore, this study aimed to devise a new network pharmacology-driven strategy to investigate the active components and network molecular mechanism of DGS. Moreover, transgenic zebrafish were used as the in vivo experimental animal model to evaluate the bioactive effects of DGS samples. Finally, this research aspired to reveal the potential components and pharmacodynamic mechanisms of DGS in the intervention of CAD.

2. Results

2.1. Identification of the Chemical Components of DGS

HPLC-qTOF-MS/MS was used to analyze the composition of DGS. Based on chromatographic retention times and fragment ion information of the molecules, 37 compounds were identified from the DGS samples (Figure 1). The compound names, retention times (min), and molecular formulae are listed in Table 1. The m/z 359.0761 in the negative ion mode corresponds to the [M-H]$^-$ ion peak of Rosmarinic acid. The primary and secondary mass spectra of Rosmarinic acid are consistent with the cleavage of primary and secondary given in the MoNA database. Furthermore, a molecular ion peak was seen at m/z 609.1415, which corresponded to the [M-H]$^-$ ion peak of rutin [20]. The secondary mass spectrum of rutin showed a mass spectral fragment of m/z 301, a [M-H-C$_{12}$H$_{22}$O$_{10}$]$^-$ mass spectral fragment formed by the loss of two sugar rings of rutin. Salvianolic acid B [21], tanshinone IIA, cryptotanshinone, and hyperoside were respectively identified by the [M-H]$^-$ ion peak at m/z 717.1220, m/z 293.1192, m/z 295.1362, and [M-H]$^-$ ion peak at m/z 463.0851 in the negative ion mode. Tanshinone IIA, cryptotanshinone, and hyperoside were identified through the NoMB database. Based on mass spectrometry information of primary and second ion fractions, nearly half of the 34 compounds identified from the DGS samples were found to be flavonoids and their glycoside derivatives and the rest were phenolic acids and tanshinones.

Table 1. Compound analysis and identification of extract of DGS.

Peak No.	Retention Time (min)	Formular	Calc. Mass	Molecular Ion (m/z)	Mass Error (ppm)	Fragment Ion	Compound Name
1	1.256	$C_7H_{12}O_6$	191.05611	191.0576	7.8	173.0375, 93.0279, 87.0043, 85.0251	Quinic acid
2	3.268	$C_{11}H_{12}O_7$	255.0510	255.0505	−2.1	165.0451, 147.0418, 131.0459	Piscidic acid
3	4.823	$C_{26}H_{28}O_{13}$	547.14571	547.1468	2.0	295.0490, 267.0542	Mirificin
4	4.908	$C_{30}H_{26}O_{12}$	577.1352	577.138	3.2	425.5437, 255.4897	Procyanidin B2
5	5.23	$C_{15}H_{14}O_6$	289.07176	289.0737	6.9	205.2053, 189.8794	(−)-Epicatechin
6	6.283	$C_{21}H_{20}O_{10}$	431.0937	431.0899	−8.8	311.0481, 283.0500	Vitexin
7	6.425	$C_{16}H_{18}O_9$	353.0878	353.08501	−7.9	191.0235, 179.0027	Chlorogenic acid
8	6.866	$C_{11}H_{12}O_6$	239.0561	239.0551	−4.2	179.0277, 177.0486, 107.0453	2-(Carboxymethyl)-4,5-dimethoxybenzoic acid
9	7.199	$C_{22}H_{22}O_{11}$	461.1035	461.1033	−0.4	253.0397	Tectoridin
10	7.55	$C_{26}H_{28}O_{14}$	563.1406	563.1428	3.9	311.0433, 283.0471, 227.0628	Schaftoside
11	8.52	$C_{21}H_{20}O_9$	415.1005	415.0963	−7.2	296.0397, 267.0549, 207.0577, 193.0550	Puerarin
12	9.137	$C_{26}H_{20}O_{10}$	491.0984	491.0985	0.3	363.1897, 351.6277, 320.0439	Salvianolic acid C
13	9.202	$C_{22}H_{22}O_{10}$	445.1140	445.1103	−8.3	283.0402	3′-Methoxypuerarin
14	10.518	$C_{15}H_{10}O_4$	253.0506	253.0487	−7.6	223.0306, 209.0538	Daidzein
15	11.318	$C_{29}H_{34}O_{14}$	605.1876	605.1836	−6.6	297.0657, 253.0771	Pueroside A
16	11.706	$C_{21}H_{20}O_{10}$	431.0984	431.0955	−6.7	269.0337	Genistin
17	12.261	$C_{27}H_{30}O_{14}$	577.1563	577.1557	−1.1	413.0696, 313.0373, 293.0346	Vitexin 2″-O-rhamnoside
18	13.302	$C_{15}H_{10}O_5$	269.0456	269.0478	8.4	252.0337	Genistein
19	13.341	$C_{27}H_{30}O_{15}$	593.1512	593.1483	−4.9	311.0412, 283.0351	Tectorigenin 7-O-xylosylglucoside
20	14.927	$C_{21}H_{20}O_{12}$	463.0882	463.0851	−6.7	303.0245	Hyperoside
21	15.64	$C_{27}H_{30}O_{16}$	609.1461	609.1415	−7.6	301.03118	Rutin
22	15.74	$C_{18}H_{16}O_8$	359.0772	359.0761	−3.2	161.0169, 135.0388, 119.4798	Rosmarinic acid
23	15.1	$C_{26}H_{22}O_{10}$	493.1140	493.1100	−8.2	295.0498, 313.0585, 197.0355, 162.0195	Salvianolic acid A
24	16.452	$C_{27}H_{22}O_{12}$	537.1038	537.0986	−9.8	313.0583, 295.0497	Lithospermic acid
25	17.822	$C_{36}H_{30}O_{16}$	717.14611	717.1420	−5.7	429.1065, 339.0366, 320.0352, 279.0567, 185.0106	Salvianolic acid B
26	19.44	$C_{27}H_{30}O_{15}$	593.1512	593.1471	−6.9	414.0734, 311.0541, 293.0322	Vitexin-4″-O-glucoside
27	20.929	$C_{36}H_{30}O_{16}$	717.1461	717.1420	−5.7	537.0832, 493.0698, 339.0386, 295.0556	Salvianolic acid L
28	23.049	$C_{29}H_{26}O_{12}$	565.1352	565.1376	4.3	339.0376, 321.0272, 293.0328	Dimethyl Lithospermate
29	24.749	$C_{26}H_{20}O_{10}$	491.0984	491.0939	−9.1	295.0532	isosalvianolic acid C
30	31.259	$C_{20}H_{28}O_2$	301.2162	301.2156	−2.1	271.6406, 259.5924	Sugiol
31	33.13	$C_{16}H_{16}O_5$	287.0925	287.0906	−6.8	269.2039, 258.1443	Shikonin
32	33.32	$C_{19}H_{18}O_3$	293.1138	293.1192	3	231.3145, 221.1474	Tanshinone IIA
33	33.845	$C_{19}H_{22}O_4$	313.14453	313.1455	3.0	227.0246, 212.0280, 267.0357	Tanshinone V
34	35.93	$C_{19}H_{20}O_3$	295.1340	295.1362	7.6	277.0811, 265.0752, 209.0584	Cryptotanshinone
35	36.129	$C_{19}H_{22}O_3$	297.1496	297.1519	7.7	270.3796, 253.1184	2-[2-(6-methoxy-3,4-dihydro-2H-naphthalen-1-ylidene)ethyl]-2-methylcyclopentane-1,3-dione
36	36.555	$C_{17}H_{14}O_6$	313.0717	313.0703	−4.4	295.1204, 283.1224, 268, 255.1049	Salvianolic acid F
37	37.433	$C_{39}H_{54}O_7$	633.3797	633.3828	4.9	617.3838, 471.358	3-O-p-Coumaroyltormentic acid

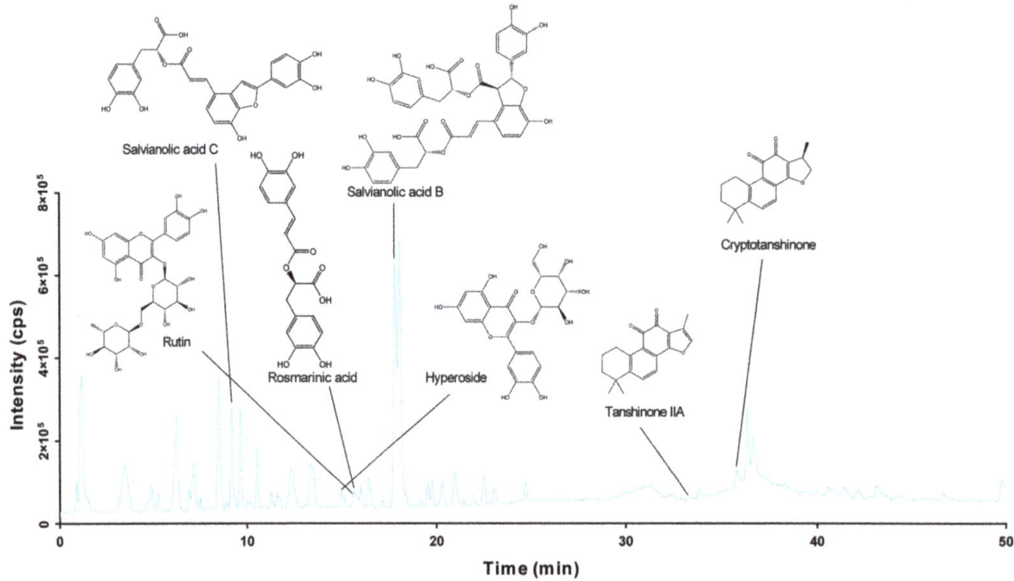

Figure 1. The HPLC-QTOF/MS total ion chromatogram of DGS in the negative ion modes.

2.2. Key Component and Target Screening and Prediction Using Network Pharmacology Analysis

A total of 35 compounds were successfully predicted to be the targets of action in the SwissTargetPrediction database. The compounds Piscidic acid and Schaftoside could not be predicted in the SwissTargetPrediction database based on two-dimensional (2D) and three-dimensional (3D) similarities. Hence, they were excluded. A total of 442 biological targets were obtained by combining these 35 compounds and removing duplicates. Moreover, 859 CAD-related targets were obtained from the Genecards database using the correlation score of >20 as the screening criterion. A Venn diagram analysis of the compound's target library against the disease's target library yielded 112 incorporative targets (Figure 2A). The top five enriched target categories were protease, enzyme, kinase, nuclear receptor, and secreted protein. As shown in Figure 2B, the top five targets in the protein–protein interaction (PPI) network in terms of degree-value were ALB, TNF, VEGFA, EGFR, and CASP3. Cytoscape 3.9.1 was employed to calculate the network topology data, and the MCODE plug-in was used to calculate the most closely linked clusters in the PPI network. A total of 4 clusters were obtained, with cluster scores of 23.667, 8.147, 3.5, and 3 (Figure 2B′). Cluster 1 with the highest clustering score was used for subsequent analysis.

In the biological process enrichment analysis, Cluster1 was imported into the David database for GO enrichment analysis. Annotations of genes were obtained by using GO analysis. The screening was performed on a $p < 0.05$ basis, and the top 30 biological processes were listed according to the magnitude of the p-value (Figure 2C). It chiefly involved cytokine-mediated signaling pathway, positive regulation of MAPK cascade, positive regulation of cell migration, positive regulation of kinase activity, protein autophosphorylation, positive regulation of MAP kinase activity, MAPK cascade, positive regulation of protein phosphorylation, positive regulation of nitric oxide biosynthetic process, positive regulation of ERK1 and ERK2 cascade, positive regulation of cell proliferation, positive regulation of phosphatidylinositol 3-kinase signaling, and negative regulation of endothelial cell apoptotic. These biological processes were mostly involved in the regulation of cytokines, cascade activation of the MAPK pathway, positive regulation of protein phosphorylation, positive regulation of NO synthesis, positive regulation of cell proliferation, anti-endothelial cell apoptosis, etc. The top 30 KEGG signaling pathways at $p < 0.05$ and sorted according to count value size. As shown in Figure 2D, KEGG pathway enrichment analysis revealed that most of the DGS targets were concerned with the regulation of cell proliferation, anti-apoptosis, cell migration, and angiogenesis.

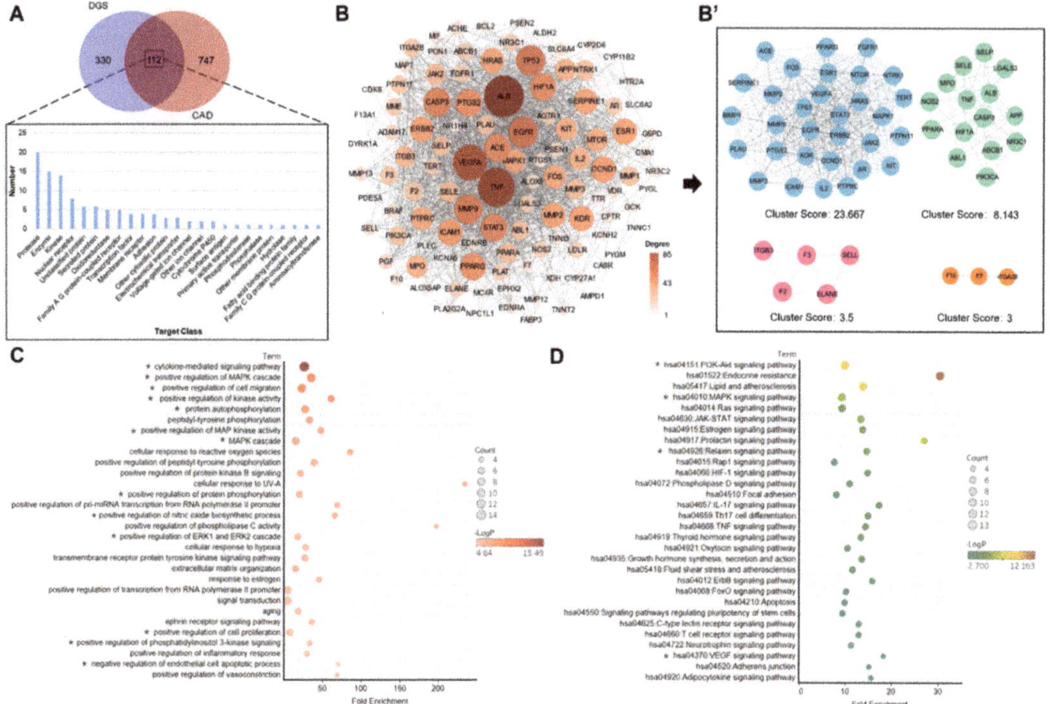

Figure 2. Network pharmacology analysis: (**A**) The Venn diagram analysis for DGS and HCD targets. (**B**) PPI network of the DGS compounds targets against CAD. (**B′**) MCODE analysis of PPI network. (**C**) Biological process analysis of PPI networks with a clustering score of 23.667. * Represents a potentially important biological process. (**D**) KEGG enrichment analysis of PPI networks with a clustering score of 23.667. * Represents a potentially important signaling pathway.

2.3. Pharmacological Mechanism Analysis in HUVECs

2.3.1. Effect of DGS on the Proliferation and Migration of HUVECs

The CCK-8 assay was performed to evaluate the cell viability after the treatment with DGS. As illustrated in Figure 3A, DGS significantly increased the viability of HUVECs. Six concentration groups of DGS were used to treat the HUVECs for 24 h. The mean cell survival rate of the 1 µg/mL DGS group was 116%, which did not show a proliferative effect compared with the control (Ctrl) group (100%). After treatment with DGS, the cell survival rates were 129.0% ($p < 0.01$), 127.4% ($p < 0.01$), 126.3% ($p < 0.01$), 128.0% ($p < 0.01$), and 124.3% ($p < 0.01$) in the five concentration groups of 2, 4, 8, 16, and 32, respectively, in comparison with the control group. These results indicate that DGS significantly promotes the proliferation of HUVECs.

Endothelial cell migration is essential for angiogenesis. Therefore, the effect of DGS on the migration of HUVECs was evaluated using wound-healing assays. As shown in Figure 3A,B, VEGF significantly increased the migration of endothelial cells to the wound healing zone after injury ($p < 0.0001$). Treatment of HUVECs with DGS revealed that DGS significantly promoted the migration of endothelial cells toward the middle of the scratch and promoted scratch healing.

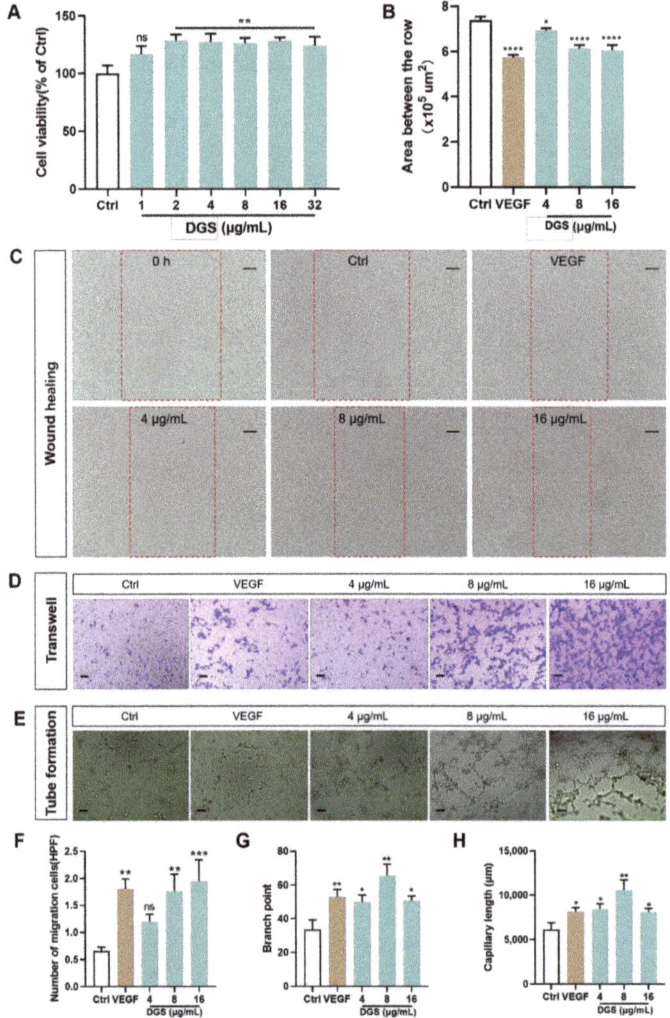

Figure 3. Effect of DGS on HUVECs cells in vitro: (**A**) An CCK-8 assay was carried out to measure HUVECs viability. (**B**) Effect of different concentrations of DGS on the migration of HUVECs cells. Results are presented as the mean ± SEM. (**C**) The healing area of the wound at 0 and 24 h were photographed by microscopy. The red dashed box represents the area counted after migration. Scale bar: 100 μm. (**D**) The migration of HUVECs in Transwell migration assays. Scale bar: 100 μm. (**E**) DGS promoted tube formation of HUVECs. Scale bar: 100 μm. (**F**) Quantification of the number of migrated cells. (**G**) Quantitative analysis of branch points for tube formation assays. (**H**) Quantitative analysis of capillary length for tube formation assays. Values are expressed as the mean ± SEM. ^{ns} $p < 0.05$ vs. Control, * $p < 0.05$ vs. Control, ** $p < 0.01$ vs. Control, *** $p < 0.001$ vs. Control, **** $p < 0.0001$ vs. Control.

The formation of blood vessels rests on the proliferation and migration of endothelial cells. Therefore, in addition to the above wound healing assay that examined the lateral migration of the endothelial cells, the transwell assay that investigated the longitudinal migration of the cells after the treatment with DGS was performed. As presented in Figure 3D,F, the migration of HUVECs was significantly increased compared with the control after treatment with VEGF-positive drugs. Moreover, DGS significantly increased the ability of HUVECs to migrate toward the backside of the membrane. Compared with the Control group (0.6568 HPF), the mean values of the number of cells migrating to the back of the membrane with 4, 8, and 16 µg/mL of DGS were 1.198 ($p > 0.05$), 1.763 ($p < 0.01$), and 1.946 ($p < 0.001$) HPF, respectively.

Angiogenesis was evaluated in terms of endothelial cell migration. Matrix gel was used to evaluate the ability of DGS to promote tube formation in HUVECs. Furthermore, the ability to promote tube formation was assessed based on the capillary length and the number of branching points. HUVECs treated with VEGF resulted in a mean number of branch points and a mean capillary length of 52.80 ($p < 0.01$) and 8144 µm ($p < 0.05$), respectively, compared with the control group; the mean value of the number of branch points for the three concentrations after the treatment of HUVECs with DGS control were 49.91, 65.75, and 50.64, respectively, and the lumen lengths were 8417, 10,568, and 8098, respectively (Figure 3E,G,H).

2.3.2. NO and VEGF Release and Protein Expression of DGS

Compared with the control group, the NO level was increased in the high concentration group of DGS (16 µg/mL) ($p < 0.05$), but there was no difference in NO secretion between 4 µg/mL DGS and 8 µg/mL DGS groups ($p > 0.05$). In ELISA, compared with control, the secretion of 8 µg/mL DGS and 16 µg/mL VEGFA treatment groups compared to the control group ($p < 0.05$, $p < 0.05$) (Figure 4A,B).

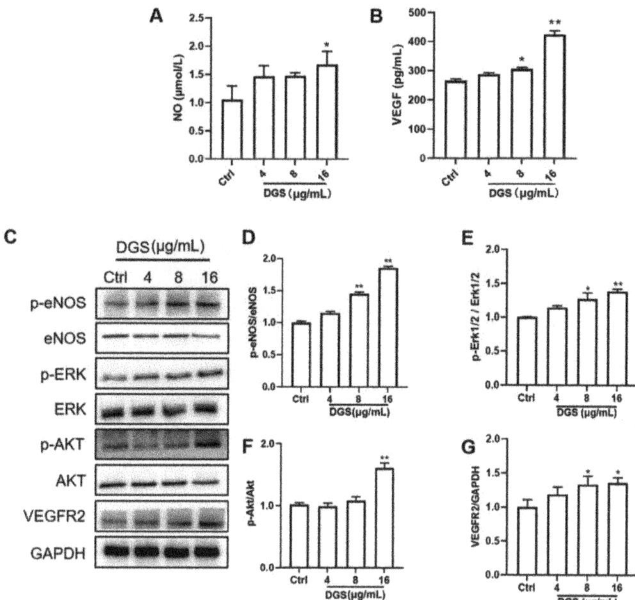

Figure 4. Regulation of NO, VEGF, and related proteins by DGS: (**A**) Effect of DGS on NO levels. (**B**) Effect of DGS on VEGFA levels. (**C**) Western blot results. (**D–G**) are the results of statistical analysis of VEGFR2/GAPDH, p-Akt/Akt, p-Erk1/2/Erk1/2, and p-eNOS/eNOS, respectively. Data are presented as the mean ± SEM from at least three independent experiments. * $p < 0.05$ vs. Control, ** $p < 0.01$ vs. Control.

The molecular mechanisms associated with the promotion of angiogenesis by DGS were elucidated. Further experiments to explore the pathways by which DGS enhanced the proliferation and migration of HUVECs. As shown in Figure 4C,D, DGS significantly upregulated VEGFR2, thereby activating the phosphorylation of proteins related to downstream signaling pathways. As illustrated in Figure 4C,E, the level of Akt phosphorylation was significantly elevated after DGS treatment, and the value of p-Akt/Akt was statistically different compared with that of the control group ($p < 0.01$). Furthermore, Erk phosphorylation level was found to be significantly elevated and statistically different after DGS treatment ($p < 0.05$, $p < 0.01$) (Figure 4C,F). Finally, vasodilation-related proteins were examined, which revealed that the phosphorylation levels of eNOS were also significantly elevated, with statistical differences ($p < 0.01$, $p < 0.01$) (Figure 4C,G). The increase in phosphorylation of VEGFR2 and eNOS was consistent with the increased secretion of VEGFA and NO in the cell cultures described above. These results suggest that DGS activates the VEGFA/VEGFR2/Akt/Erk/eNOS signaling pathway to promote angiogenesis and vasodilation.

2.4. Bioactivity Evaluation of DGS Based on Zebrafish Model In Vivo

A zebrafish assay was performed according to the schematic diagram (Figure 5A). As shown in Figure 5B, zebrafish treated with DGS for 3 days demonstrated mortality and abnormalities at concentrations $\geq$ 100 µg/mL. The LD50 of DGS was 753.7 µg/mL and the 5% lethality was 478 µg/mL. These results suggest that efficacy assessment of DGS at concentrations $\leq$ 100 µg/mL can be used for the evaluation of the efficacy of angiogenesis.

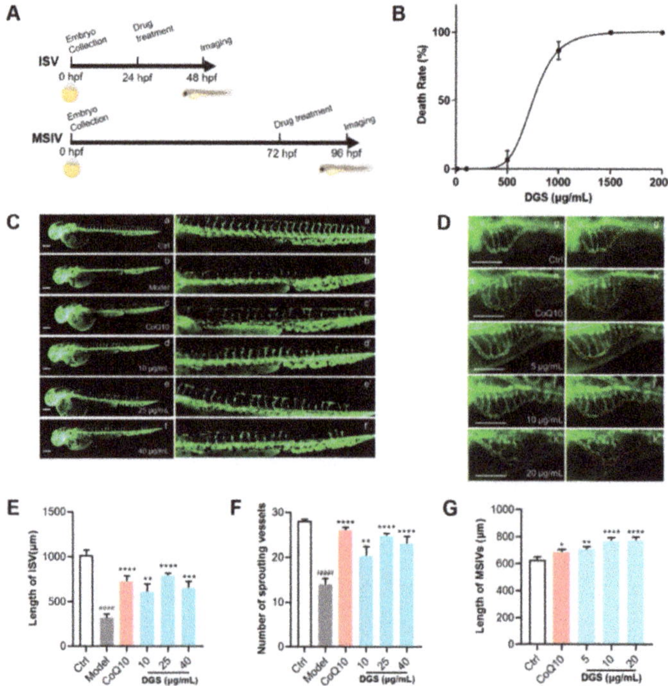

Figure 5. DGS promoted the angiogenesis of zebrafish: (**A**) Schematic diagram of the zebrafish experiment. (**B**) The lethal curve of DGS (**C**) Fluorescent images of the ISV of the zebrafish. The images of a′–f′ were partial enlargements of images a–f. Scale bar: 200 µm. (**D**) Fluorescent image of the MSIV of the zebrafish. The images of g′–k′ were partial enlargements of images g–k. Scale bar: 200 µm. (**E**) Effect of DGS on the length of ISV in zebrafish. (**F**) The effect of DGS on the sprouting of SIV in zebrafish. (**G**) Effect of DGS on the growth of the MSIV in zebrafish. Values are expressed as the mean ± SEM (n = 10). #### $p < 0.0001$ vs. Control, * $p < 0.05$ vs. Model, ** $p < 0.01$ vs. Control, *** $p < 0.01$ vs. Control, **** $p < 0.0001$ vs. Control.

Transgenic Tg (flk1a: EGFP) zebrafish embryos were used to explore the proangiogenic effects of DGS. The findings indicated that PTK787 significantly inhibited the formation of intersegment vessels (ISVs). Most vessels were incomplete and discontinuous after PTK787 treatment (Figure 5B,D). The ISVs length in the model was 323.3 ± 38.42 μm, which was significantly lower than that of the control (1017 ± 58.96 μm; $p < 0.0001$). After treatment with CoQ10 as a positive drug, the total ISV length reached 730.0 ± 59.77 μm, which was significantly higher than that of the model group ($p < 0.0001$). The lengths of ISVs after treatment with DGS at concentrations of 10, 25, and 40 μg/mL were 610.0 ± 85.08 μm ($p < 0.01$), 793.7 ± 23.71 μm ($p < 0.0001$), and 654.3 ± 70.81 μm ($p < 0.0001$), respectively. Concerning the number of vascular sprouting, the number was significantly increased in the CoQ10 and DGS-treated groups. In conclusion, DGS was able to significantly promote the growth of the inhibited vessels.

To further investigate the proangiogenic effect of DGS on normal blood vessels, the effects were assessed using the main subintestinal veins (MSIV) of the zebrafish. As depicted in Figure 5C,F, the length of MSIV in the zebrafish treated with CoQ10 for 24 h was 688.2 ± 22.13 μm, which was significantly higher compared with that of the control (628.0 ± 22.13 μm; $p < 0.05$). After treatment with DGS at concentrations of 5, 10, and 20 μg/mL, the lengths of MSIV were 708.0 ± 18.22 μm ($p < 0.01$), 771.5 ± 22.72 μm ($p < 0.0001$), and 775.5 ± 23.62 μm ($p < 0.0001$), respectively.

2.5. Dynamic Molecular Docking of Active Components of DGS with Key Targets

The binding ability of compounds from DGS to VEGFR2 was validated in the present study using the same molecular docking to provide evidence for the proangiogenic effect of DGS with a virtual approach. All 37 components of DGS were docked to the binding pocket of VEGFR2, and an agonist of VEGFR2 was used as the positive control. All compounds were docked into the active pocket of VEGFR2, and all exhibited binding energies were greater than -5 kcal/mol. When the binding energy of agnuside was used as a control [22], nine compounds had binding energies greater than that of agnuside, as shown in Table 2. The binding modes of agnuside and salvianolic acid C with VEGFR2 were visualized separately using Pymol, which showed that the binding of agnuside to VEGFR2 was mainly via hydrogen bonding and pi-sigma interactions (Figure 6A). On the contrary, salvianolic acid C displayed pi–pi stacked interactions in addition to hydrogen bonding and pi–sigma interactions (Figure 6B).

Based on the docking results, salvianolic acid C and agnuside with VEGFR2 were further selected for MD simulation to examine their stability in the binding pocket. Root mean square deviation (RMSD) serves as an important basis for measuring system stability. In this study, the mean RMSD of the salvianolic acid C–VEGFR2 system after balance was 2.63 ± 0.323 Å (Figure 7A). Meanwhile, the mean RMSD of the agnuside–VEGFR2 system after balance was 2.36 ± 0.44 Å (Figure 7A). The mean RMSD of the two systems is <3 Å, which is completely acceptable in the protein system. Subsequently, the flexibility changes and root mean square fluctuation (RMSF) values of amino acid residues in VEGFR2 were evaluated. Figure 7B,F, the amino acid residues that interact with the ligand for more than 30% in the RMSF diagram. The RMSF values of most of the amino acid residues involved in the interaction were small, thereby indicating that the stability of the entire system was increased after salvianolic acid C and agnuside are combined with VEGFR2. The 2D visualization analysis of the salvianolic acid C–VEGFR2 system showed that salvianolic acid C formed hydrogen bonds with VEGFR2 via LYS-920, ASN-923, ARG-842, CYS-919, GLU-917, and GLU-885; furthermore, it forms Pi–Pi stacking interaction with PHE-1047 (Figure 7C,D). The 2-D visual analysis of the agnuside–VEGFR2 system showed that agnuside formed hydrogen bonds with VEGFR2 through VAL-914, THR-916, GLU-885, VAL-899, ASP-1046, and ILE-1025; moreover, it formed π–cation stacking with LYS-868 (Figure 7G,H). All the interaction durations exceed 30% of the whole simulation time.

Table 2. Docking information of VEGFR2 with the corresponding compounds.

Ligand	Binding Affinity (kcal/mol)	Type of Interaction
Agnuside	−8.8	Hydrogen bonding: PHE-1047, ILE-1044, ASP-046, LYS-868, VAL-914, ALA-866; π–Sigma: THR-916
Salvianolic Acid C	−10.7	Hydrogen bonding: LYS-920, CYS-919, ASN-923, ASP-1046, GLU-885, THR-916; π–Sigma: LEU-840, LEU-1035, THR-916; π–π stacked: PHE-918
Isosalvianolic Acid C	−9.7	Hydrogen bonding: GLU-917, CYS-919, ASN-923; π–Sigma: PHE-1047; π–π stacked: PHE-918, PHE-1047
Genistin	−9.6	Hydrogen bonding: GLU-885, THR-916, ASN-923; π–Sigma: THR-916, VAL-848, LEU-1035; π–π stacked: PHE-918
3-O-p-Coumaroyltormentic acid	-9.4	Hydrogen bonding: ILE-1025; π–π stacked: PHE-845
Tanshinone IIA	−9.2	π–Sigma: PHE-1047, PHE-845; π–π stacked: PHE-845
2-[2-(6-Methoxy-3,4-Dihydro-2H-Naphthalen-1-Ylidene)Ethyl]-2-Methylcyclopentane-1,3-Dione	−9.2	π–Sigma: LEU-840, π–π stacked: PHE-845
Daidzein	−9.1	π–Sigma: THR-916, LEU-1035, LEU-840; π–π stacked: PHE-918
Rosmarinic Acid	−9	Hydrogen bonding: GLU-917, CYS-919, ASP-1046, LYS-868; π–Sigma: LEU-889, LEU-1035
Lithospermic Acid	−9	Hydrogen bonding: GLU-885, MET-869, ASP-1046; π–π stacked: PHE-845

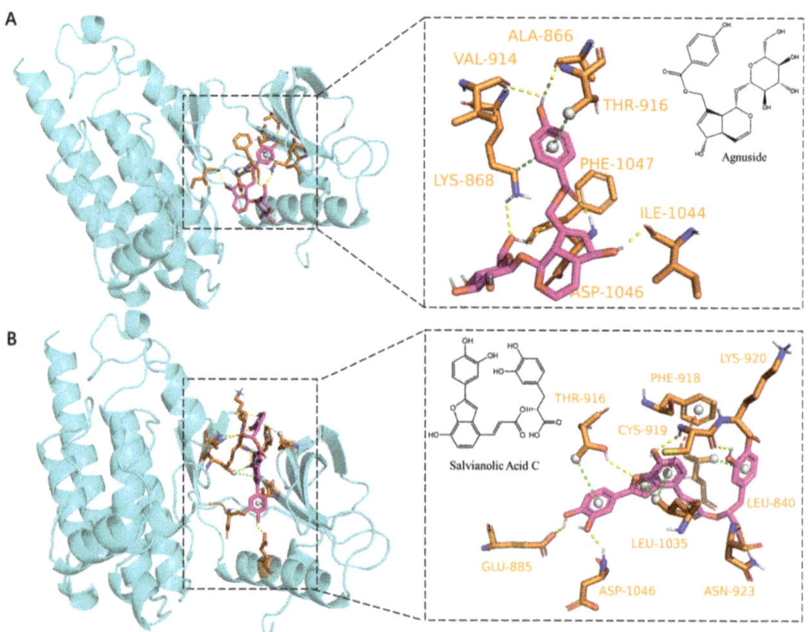

Figure 6. Molecular docking of Agnuside and Salvianolic Acid C to VEGFR2 protein: (**A**) Agnuside binding model with VEGFR2, yellow dashed lines represent hydrogen bonding interactions, green dashed lines represent π–Sigma interactions. (**B**) Salvianolic Acid C binding model with VEGFR2, yellow dashed lines represent hydrogen bonding interactions, green dashed lines represent π–Sigma interactions and red dashed lines represent π–π stacked interactions.

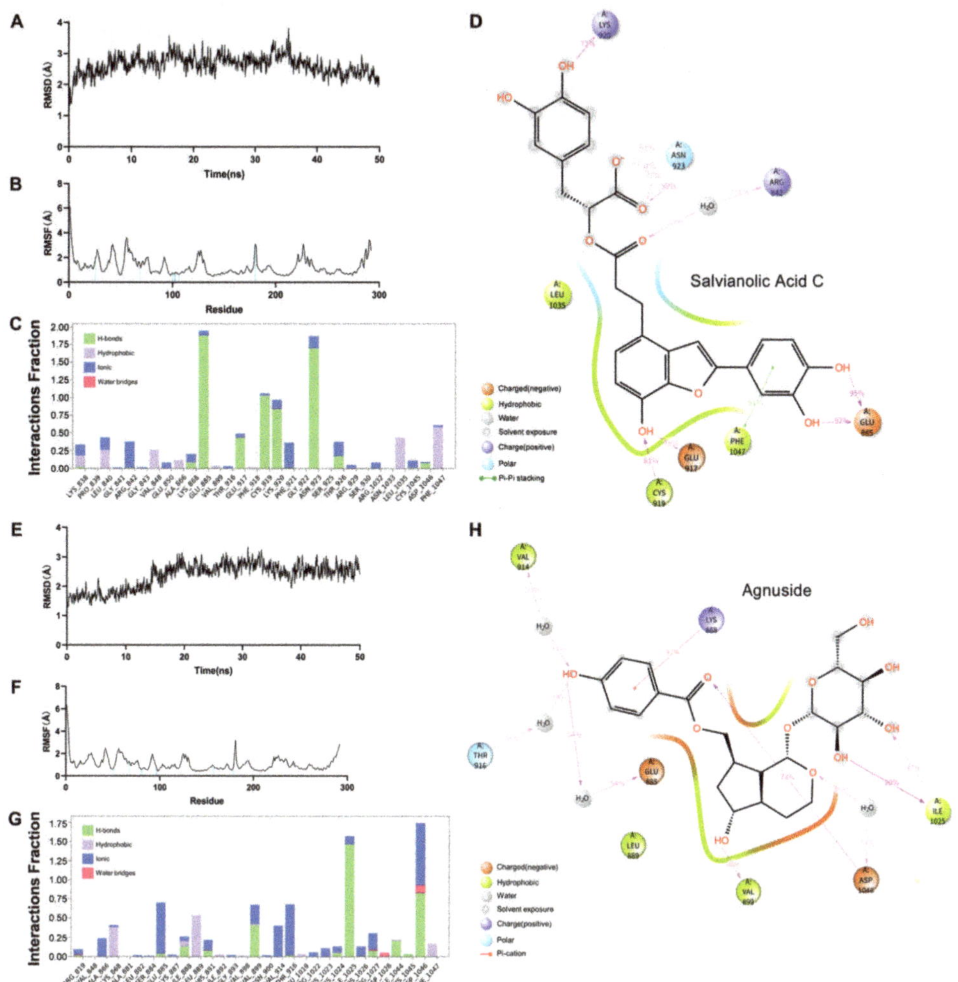

Figure 7. Dynamics molecular docking of two protein–ligand complexes, Salvianolic Acid C–VEGFR2 (**A–D**) and Agnuside–VEGFR2 (**E–H**): (**A**) RMSD of Salvianolic Acid C–VEGFR2. (**B**) RMSF of Salvianolic Acid C–VEGFR2. (**C**) Protein–Ligand Contacts Diagram of Salvianolic Acid C–VEGFR2; Y axis suggests that percentage of the simulation time the specific interaction is maintained; Values over 1.0 are possible as some protein residue may make multiple contacts of the same subtype with the ligand. (**D**) A schematic of detailed Salvianolic Acid C atom interactions with the VEGFR2 residues. (**E**) RMSD of Agnuside–VEGFR2. (**F**) RMSF of Agnuside–VEGFR2. (**G**) Protein–Ligand Contacts Diagram of Agnuside–VEGFR2. (**H**) A schematic of detailed Agnuside atom interactions with the VEGFR2 residues.

3. Discussion

It is well known that TCM has complex component systems, and their mechanisms are often difficult to explain using existing techniques. This difficulty greatly limits the globalization of traditional Chinese medicine. Effective study of the holistic effects of complex component systems has become a huge challenge. However, "Network pharmacology" has offered a new way to solve this problem [23,24]. Using system biology and computer technology, network pharmacology builds a "disease–gene–target–drug" interaction network, and systematically displays the impacts of drugs on the disease network [25]. Therefore, compared with the conventional single-target mode of action, network pharmacology is more suitable for studying the complex effects of TCM and has been rightly called the "next-generation drug development model". However, most of the network pharmacology studies in TCM are based on the results of network pharmacological analysis to explain the complexity of Chinese medicine, and there is a lack of mature theoretical guidance for subsequent experimental studies. Hence, this study proposes a Network pharmacology-driven strategy to investigate the molecular mechanism of TCM, with the hope of enhancing the precision of experimental research.

In this study, most of the components of DGS, including phenols, tanshinones, and flavonoids, were identified using HPLC-qTOF/MS. The phenolic compounds in Danshen have been reported to promote myocardial ischemic angiogenesis [26]. Mass spectrometry asserted the presence of salvianolic acid A, B, C, L, and other phenolic acids in DGS. Flavonoids are present in almost all plants and studies have shown that they can inhibit cardiac injury through a variety of mechanisms [27]. Mass spectrometric detection revealed the presence of rutin, hyperoside, and other flavonoids in DGS. The mechanisms by which these components acted in the treatment of CAD were investigated. Network pharmacology was used to analyze the action network of DGS, which indicated that the targets of DGS action are mainly enriched in biological processes related to cytokine regulation, protein phosphorylation, activation of the MAPK signaling pathway, and cell proliferation. Subsequently, KEGG pathway enrichment analysis revealed that the highest enrichment of targets was in the PI3K-Akt signaling pathway. A study has reported that the PI3K-Akt pathway plays a key role in the emergence, progression, and treatment of CAD, thereby activating downstream pathways that control cell survival, proliferation, migration, and other biological processes after receiving intracellular and extracellular feedback [28]. As an effector downstream of Akt, eNOS can be activated to control endothelial cell survival, proliferation, apoptosis, and intravascular environment stability after the development of coronary lesions in the heart [29,30]. Additionally, activation of downstream eNOS contributes to myocardial angiogenesis, which is similar to the effect of VEGF in promoting therapeutic angiogenesis in CAD [31]. It has been reported that Danhong injection can promote the growth of vessels in the myocardial infarct site by activating the ERK signaling pathway [32]. Therefore, the phenotypic and signaling pathway mechanisms were validated in subsequent studies.

The zebrafish is an emerging internationally recognized model animal. During zebrafish embryonic development, angiogenesis can be visualized under a microscope and is currently considered one of the best models for studies on angiogenesis [33,34]. Specifically, the transgenic Tg (flk1a:EGFP) zebrafish with green fluorescence of the vascular can be observed via video and imaged under a fluorescence microscope with great convenience to assess the effect of the drug on vessel growth [35]. All mammalian and zebrafish endothelial cells are extremely plastic and retain their plasticity even after reaching adulthood. After enduring cardiac injury, mice can form new coronary arteries from pre-existing endothelial cells [36,37]. The zebrafish vascular growth model is ideally suited as it mimics human vascular disease models for new drug screening and evaluation [38]. Our experimental results established that DGS can promote vascular growth in vivo by increasing the vessel length and the number of vascular outgrowths, without any effects on zebrafish development, at the effective dose.

The findings from network pharmacology and zebrafish experiments suggest that effective angiogenesis can enhance blood perfusion in the case of CAD and alleviate cardiac injury. Therefore, the angiogenic mechanisms of DGS in vitro were further investigated. Endothelial cells are involved in angiogenesis at the lesion site mainly via two pathways: proliferation and migration [39]. The CCk-8 assay demonstrated that DGS significantly increased the viability of HUVECs, which established that DGS could promote the proliferation of HUVECs. The ability of DGS to promote the migration of HUVECs was then assessed using the wound healing assay [40], transwell assay [41], and tube formation assay [42]. The results showed that DGS significantly stimulated the migration of HUVECs towards the scratch region and the lower transwell compartment and significantly stimulated the tube formation ability of HUVECs.

Furthermore, the levels of VEGFA and NO secretion in HUVECs were measured, and angiogenic and diastolic-associated proteins were quantified using Western blot. NO is a key cell-signaling molecule that is produced by activated eNOS catalyzing L-arginine[43]. Activation of eNOS and release of NO will induce the vasodilation of vessels, and NO is necessary to maintain endothelial cell function [44]. In this study, the secretion of NO and the phosphorylation level of eNOS in HUVECs were measured, which signified that DGS significantly upregulated p-eNOS and stimulated the secretion of NO. Moreover, NO is one of the main mechanisms that promote angiogenesis [45]. Our results suggest that DGS can stimulate the secretion of VEGFA to upregulate VEGFR2, thereby accentuating the binding of VEGFA to VEGFR2. ERK and Akt are located downstream of the VEGF/VEGFR2 pathway, and it has been shown that the phosphorylation of Erk and Akt can further mediate the proliferation and migration of HUVECs, thereby promoting angiogenesis [46]. The results allude that DGS may promote angiogenesis and vasodilation by promoting the phosphorylation of Akt and Erk. As shown in Figure 8, DGS activates the downstream Akt/Erk/eNOS pathway by promoting the binding of VEGFA to VEGFR2. This activation ultimately releases NO to expand blood vessels and mediates the proliferation, migration, and angiogenesis of HUVECs. The dynamic molecular docking results showed that the key compounds in DGS exhibit a good binding capacity for VEGFR2, specifically salvianolic acid C, which may be an important agonist of VEGFR2 and activate downstream signaling pathways.

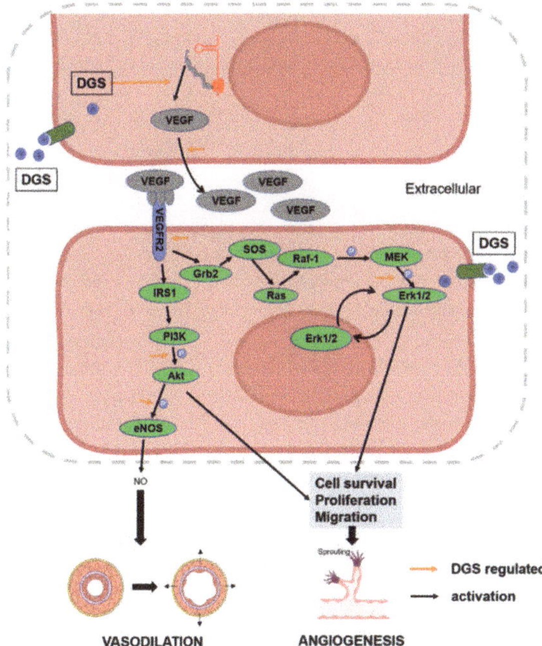

Figure 8. DGS participates in the overall regulatory network of CAD inhibition through angiogenesis and vasodilation.

4. Materials and Methods

4.1. Cell and Reagents

HUVECs were purchased from Jinan Dinguo Changsheng Biotechnology Co., Ltd. (Jinan, China). PTK787 was purchased from the AbMole BioScience (Houston, TX, USA). Pronase E and dimethyl sulfoxide (DMSO) were purchased from Sigma-Aldrich (Mannheim, Germany). Coenzyme Q10(CoQ10) was purchased from Yuanye (≥98% HPLC, Shanghai, China). Methyl alcohol and acetonitrile were purchased from TEDIA (Anhui, China). Cell Counting Kit-8 (CCK-8) and bovine serum albumin (BSA) were purchased from GeneView (Houston, TX, USA). Nitric oxide (NO) assay kit was purchased from Nanjing Jiancheng Bioengineering Institute (Nanjing, China). A Human VEGF ELISA kit was purchased from ABclonal (Boston, MA, USA). We used the following primary antibodies: rabbit anti-GAPDH (Cell Signaling; Danvers, MA, USA), rabbit anti-VEGFR2 (Cell Signaling), rabbit anti-Akt (Cell Signaling), rabbit anti-p-Akt (Cell Signaling), rabbit anti-Erk1/2 (Cell Signaling), rabbit anti-p- Erk1/2 (Cell Signaling), rabbit anti-eNOS (Cell Signaling), and rabbit anti-p-eNOS (Cell Signaling). We used the following secondary antibody for the experiment: HRP-conjugated goat anti-rabbit IgG (ABclonal).

4.2. Plant Material and Extraction

The DGS is composed of Danshen/Gegen/Hawthorn leaf = (1:1:1. Ratio). The medicinal material samples of Danshen, Gegen, and Hawthorn leaves were collected from Hongji Tang Herbal Drug Store (Qingdao, China) and identified as Salviae Miltiorrhizae Radix et Rhizoma (the root of Salvia miltiorrhiza Bge.), Puerariae Lobatae Radix (The root of Pueraria lobata (Willd.) Ohwi), and Crataegi Folium (the leaf of Crataegus pinnatifida Bge.) by Dr. Liwen Han, Shandong First Medical University and Shandong Academy of Medical Sciences. The sample preparation method was as follows: 1000 g of crushed Danshen was extracted twice with 10 L of 80% ethanol at reflux for 2 h each time, filtered, and then combined with the extract, and concentrated and dried under reduced pressure to obtain the Danshen extract. Then, 1000 g of Gegen was crushed and extracted thrice with 10 L of 70% ethanol for 2 h each time; the extracts were filtered, concentrated, and then dried under reduced pressure. The Hawthorn leaves were then crushed and extracted by refluxing twice with 70% ethanol, each time for 2 h; the obtained extract was filtered, combined, concentrated, and then dried under reduced pressure to obtain the Hawthorn leaf extract. Three extracts were blended to obtain the sample of DGS.

4.3. Chromatographic and Mass Spectrometry Conditions

The DGS powder sample was dissolved in methanol at the concentration of 6.6 mg/mL methanol solution and filtered through a 0.22 μm microporous membrane. SCIEX ExionLC^{TM} AD ultra-performance liquid chromatography system, and SCIEX X500R Quadrupole time-of-flight mass spectrometer (SCIEX, Framingham, MA, USA) with SCIEX OS software were used under the chromatographic condition: Shimadzu Shim-pack GIST-C18 column (2 um, 3.0 × 100 mm), column temperature 40 °C; mobile phase composed of 0.1% formic acid in water and methanol in B; the gradient elution program 0 min, 18% B; 1 min, 18% B; 10 min, 35% B; 27 min, 55% B; 35 min, 80% B; 48 min, 82% B; 49 min, 18% B. The flow rate was set to 0.4 mL/min, the injector temperature was set to 15 °C, and the injection volume was 2 μL.

Mass spectrum condition: scanning was performed in the negative ionization mode using an electrospray ionization (ESI) source. The data acquisition method was information dependent acquisition (IDA). The scan range was 50–2000 Da, the ion source temperature was 550 °C, the declustering voltage was 5500 V and the accumulation time was 0.2 s. The settings for Gas 1, Gas 2, Curtain Gas, and CAD Gas were 50, 55, 35, and 7 psi, respectively. The data were processed using the SCIEX OS software. Compounds were characterized using primary and secondary mass spectra plotted by comparison with literature or the MassBank of North America Database (MoNA) (https://mona.fiehnlab.ucdavis.edu/, accessed on 10 November 2021).

4.4. Collection of DGS-Related Targets and CAD-Related Targets

The compounds' structure obtained by mass spectrometry was input in the SMILES format using the Pubchem (https://pubchem.ncbi.nlm.nih.gov/, accessed on 10 December 2021) database. The compounds were imported into the SwissTargetPrediction (http://www.swisstargetprediction.ch/, accessed on 10 December 2021) database [47] in SMILES format, and the targets of Homo Sapiens species were selected. The results of the predicted targets were screened with a probability value > 0. Finally, the predicted targets were combined and de-duplicated to build a compound target library. All targets for the disease were obtained in the GeneCards (https://www.genecards.org/, accessed on 10 December 2021) database [48] using the keyword "Coronary Artery Disease" and the targets were filtered according to the Relevance score > 20.

4.5. Protein–Protein Interaction (PPI) Network

The intersecting targets were obtained by combining the DGS target library with the CAD target library using the online Venn Analysis tool (http://bioinformatics.psb.ugent.be/webtools/Venn/, accessed on 10 December 2021). All intersecting targets were imported into the String database to obtain protein interaction network data for the intersecting targets. The obtained protein data network was imported into Cytoscape 3.9.1 for PPI visualization and topological analysis. MOCODE, a data connectivity-based plug-in for finding dense regions of the protein interaction networks, was employed to uncover the most closely related protein interaction networks in a complex network [49]. The PPI network was further analyzed using the MCODE plugin in Cytoscape 3.9.1 to identify small, closely related networks in the entire interactions, and the targets in the interactions with the highest scores were selected for subsequent enrichment analyses based on the MCODE clustering scores.

4.6. Gene Ontology (GO) and KEGG Pathway Enrichment Analysis

The targets obtained from the MCODE plugin screening were imported into the David (https://david.ncifcrf.gov/ accessed on 20 June 2022) database [50,51] for GO enrichment analysis and the KEGG pathway enrichment analysis. The top 30 enrichment results were visualized using Tableau2019 using either p-value or Count value; $p < 0.05$ items were retained for the screening results.

4.7. Cell Culture and Treatment

HUVECs were cultured at 37 °C in DMEM (Gibco; Thermo Fisher Scientific, Inc., Waltham, MA, USA) with 10% fetal bovine serum (FBS, Gibco; Thermo Fisher Scientific, Inc., Waltham, MA, USA), 1% penicillin–streptomycin (Gibco), and 1% endothelial cell growth supplement (ECGS). The culture medium was refreshed every other day unless otherwise stated. The drugs used to treat HUVECs cells were formulated with DMSO, and the amount of DMSO in all cell cultures was required to not exceed 5‰ of the total volume.

4.8. CCK-8 Assay

CCK-8 was used to test the viability of cells after treatment with different concentrations of DGS. The HUVECs were seeded into a 96-well plate at a density of 1×10^4 cells per well. The microplate was placed in a CO_2 incubator at 37 °C for 24 h, after which 10 μL of the CCK-8 solution was added to each well and the plate was incubated for 2 h. The absorbance was measured using an enzyme marker (Spectramax id5) at 450 nm.

4.9. Wound Healing Assay

HUVECs cells were seeded in a 6-well plate and cultured overnight until a confluent monocytic layer was formed and a straight cell scratch was made on the monolayer with the tip of a 200 μL pipette tip. The cells were treated with VEGF or different concentrations of DGS. The scratches were imaged separately using an Olympus IX83 inverted microscope

(10×, magnification) (Olympus, Tokyo, Japan). The scratch area was quantified using Image Pro Plus.

4.10. Transwell Assay

In this study, the migration and invasive abilities of cultured HUVECs were assayed using the Transwell culture system (Corning®). To the lower chamber of the Transwell, DMEM medium containing the drug but without FBS. HUVECs cells were placed in the upper chamber of the DMEM medium without FBS at 37 °C for 24 h. After treatment, the cells were fixed with 4% paraformaldehyde for 15 min, and stained with 0.5% crystal violet for 10 min. Cell migration was imaged and quantified using an Olympus IX83 inverted microscope.

4.11. Tube Formation Assay

The angiogenic potential of HUVECs was assessed by in vitro tube formation assays. Briefly, a 96-well plate was pre-coated with Matrigel (Corning®) substrate (50 μL/well) and then polymerized at 37 °C for 60 min. Approximately 3×10^4 HUVECs were seeded in each well and the plate was incubated in a CO_2 incubator for 6–8 h at 37 °C. Finally, the tubular structures formed by HUVECs in the 96-well plate were observed under an inverted microscope.

4.12. VEGF and NO Level Detection

Cultures of drug-treated HUVECs cells were collected separately, and the supernatant was removed through centrifugation at 10,000 rpm and then stored at −20 °C. The VEGFA expression was assessed using an ELISA kit (R&D), as per the manufacturer's institutions. The amount of NO secreted was enumerated according to the manufacturer's instructions.

4.13. Western Blotting

Cells in the exponential differentiation phase were diluted to a cell suspension of 1×10^5 cell/mL using complete medium. Cells were seeded into a 6-well plate at 2 mL/well in an incubator at 37 °C and 5% CO_2. When the cells spread across 70% of the bottom of the 6-well plate, the complete medium was replaced with serum-free medium with or without DGS drug for 24 h.

HUVECs cells were washed and scraped out after centrifugation at 4 °C, followed by washing in RIPA buffer (150 mM NaCl, 50 mM Tris-HCl, 1% Triton, 0.5% NP40, 1 nM PMSF). After centrifugation at 10,000 rpm for 30 min, the supernatant was collected. The protein concentrations were determined using the BCA Protein Assay Kit (Beyotime Biotechnology, Shanghai, China). Each lane was loaded with 20 μg of the protein and separated. The proteins were then loaded on each lane and separated on SDS-PAGE gels, followed by electrophoretic transfer to nitrocellulose (NC) membranes (MilliPore, Boston, MA, USA). The membranes were blocked with 5% BSA for 2 h and then incubated with primary antibodies at 4 °C. The primary antibodies were diluted in the BSA buffer, and secondary antibodies were diluted in TBST. The antibody-reactive bands were revealed by chemiluminescence. The images were scanned and band intensities were analyzed with Image J software.

4.14. Animals

Adult wild-type (AB) zebrafish and Tg (flk1a:EGFP) transgenic zebrafish ((labeled green fluorescent protein at endothelial growth factor receptor)) were reared and managed at the Zebrafish Research Center in the School of Pharmacy and Pharmaceutical Sciences, Shandong First Medical University. The fish were kept in a recirculating system (Shanghai Haisheng Biological Experimental Equipment Co., Ltd., Shanghai, China) at a temperature of 28 °C under a mixed light schedule of 12 h light/dark. The water systems were monitored for nitrite (<0.2 ppm), nitrate (<50 ppm), and ammonia nitrogen (0.01–0.1 ppm). Conductivity and pH were maintained at 500 μS cm^{-1} and 7, respectively. The fish were fed with fungus shrimp twice daily and fasted for a day before spawning. The day before spawning the fish were placed in spawning tanks in a 1:1 ratio of male to female; the next day the isolation plates were removed after light stimulation for natural spawning, and all eggs were collected and placed in the E3 culture water (5 mM NaCl, 0.17 mM KCl, 0.4 mM $CaCl_2$, 0.16 mM $MgSO_4$) for incubation.

4.15. Animal Grouping and Administration

Transgenic Tg (flk1a: EGFP) zebrafish embryos (24 hpf) were stripped of their egg membranes in Petri dishes containing Pronase E (1 mg/mL) and randomly distributed in 24-well plates with 10 strips/well, using 3 replicates set up for each concentration group. The plates were then treated with different concentrations of DGS, the specific concentrations administered are illustrated in the corresponding results. After 72 h of treatment at 28.5 °C, the development and mortality of each group of zebrafish were enumerated.

Transgenic Tg (flk1a: EGFP) zebrafish embryos (24 hpf) were stripped of their egg membranes with Pronase E (1 mg/mL) in Petri dishes, which were randomly distributed in 6-well plates and then treated with different concentrations of DGS as indicated in the corresponding results. After treatment at 28.5 °C for 24 h, the zebrafish larvae were rinsed several times with water and then transferred to a 96-well plate. The total length of the intersegment vessels (ISVs) of zebrafish larvae was observed and imaged under a fluorescent microscope. The total ISV length was quantified for each zebrafish using the Image Pro Plus and the average ISVs length was calculated for each group of zebrafish larvae.

Transgenic Tg (flk1a: EGFP) zebrafish larvae (72 hpf) were randomly distributed in 6-well plates and then treated with different concentrations of DGS, as indicated in the corresponding results. After treatment at 28.5 °C for 24 h, the zebrafish larvae were rinsed several times with water and then transferred into a 96-well plate with the appropriate amount of anesthetic. The total length of the main subintestinal vein (MSIV) of zebrafish larvae was determined under a fluorescent microscope. The total MSIV length of each zebrafish was quantified using Image Pro Plus, and the mean MSIV length was calculated for each group of zebrafish larvae.

4.16. Dynamics Molecular Docking

The molecular docking study was conducted with the PyRx v0.8 [52]. VEGFR2 (PDB ID:3B8Q) was obtained from the RCSB Protein Data Bank (https://www.rcsb.org/, accessed on 1 March 2022). All polar hydrogens of the protein crystals were added, and the solvent water molecules were removed and converted into the pdbqt format. All compounds were obtained from the PubChem database as the sdf format files, and saved in the pdbqt format after conversion to an energy-minimized form using Pthe yRx v0.8. Docking was performed in the PyRx v0.8 with the following docking box parameters center_x 39.0 center_y 33.4 center_z 14.5 size_x 19.4 size_y 27.7 size_z 17.4. The results were visualized with the Pymol v2.4.1.

Desmond software package (developed at D. E. Shaw Research) was used to investigate the molecular interactions. In the molecular dynamics simulations, the complexes were placed into an automatically calculated cube box in which the complexes were modeled separately using transferable interatomic potential with three points model (TIP3P). The optimization of the models was further accomplished by optimized potentials for liquid simulations 4 (OPLS4). The system was neutralized by the addition of NaCl to make the system isotonic. A Nosehoover thermostat was used to provide a temperature of 300 k. The Martyna–Tobias–Klien barostat was used to maintain a pressure of 1.01325 bars. The total time for the molecular dynamics simulation was 50 ns. Ligand–protein interactions were simulated using the Interaction Diagram tool in the Desmond package. The Desmond package was used to generate the RMSD and RMSF of the proteins, and the interactions were further analyzed.

4.17. Statistical Analysis

Data are presented as the mean $\pm$ SEM. All statistical analyses were performed using the GraphPad Prism 9.0 software. The Student's two-tailed *t*-test was performed to compare the two study groups, while one-way ANOVA was applied to compare multiple groups. $p < 0.05$ was considered to indicate statistical significance.

5. Conclusions

We explored the potential mechanisms of DGS as a phytotherapy for CAD by employing a network pharmacology-driven strategy. Our findings suggest that DGS may exert proangiogenic and vasodilatory effects through the activation of the VEGF/VEGFR2/Akt/Erk/eNOS signaling pathway. Molecular docking and molecular dynamics suggest that salvianolic acid C may be a key component in exerting angiogenic and vasodilatory effects. Furthermore, our study results can serve as a reference for the mechanism of DGS as a natural product for phytotherapy.

Author Contributions: Conception and design of the work, Y.-G.Z., X.-X.L., Z.-M.L. and L.-W.H.; data collection, Y.-G.Z., X.-X.L., J.-C.Z., Y.-T.-J.Z., R.D., N.W. and Z.-H.M.; analysis and interpretation of the data, Y.-G.Z., X.-X.L., J.-C.Z., Y.-T.-J.Z., R.D., N.W. and Z.-H.M.; statistical analysis, Y.-G.Z., X.-X.L., J.-C.Z., R.D., N.W. and Z.-H.M.; drafting of the manuscript, Y.G-.Z., X.-X.L., J.-C.Z., R.D., N.W. and Z.-H.M.; critical revision of the manuscript, L.L., S.-L.W., Y.-L.M., S.-S.W., Z.-H.M. and L.-W.H.; project administration, Z.-M.L. and L.-W.H. All authors have read and agreed to the published version of the manuscript.

Funding: This work was funded by the Funding of the Key Project at Central Government Level: The ability establishment of sustainable use for valuable Chinese medicine resources (2060302), Natural Science Foundation of Shandong Province (ZR2019MH037), and Academic Promotion Programe of Shandong First Medical University (No. 2019LJ003).

Institutional Review Board Statement: Not applicable.

Informed Consent Statement: Not applicable.

Data Availability Statement: The data presented in this study are available on request from the corresponding author.

Acknowledgments: Thanks to the Zebrafish Center of School of Pharmacy and Pharmaceutical Science, Shandong First Medical University and Shandong Academy of Medical Sciences for providing zebrafish.

Conflicts of Interest: The authors declare that they have no conflict of interest to disclose.

Sample Availability: Not available.

References

1. Juan-Salvadores, P.; Jiménez Díaz, V.A.; Iglesia Carreño, C.; Guitián González, A.; Veiga, C.; Martínez Reglero, C.; Baz Alonso, J.A.; Caamaño Isorna, F.; Iñiguez Romo, A. Coronary Artery Disease in Very Young Patients: Analysis of Risk Factors and Long-Term Follow-Up. *J. Cardiovasc. Dev. Dis.* **2022**, *9*, 82. [CrossRef] [PubMed]
2. Rocker, A.J.; Lee, D.J.; Shandas, R.; Park, D. Injectable Polymeric Delivery System for Spatiotemporal and Sequential Release of Therapeutic Proteins To Promote Therapeutic Angiogenesis and Reduce Inflammation. *ACS Biomater. Sci. Eng.* **2020**, *6*, 1217–1227. [CrossRef] [PubMed]
3. Gautam, N.; Saluja, P.; Malkawi, A.; Rabbat, M.G.; Al-Mallah, M.H.; Pontone, G.; Zhang, Y.; Lee, B.C.; Al'Aref, S.J. Current and Future Applications of Artificial Intelligence in Coronary Artery Disease. *Healthcare* **2022**, *10*, 232. [CrossRef] [PubMed]
4. Khan, M.A.; Hashim, M.J.; Mustafa, H.; Baniyas, M.Y.; Al Suwaidi, S.K.B.M.; AlKatheeri, R.; Alblooshi, F.M.K.; Almatrooshi, M.E.A.H.; Alzaabi, M.E.H.; Al Darmaki, R.S.; et al. Global Epidemiology of Ischemic Heart Disease: Results from the Global Burden of Disease Study. *Cureus* **2020**, *12*, e9349. [CrossRef] [PubMed]
5. Sen, A.; Singh, A.; Roy, A.; Mohanty, S.; Naik, N.; Kalaivani, M.; Ramakrishnan, L. Role of Endothelial Colony Forming Cells (ECFCs) Tetrahydrobiopterin (BH4) in Determining ECFCs Functionality in Coronary Artery Disease (CAD) Patients. *Sci. Rep.* **2022**, *12*, 1–12. [CrossRef]
6. Jamaiyar, A.; Juguilon, C.; Wan, W.; Richardson, D.; Chinchilla, S.; Gadd, J.; Enrick, M.; Wang, T.; McCabe, C.; Wang, Y.; et al. The Essential Role for Endothelial Cell Sprouting in Coronary Collateral Growth. *J. Mol. Cell. Cardiol.* **2022**, *165*, 158–171. [CrossRef]

7. Brendler, T.; Al-Harrasi, A.; Bauer, R.; Gafner, S.; Hardy, M.L.; Heinrich, M.; Hosseinzadeh, H.; Izzo, A.A.; Michaelis, M.; Nassiri-Asl, M.; et al. Botanical Drugs and Supplements Affecting the Immune Response in the Time of COVID-19: Implications for Research and Clinical Practice. *Phytother. Res.* **2021**, *35*, 3013–3031. [CrossRef]
8. Orekhov, A.N.; Ivanova, E.A. Cellular Models of Atherosclerosis and Their Implication for Testing Natural Substances with Anti-Atherosclerotic Potential. *Phytomedicine* **2016**, *23*, 1190–1197. [CrossRef]
9. Ling, Y.; Shi, J.; Ma, Q.; Yang, Q.; Rong, Y.; He, J.; Chen, M. Vasodilatory Effect of Guanxinning Tablet on Rabbit Thoracic Aorta Is Modulated by Both Endothelium-Dependent and -Independent Mechanism. *Front. Pharmacol.* **2021**, *12*, 754527. [CrossRef]
10. Wang, X.-P.; Wang, P.-F.; Bai, J.-Q.; Gao, S.; Wang, Y.-H.; Quan, L.-N.; Wang, F.; Wang, X.-T.; Wang, J.; Xie, Y.-D. Investigating the Effects and Possible Mechanisms of Danshen-Honghua Herb Pair on Acute Myocardial Ischemia Induced by Isoproterenol in Rats. *Biomed. Pharmacother.* **2019**, *118*, 109268. [CrossRef]
11. Wu, W.; Wang, Y. Pharmacological Actions and Therapeutic Applications of Salvia Miltiorrhiza Depside Salt and Its Active Components. *Acta Pharmacol. Sin.* **2012**, *33*, 1119–1130. [CrossRef] [PubMed]
12. Tseng, Y.-J.; Hung, Y.-C.; Kuo, C.-E.; Chung, C.-J.; Hsu, C.Y.; Muo, C.-H.; Hsu, S.-F.; Hu, W.-L. Prescription of Radix Salvia Miltiorrhiza in Taiwan: A Population-Based Study Using the National Health Insurance Research Database. *Front. Pharmacol.* **2021**, *12*, 719519. [CrossRef] [PubMed]
13. Cicero, A.F.G.; Colletti, A.; Bellentani, S. Nutraceutical Approach to Non-Alcoholic Fatty Liver Disease (NAFLD): The Available Clinical Evidence. *Nutrients* **2018**, *10*, E1153. [CrossRef] [PubMed]
14. Li, Q.; Cui, Y.; Xu, B.; Wang, Y.; Lv, F.; Li, Z.; Li, H.; Chen, X.; Peng, X.; Chen, Y.; et al. Main Active Components of Jiawei Gegen Qinlian Decoction Protects against Ulcerative Colitis under Different Dietary Environments in a Gut Microbiota-Dependent Manner. *Pharmacol. Res.* **2021**, *170*, 105694. [CrossRef] [PubMed]
15. Dai, H.; Lv, Z.; Huang, Z.; Ye, N.; Li, S.; Jiang, J.; Cheng, Y.; Shi, F. Dietary Hawthorn-Leaves Flavonoids Improves Ovarian Function and Liver Lipid Metabolism in Aged Breeder Hens. *Poult. Sci.* **2021**, *100*, 101499. [CrossRef]
16. Hu, Y.; Li, Q.; Pan, Y.; Xu, L. Sal B Alleviates Myocardial Ischemic Injury by Inhibiting TLR4 and the Priming Phase of NLRP3 Inflammasome. *Molecules* **2019**, *24*, 4416. [CrossRef]
17. Huo, Y.; Yi, B.; Chen, M.; Wang, N.; Chen, P.; Guo, C.; Sun, J. Induction of Nur77 by Hyperoside Inhibits Vascular Smooth Muscle Cell Proliferation and Neointimal Formation. *Biochem. Pharmacol.* **2014**, *92*, 590–598. [CrossRef]
18. Yang, H.; Cheung, M.-K.; Yue, G.G.-L.; Leung, P.-C.; Wong, C.-K.; Lau, C.B.-S. Integrated Network Pharmacology Analysis and In Vitro Validation Revealed the Potential Active Components and Underlying Mechanistic Pathways of Herba Patriniae in Colorectal Cancer. *Molecules* **2021**, *26*, 6032. [CrossRef]
19. Zhang, R.; Zhu, X.; Bai, H.; Ning, K. Network Pharmacology Databases for Traditional Chinese Medicine: Review and Assessment. *Front. Pharmacol.* **2019**, *10*, 123. [CrossRef]
20. Li, S.; Zhang, W.; Wang, R.; Li, C.; Lin, X.; Wang, L. Screening and Identification of Natural α-Glucosidase and α-Amylase Inhibitors from Partridge Tea (Mallotus Furetianus Muell-Arg) and in Silico Analysis. *Food Chem.* **2022**, *388*, 133004. [CrossRef]
21. Chen, Y.; Zhang, N.; Ma, J.; Zhu, Y.; Wang, M.; Wang, X.; Zhang, P. A Platelet/CMC Coupled with Offline UPLC-QTOF-MS/MS for Screening Antiplatelet Activity Components from Aqueous Extract of Danshen. *J. Pharm. Biomed. Anal.* **2016**, *117*, 178–183. [CrossRef] [PubMed]
22. Pillarisetti, P.; Myers, K.A. Identification and Characterization of Agnuside, a Natural Proangiogenic Small Molecule. *Eur. J. Med. Chem.* **2018**, *160*, 193–206. [CrossRef] [PubMed]
23. Xiang, W.; Suo, T.-C.; Yu, H.; Li, A.-P.; Zhang, S.-Q.; Wang, C.-H.; Zhu, Y.; Li, Z. A New Strategy for Choosing "Q-Markers" via Network Pharmacology, Application to the Quality Control of a Chinese Medical Preparation. *J. Food Drug Anal.* **2018**, *26*, 858–868. [CrossRef] [PubMed]
24. Li, S.; Zhang, B.; Jiang, D.; Wei, Y.; Zhang, N. Herb Network Construction and Co-Module Analysis for Uncovering the Combination Rule of Traditional Chinese Herbal Formulae. *BMC Bioinform.* **2010**, *11*, S6. [CrossRef] [PubMed]
25. Li, S.; Zhang, B. Traditional Chinese Medicine Network Pharmacology: Theory, Methodology and Application. *Chin. J. Nat. Med.* **2013**, *11*, 110–120. [CrossRef]
26. Yu, L.-J.; Zhang, K.-J.; Zhu, J.-Z.; Zheng, Q.; Bao, X.-Y.; Thapa, S.; Wang, Y.; Chu, M.-P. Salvianolic Acid Exerts Cardioprotection through Promoting Angiogenesis in Animal Models of Acute Myocardial Infarction: Preclinical Evidence. *Oxidative Med. Cell. Longev.* **2017**, *2017*, 8192383. [CrossRef]
27. Syahputra, R.A.; Harahap, U.; Dalimunthe, A.; Nasution, M.P.; Satria, D. The Role of Flavonoids as a Cardioprotective Strategy against Doxorubicin-Induced Cardiotoxicity: A Review. *Molecules* **2022**, *27*, 1320. [CrossRef]
28. Ghigo, A.; Laffargue, M.; Li, M.; Hirsch, E. PI3K and Calcium Signaling in Cardiovascular Disease. *Circ. Res.* **2017**, *121*, 282–292. [CrossRef]
29. Kazakov, A.; Hall, R.; Jagoda, P.; Bachelier, K.; Müller-Best, P.; Semenov, A.; Lammert, F.; Böhm, M.; Laufs, U. Inhibition of Endothelial Nitric Oxide Synthase Induces and Enhances Myocardial Fibrosis. *Cardiovasc. Res.* **2013**, *100*, 211–221. [CrossRef]
30. Apte, R.S.; Chen, D.S.; Ferrara, N. VEGF in Signaling and Disease: Beyond Discovery and Development. *Cell* **2019**, *176*, 1248–1264. [CrossRef]
31. Li, J.; Zhang, Y.; Li, C.; Xie, J.; Liu, Y.; Zhu, W.; Zhang, X.; Jiang, S.; Liu, L.; Ding, Z. HSPA12B Attenuates Cardiac Dysfunction and Remodelling after Myocardial Infarction through an ENOS-Dependent Mechanism. *Cardiovasc. Res.* **2013**, *99*, 674–684. [CrossRef] [PubMed]

32. Li, S.-N.; Li, P.; Liu, W.-H.; Shang, J.-J.; Qiu, S.-L.; Zhou, M.-X.; Liu, H.-X. Danhong Injection Enhances Angiogenesis after Myocardial Infarction by Activating MiR-126/ERK/VEGF Pathway. *Biomed. Pharmacother.* **2019**, *120*, 109538. [CrossRef] [PubMed]
33. Zhang, X.; Shi, Y.; Wang, L.; Li, X.; Zhang, S.; Wang, X.; Jin, M.; Hsiao, C.-D.; Lin, H.; Han, L.; et al. Metabolomics for Biomarker Discovery in Fermented Black Garlic and Potential Bioprotective Responses against Cardiovascular Diseases. *J. Agric. Food Chem.* **2019**, *67*, 12191–12198. [CrossRef] [PubMed]
34. Dong, R.; Tian, Q.; Shi, Y.; Chen, S.; Zhang, Y.; Deng, Z.; Wang, X.; Yao, Q.; Han, L. An Integrated Strategy for Rapid Discovery and Identification of Quality Markers in Gardenia Fructus Using an Omics Discrimination-Grey Correlation-Biological Verification Method. *Front. Pharmacol.* **2021**, *12*, 705498. [CrossRef]
35. Lawson, N.D.; Weinstein, B.M. In Vivo Imaging of Embryonic Vascular Development Using Transgenic Zebrafish. *Dev. Biol.* **2002**, *248*, 307–318. [CrossRef]
36. He, L.; Huang, X.; Kanisicak, O.; Li, Y.; Wang, Y.; Li, Y.; Pu, W.; Liu, Q.; Zhang, H.; Tian, X.; et al. Preexisting Endothelial Cells Mediate Cardiac Neovascularization after Injury. *J. Clin. Investig.* **2017**, *127*, 2968–2981. [CrossRef]
37. Tang, J.; Zhang, H.; He, L.; Huang, X.; Li, Y.; Pu, W.; Yu, W.; Zhang, L.; Cai, D.; Lui, K.O.; et al. Genetic Fate Mapping Defines the Vascular Potential of Endocardial Cells in the Adult Heart. *Circ. Res.* **2018**, *122*, 984–993. [CrossRef]
38. Greenspan, L.J.; Weinstein, B.M. To Be or Not to Be: Endothelial Cell Plasticity in Development, Repair, and Disease. *Angiogenesis* **2021**, *24*, 251–269. [CrossRef]
39. Xiang, Y.; Yao, X.; Wang, X.; Zhao, H.; Zou, H.; Wang, L.; Zhang, Q.-X. Houshiheisan Promotes Angiogenesis via HIF-1α/VEGF and SDF-1/CXCR4 Pathways: In Vivo and in Vitro. *Biosci. Rep.* **2019**, *39*, 1–12. [CrossRef]
40. Monika, P.; Chandraprabha, M.N.; Rangarajan, A.; Waiker, P.V.; Chidambara Murthy, K.N. Challenges in Healing Wound: Role of Complementary and Alternative Medicine. *Front. Nutr.* **2021**, *8*, 791899. [CrossRef]
41. Salo, T.; Dourado, M.R.; Sundquist, E.; Apu, E.H.; Alahuhta, I.; Tuomainen, K.; Vasara, J.; Al-Samadi, A. Organotypic Three-Dimensional Assays Based on Human Leiomyoma-Derived Matrices. *Philos. Trans. R. Soc. Lond. B Biol. Sci.* **2018**, *373*, 20160482. [CrossRef] [PubMed]
42. Quesada-Gómez, J.M.; Santiago-Mora, R.; Navarro-Valverde, C.; Dorado, G.; Casado-Díaz, A. Stimulation of In-Vitro Angiogenesis by Low Concentrations of Risedronate Is Mitigated by 1,25-Dihydroxyvitamin D3 or 24,25-Dihydroxyvitamin D3. *J. Steroid Biochem. Mol. Biol.* **2015**, *148*, 214–218. [CrossRef] [PubMed]
43. Lucas-Herald, A.K.; Montezano, A.C.; Alves-Lopes, R.; Haddow, L.; Alimussina, M.; O'Toole, S.; Flett, M.; Lee, B.; Amjad, S.B.; Steven, M.; et al. Vascular Dysfunction and Increased Cardiovascular Risk in Hypospadias. *Eur. Heart J.* **2022**, *43*, 1832–1845. [CrossRef]
44. Zullino, S.; Buzzella, F.; Simoncini, T. Nitric Oxide and the Biology of Pregnancy. *Vasc. Pharmacol.* **2018**, *110*, 71–74. [CrossRef] [PubMed]
45. Ambari, A.M.; Lilihata, G.; Zuhri, E.; Ekawati, E.; Wijaya, S.A.; Dwiputra, B.; Sukmawan, R.; Radi, B.; Haryana, S.M.; Adiarto, S.; et al. External Counterpulsation Improves Angiogenesis by Preserving Vascular Endothelial Growth Factor-A and Vascular Endothelial Growth Factor Receptor-2 but Not Regulating MicroRNA-92a Expression in Patients With Refractory Angina. *Front. Cardiovasc. Med.* **2021**, *8*, 761112. [CrossRef]
46. Li, J.-Z.; Zhou, X.-X.; Wu, W.-Y.; Qiang, H.-F.; Xiao, G.-S.; Wang, Y.; Li, G. Concanavalin A Promotes Angiogenesis and Proliferation in Endothelial Cells through the Akt/ERK/Cyclin D1 Axis. *Pharm. Biol.* **2022**, *60*, 65–74. [CrossRef]
47. Gfeller, D.; Michielin, O.; Zoete, V. Shaping the Interaction Landscape of Bioactive Molecules. *Bioinformatics* **2013**, *29*, 3073–3079. [CrossRef]
48. Safran, M.; Rosen, N.; Twik, M.; BarShir, R.; Stein, T.I.; Dahary, D.; Fishilevich, S.; Lancet, D. The GeneCards Suite. In *Practical Guide to Life Science Databases*; Abugessaisa, I., Kasukawa, T., Eds.; Springer Nature Singapore: Singapore, 2021; pp. 27–56. ISBN 978-981-16-5812-9.
49. Bader, G.D.; Hogue, C.W.V. An Automated Method for Finding Molecular Complexes in Large Protein Interaction Networks. *BMC Bioinform.* **2003**, *4*, 2–27. [CrossRef]
50. Sherman, B.T.; Hao, M.; Qiu, J.; Jiao, X.; Baseler, M.W.; Lane, H.C.; Imamichi, T.; Chang, W. DAVID: A Web Server for Functional Enrichment Analysis and Functional Annotation of Gene Lists (2021 Update). *Nucleic Acids Res.* **2022**, *10*, 1–6. [CrossRef]
51. Huang, D.W.; Sherman, B.T.; Lempicki, R.A. Systematic and Integrative Analysis of Large Gene Lists Using DAVID Bioinformatics Resources. *Nat. Protoc.* **2009**, *4*, 44–57. [CrossRef]
52. Dallakyan, S.; Olson, A.J. Small-Molecule Library Screening by Docking with PyRx. *Methods Mol. Biol.* **2015**, *1263*, 243–250. [CrossRef] [PubMed]

Article

Binding Studies of Caffeic and p-Coumaric Acid with α-Amylase: Multispectroscopic and Computational Approaches Deciphering the Effect on Advanced Glycation End Products (AGEs)

Mohd Shahnawaz Khan *, Majed S. Alokail, Amal Majed H. Alenad, Nojood Altwaijry, Nouf Omar Alafaleq, Abdulaziz Mohammed Alamri and Mubarak Ali Zawba

Department of Biochemistry, College of Science, King Saud University, Riyadh 11451, Saudi Arabia; malokail@ksu.edu.sa (M.S.A.); aalenad@ksu.edu.sa (A.M.H.A.); nojood@ksu.edu.sa (N.A.); nalafaleg@ksu.edu.sa (N.O.A.); abalamri@ksu.edu.sa (A.M.A.); 436160113@student.ksu.edu.sa (M.A.Z.)
* Correspondence: moskhan@ksu.edu.sa

Abstract: Alpha-amylase (α-amylase) is a key player in the management of diabetes and its related complications. This study was intended to have an insight into the binding of caffeic acid and coumaric acid with α-amylase and analyze the effect of these compounds on the formation of advanced glycation end-products (AGEs). Fluorescence quenching studies suggested that both the compounds showed an appreciable binding affinity towards α-amylase. The evaluation of thermodynamic parameters (ΔH and ΔS) suggested that the α-amylase-caffeic/coumaric acid complex formation is driven by van der Waals force and hydrogen bonding, and thus complexation process is seemingly specific. Moreover, glycation and oxidation studies were also performed to explore the multitarget to manage diabetes complications. Caffeic and coumaric acid both inhibited fructosamine content and AGE fluorescence, suggesting their role in the inhibition of early and advanced glycation end-products (AGEs). However, the glycation inhibitory potential of caffeic acid was more in comparison to p-coumaric acid. This high antiglycative potential can be attributed to its additional –OH group and high antioxidant activity. There was a significant recovery of 84.5% in free thiol groups in the presence of caffeic acid, while coumaric attenuated the slow recovery of 29.4% of thiol groups. In vitro studies were further entrenched by in silico studies. Molecular docking studies revealed that caffeic acid formed six hydrogen bonds (Trp 59, Gln 63, Arg 195, Arg 195, Asp 197 and Asp 197) while coumaric acid formed four H-bonds with Trp 59, Gln 63, Arg 195 and Asp 300. Our studies highlighted the role of hydrogen bonding, and the ligands such as caffeic or coumaric acid could be exploited to design antidiabetic drugs.

Keywords: α-amylase; advanced glycation end-products; caffeic acid; coumaric acid

Citation: Khan, M.S.; Alokail, M.S.; Alenad, A.M.H.; Altwaijry, N.; Alafaleq, N.O.; Alamri, A.M.; Zawba, M.A. Binding Studies of Caffeic and p-Coumaric Acid with α-Amylase: Multispectroscopic and Computational Approaches Deciphering the Effect on Advanced Glycation End Products (AGEs). *Molecules* **2022**, *27*, 3992. https://doi.org/10.3390/molecules27133992

Academic Editor: Michael Assfalg

Received: 8 April 2022
Accepted: 27 May 2022
Published: 21 June 2022

Publisher's Note: MDPI stays neutral with regard to jurisdictional claims in published maps and institutional affiliations.

1. Introduction

In the past decades, rigorous research and management strategies have been implicated in tackling the menace of diabetes mellitus (DM) [1]. According to the International Diabetes Federation, the number of diabetic patients reached 463 million in 2019, and the numbers are expected to rise by 66% by 2045 [2]. DM is a metabolic syndrome characterized by a loss of capability to oxidize carbohydrates by individuals due to altered insulin production, low insulin production and desensitized insulin receptors [3]. The inefficiency in metabolizing carbohydrates causes a surge in glucose levels in the patients' blood. Monomeric carbohydrates can undergo non-enzymatic glycation towards proteins, nucleic acids and lipids [4].

Glycation is a non-enzymatic and spontaneous reaction. It is commenced by the reaction between reducing sugars carbonyl group with a free amino group, generally lysine's ε-amino group and α-amino group of the protein. The process forms complex adducts referred to as Schiff base [5]. The base undergoes Amadori rearrangement to form various complex compounds known as advanced glycation end products (AGEs) [6,7]. There are a series of events that results in cellular malfunctioning as a result of high glucose, and these are not fully elucidated, thereby attracting researchers across the globe. Amongst these, a major event is the formation of AGEs that can contribute to diabetic complications in different ways. Glycation of proteins results in AGEs formation, leading to many diabetic complications. Protein glycation alters their normal functioning in various ways, viz. disruption of molecular conformation, alteration of enzymatic activity, reducing degradation capacity and others. AGEs interaction with their cellular receptors plays a crucial role in the pathogenesis of diabetic complications [8]. AGEs formation further aids the formation of reactive oxygen species (ROS), which further damages the cellular antioxidant defense mechanism and increases the oxidative load [9]. AGE formation is augmented in diabetic conditions, especially in Type 2 DM [10,11]. Glycation contributes greatly to AGE formation; highly reactive C=O species such as 3-deoxyglucosone and methylglyoxal are generated in glycation, lipid peroxidation and polyol pathway of DM type II [12]. Accumulation of AGEs contributes to many chronic diseases such as diabetes, Alzheimer's disease, heart failure and many other life-threatening conditions [13]. Further AGEs interaction with RAGEs, their specific receptors causes an onset of an inflammation cascade and further increases the oxidative load in the cells. The cascade of inflammation involves the activation of several pathways involved in diabetes, such as the MAP kinase pathway, TNF-α, NF-Kb, etc. [14]. The toxic accumulations can further contribute to pathogenesis in diabetic conditions such as decreased ligand binding, altered enzymatic activities and immunogenicity [15].

α-Amylase is an essential enzyme in carbohydrate degradation and is linked with post-prandial hyperglycemia in Type-II DM patients. The enzyme hydrolyzes carbohydrates and creates a glycemic environment in the bloodstream, aiding AGEs formation and thrusting diabetic complications [16]. Constraining these enzymes' activity can subdue blood glucose levels. Investigations have shown that α-amylase inhibition aids in regulating glucose levels [17]. Polyphenols, their derivatives and other naturally derived compounds show therapeutic potential against various diseases such as cancer [18–21], neurodegenerative disorders [22,23], diabetes [24,25] and many more.

Caffeic acid (3,4-dihydroxycinnamic) is one of the most abundant hydroxycinnamate present in plant tissues. Phenolic acid caffeine is prevalent in various food sources such as apple, cider, blueberries, etc., and beverages such as tea and coffee [26]. Caffeic acid is known to cross the brain barrier and is classified as an antioxidant, antibacterial and anti-cancer compound [27,28]. Similarly, Coumaric acid is an essential polyphenol that governs the synthesis of some other important polyphenolic compounds such as sinapinic, ferulic and caffeic acid [29,30]. CA also plays an important role in lignin synthesis and is an abundant dietary polyphenol present in apples, maize, tomato and wheat [31]. The present study was designed to explore the binding mechanism of polyphenols (caffeic/coumaric acid) to α-amylase. Our study was also extended to reveal the anti-glycation activities of these structures and their modes of action. Various spectroscopic, biochemicals and computational approaches were employed.

2. Materials and Methods

2.1. Materials

Human serum albumin (HSA), methylglyoxal (MG), nitro blue tetrazolium (NBT), 2,4-Dinitrophenylhydrazine (DNPH), caffeic acid, p-coumaric acid and porcine pancreatic α-amylase were purchased from Sigma-Aldrich, Milwaukee, WI, USA. All other chemicals, unless stated otherwise, were high grade and purchased from local vendors. All the blanks were run replicating the same conditions and acted as a control. All the buffers were filtered before use.

2.2. Steady-State Fluorescence

The inhibitory effect of caffeic and coumaric acid on α-amylase was studied by exploiting the fluorescence ability of the protein. Quenching studies were carried out against the protein in the presence of both the natural compounds, using a spectrofluorometer (Jasco FP-750, JASCO Corporation, Tokyo, Japan). The protein (5 μM α-amylase) was excited at 295 nm, and the emission spectra were recorded at higher wavelengths between 300–400nm [32–35]. Caffeic acid (0–40 μM) and coumaric acid (0–30 μM) were titrated with the α-amylase at three different temperatures (25, 30 and 35 °C). All these binding experiments were conducted at physiological pH of 7.4. The obtained quenching data were put into Stern–Volmer (Equation (1)) and modified Stern–Volmer equation (Equation (2)) as per earlier published literature [20,21,36] to find various binding parameters. The binding experiment was performed at physiological pH of 7.4.

F_0 corresponds to the maximum fluorescence intensity of free protein, *F shows the fluorescence intensity of complex, *K corresponds to the binding site, n depicts the binding sites and *C refers to the concentration of ligand.

$$F_0/F = 1 + Ksv\,[C] \tag{1}$$

$$log\,[(F_0 - F)/F] = log\,K + nlog[C] \tag{2}$$

Using the van't Hoff equation, (Equation (3)) [37,38], thermodynamic parameters for ligand–protein interaction such as Gibbs free energy change (ΔG^0), enthalpy change (ΔH^0) and entropy change (ΔS^0) can be calculated.

$$\Delta G^0 = -RTLnK = \Delta H^0 - T\Delta S^0 \tag{3}$$

The mode of quenching was further confirmed from the value of the bimolecular quenching rate constant, K_q, which was calculated as per Equation (4)

$$Kq = Ksv/\tau_0 \tag{4}$$

τ_0 refers to the average integral fluorescence lifetime of tryptophan and is reported to be 10^{-8}.

2.3. Synchronous Fluorescence

Synchronous fluorescence spectrometry is used as a common tool to inspect conformational changes in proteins in the presence of other molecules, also termed quenchers [39]. Synchronous fluorescence analysis of α-amylase was carried out in the absence and presence of caffeic acid and coumaric acid (0–20 μM) to obtain the spectra. A 15 nm difference was kept for Tyr residues between the excitation (245 nm) and emission spectra (260–340 nm). In a similar manner for Trp, the difference was maintained at 60 nm, the excitation was fixed at 220 nm and the emission was between 280–400 nm.

2.4. Non-Enzymatic Glycation

HSA was glycated using methylglyoxal (MG) as an inducer, as reported earlier [36,40]. Briefly, HSA was taken at the concentration of 10 mg/mL and incubated along with MG (3 mM) in the presence of caffeic and coumaric acid (0–200 μM) under sterile conditions using 0.02% sodium azide. HSA alone and in the presence of MG was also incubated under similar conditions as negative and positive control samples, respectively. Samples were further dialyzed in 20 mM sodium phosphate buffer with successive changes at regular intervals for 24 h. Protein concentration was determined using the Bradford method [41] and stored at −20 °C.

2.5. Determination of Advanced Glycation End-Products (AGEs)

AGEs were estimated for all the samples using fluorescence spectroscopy. A dilution factor of 10 was applied to all the samples, and then, the samples incubated were excited at

340 nm, and the emission was recorded from 350 to 500 nm [40]. The inhibitory effect of the ligand on the AGEs formation was calculated by the given Equation (5):

$$Inhibition\ (\%) = (F_g - F_t / F_g - F_c) \times 100 \tag{5}$$

2.6. Detection of Early Glycation (Amadori) Products: Quantification of Fructosamine

NBT assay was used to determine fructosamine content; the previously used protocol was used [42]. Briefly, 0.5 mM NBT was mixed with samples (0.5 mg/mL) and incubated in 100 mM sodium carbonate buffer of pH 10.4. The reaction mixture was incubated for 2 h at 37 °C, and reading was taken at 530 nm. The concentration of fructosamine was evaluated using its molar extinction coefficient value, i.e., 12,640 M^{-1} cm^{-1} [43].

2.7. Protein Oxidation Measurement: Carbonyl and Free Thiol (SH) Content

Carbonyl content was estimated to calculate the level of protein oxidation [42]. Briefly, aliquoted protein samples (100 µL) were mixed with 400 µL DNPH (10 mM). After thorough mixing, 500 µL of TCA (20% *w/v*) was added and centrifuged at 10,000× *g* for 10 min. The pellet was washed further with a 1 mL ethanol/ethyl acetate (1:1) mixture and resuspended in 1 mL of 6 M guanidine hydrochloride. The absorbance of the sample was recorded at 370 nm, and the concentration expressed as nanomoles of carbonyls per milligram of protein was determined using 22,000 M^{-1} cm^{-1} as molar absorptivity.

Ellman's reagent was used to calculate the free thiol content [44]. Native and glycated samples in the absence and presence of caffeic/coumaric acid (250 µL) were incubated with 750 µL of DTNB (0·5 mM) for 15 min, and the absorbance was measured at 412 nm. The concentration of free thiol groups was calculated using a standard curve of L-cysteine and expressed as nanomoles of L-cysteine per milligram of protein.

2.8. Molecular Docking

The interaction between pancreatic α-amylase and both caffeic acid and p-coumaric acid was performed using Autodock-4.2.6 and Discovery. The three-dimensional coordinates of pancreatic α-amylase (PDB ID: 1hny) were retrieved from the protein data bank (www.rcsb.org, accessed on 22 February 2022). The X-ray structure was 1.8 Å [45]. The enzyme structure was pre-processed by adding polar hydrogen atoms, deleting unessential water molecules, and adding Kollman charges through Autodock. Similarly, the two-dimensional structures of both caffeic acid and p-coumaric acid were downloaded from the PDB website (CID: 689043 and CID: 637542, respectively). The ligands were prepared through Autodock, and Gasteiger charges were added. Molecular docking was performed through Autodock using a genetic algorithm, and the output was set to Lamarckian GA (4.2).

3. Results and Discussion

3.1. Intrinsic Fluorescence

The pattern of changes exhibited in the protein fluorescence after adding a ligand provides information with structure and hence the protein's function [46]. Steady-state fluorescence quenching has been used to determine various binding parameters of ligand–protein interactions. Trp residue has the strongest fluorescence intensity and is the most sensitive to the changes in the micro-environment [47], and its emission wavelength is more sensitive to the microenvironment, indicating the protein conformational changes after binding with drugs [48]. Generally, the quenching mechanism can be classified into three categories: (1) dynamic quenching, which is caused by the collision between molecules in the transition to the excited state; (2) static quenching, which is caused by the formation of a complex between the fluorophore and the quencher; (3) the combined static and dynamic quenching [49]. The quenching was carried out with increasing concentrations of the ligands against α-amylase at different temperatures (Figures 1A and 2A). As indicated by the spectra, quenching was observed with increasing concentration of both the ligands caffeic and coumaric acid. This decrease in the fluorescence intensity amid the addition of caffeic acid and coumaric acid indicates that the ligands interact in the microenvironment of aromatic residues of α-amylase and mask the internal optimal fluorescence intensity. Previous investigations have shown that many other polyphenols exhibited similar behavior, thereby decreasing the fluorescence intensity of α-amylase [29,37,38]. Fluorescence quenching data were evaluated mathematically employing Stern–Volmer, modified Stern–Volmer, and van't Hoff equations to calculate the binding and thermodynamic parameters [26,28].

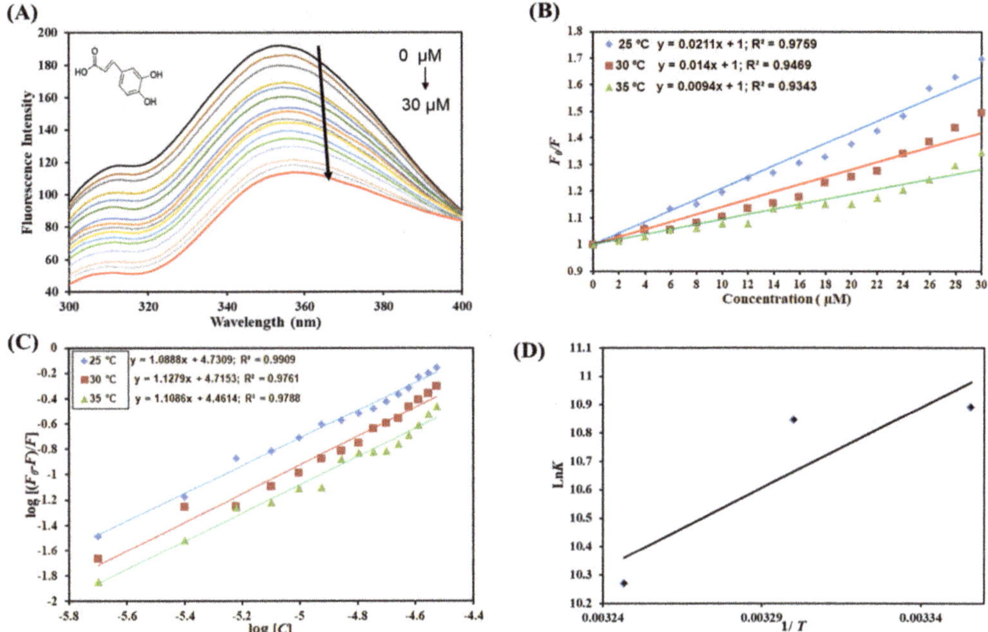

Figure 1. Binding between caffeic acid and α-amylase. (**A**) Quenching in fluorescence intensity of α-amylase (5 μM) in the absence and presence of varying caffeic acid concentration (0–30 μM) at 298 K, (**B**) Stern–Volmer plot at different temperatures, (**C**) modified Stern–Volmer plot at different temperatures, and (**D**) van't Hoff thermodynamics plot at three different temperatures.

In the SV plot of F_0/F vs. caffeic acid and F_0/F vs. [coumaric acid] (Figures 1B and 2B), the slope gives the value of the Stern–Volmer constant (K_{sv}) [40]. Figures 1B and 2B apparently show linear SV plots for caffeic acid–α-amylase and coumaric acid–α-amylase, respectively. Fluorescence quenching can be static or dynamic or a combination of both [34,35]. Temperature dependency of K_{sv} reveals the type of quenching operative for a particular interaction, i.e., protein–ligand complex formation is driven by static or dynamic quenching. The K_{sv} value decreases with increasing temperature for static quenching due to complex formation, which undergoes dissociation on increasing the temperature. In contrast, the opposite effect occurs for dynamic quenching, where K_{sv} increases with temperature. Hence, the values of K_{sv} were estimated at three different temperatures for α-amylase-caffeic acid and α-amylase-coumaric acid and are enumerated in Tables 1 and 2, respectively. K_{sv} increases with increasing temperature for α-amylase–caffeic acid interaction, implying dynamic quenching. On the contrary, K_{sv} values were found to decrease with increasing temperature for α-amylase–coumaric acid, suggesting the presence of static quenching. These observations can corroborate previously reported results [30,40]. Further, the quenching mode was confirmed by finding the biomolecular quenching rate constant (K_q) using the equation $K_q = K_{sv}/\tau_0$ ($\tau_0 = 2.7 \times 10^{-9}$ s). The values of K_q for α-amylase–caffeic acid

(Table 1) and α-amylase–coumaric acid (Table 2) were found to be higher than the maximum dynamic quenching constant (nearly 10^{10} M^{-1} s^{-1}) [49]. Thus, it can be concluded that α-amylase–caffeic acid complex formation was driven by dynamic quenching while a combination of static and dynamic directs α-amylase–coumaric acid complex formation, while α-amylase–coumaric acid interaction was driven by static quenching. Additionally, the binding constant (K) was also calculated, revealing the binding affinity for protein (Table 2). Fluorescence quenching data were fitted into a modified Stern–Volmer equation with the intercept of the plot giving the value of binding constant (K) for both the ligands (Figures 1C and 2C). It was found to be of the order of 10^4 M^{-1} for α-amylase–caffeic acid complex, while for α-amylase–coumaric acid, K was of the order of 10^4 M^{-1}, but lesser than caffeic acid, suggesting that caffeic acid binds to α-amylase with a higher affinity as compared to coumaric acid. Table 1 depicts the values of K at different temperatures for α-amylase–caffeic acid complex, which was found to decrease with increasing temperature, implying that a more stable complex is formed at lower temperatures. Table 2 depicts the values of K obtained for α-amylase–coumaric acid complex.

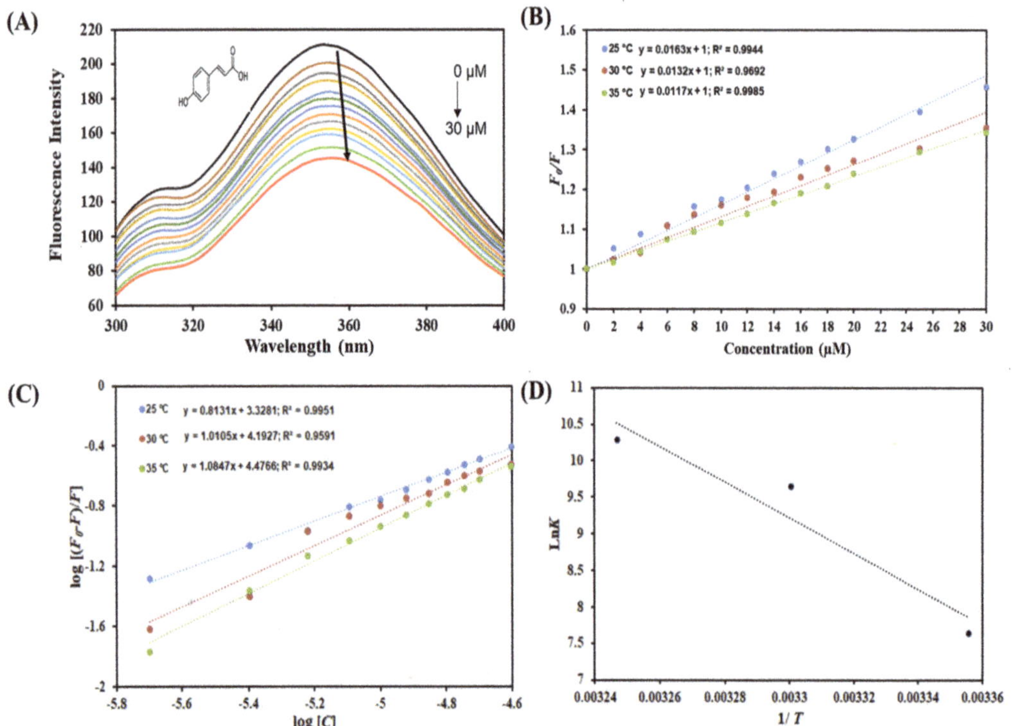

Figure 2. (**A**) Binding between p-coumaric acid and α-amylase. Fluorescence quenching intensity of α-amylase (5 µM) in the absence and presence of varying p-coumaric acid concentration (0–30 µM) at 298 K, (**B**) Stern–Volmer plot at different temperatures, (**C**) modified Stern–Volmer plot at different temperatures and (**D**) van't Hoff thermodynamics plot at three different temperatures.

Table 1. Binding and thermodynamic parameters obtained for caffeic acid–α- amylase interaction obtained through fluorescence spectroscopy.

Temp. (°C)	K_{sv} (10^4 M^{-1})	K_q (10^{13} M^{-1} s^{-1})	K (10^4 M^{-1})	ΔG (kcal mol^{-1})	ΔS (cal.mol^{-1}K^{-1})	ΔH (kcal mol^{-1})	TΔS (kcal mol^{-1})
25	2.1	0.77	5.37	−6.49			−4.78
30	1.4	0.51	5.12	−6.41	−16.04	−11.28	−4.86
35	0.9	0.33	2.88	−6.33			−4.94

Table 2. Binding and thermodynamic parameters obtained for coumaric acid–α- amylase interaction obtained through fluorescence spectroscopy.

Temp. (°C)	K_{sv} (10^4 M^{-1})	K_q (10^{13} M^{-1} s^{-1})	K (10^4 M^{-1})	ΔG (kcal mol^{-1})	ΔS (cal.mol^{-1}K^{-1})	ΔH (kcal mol^{-1})	TΔS (kcal mol^{-1})
25	1.6	0.59	0.2	−4.65			53.07
30	1.3	0.48	1.5	−5.54	178.11	48.42	53.96
35	1.1	0.40	2.9	−6.43			54.85

The data obtained at different temperatures are fitted into the van't Hoff equation to find the thermodynamic parameters of the system for both the ligands (Figures 1D and 2D). The slope of this plot gives $-\Delta H/R$, and the intercept gives the value of $\Delta S/R$. The magnitude and the sign of the thermodynamic parameters (ΔH, ΔS and ΔG) offer a clue about the forces that drive the reaction. Table 1 shows the various thermodynamic parameters obtained for the α-amylase–caffeic acid system, while Table 2 shows the thermodynamic parameters obtained for the α-amylase–coumaric acid system. We obtained negative ΔH and ΔS for the α-amylase–caffeic acid system, revealing the existence of van der Waals force and hydrogen bonding, while positive ΔH and ΔS were obtained for α-amylase–coumaric acid, implying the reaction to be driven predominantly by hydrophobic interaction [50,51]. Additionally, negative ΔG for both the systems suggested the spontaneous nature of the reaction.

3.2. Synchronous Fluorescence

Synchronous fluorescence spectroscopy is used to have deeper insights into the conformational changes in the proteins microenvironment comprising tyrosine and tryptophan residues. Synchronous fluorescence provides information on conformational changes in the molecular environment of fluorophores of proteins once the ligands bind to them [52]. When $\Delta\lambda$ ($\lambda_{em} - \lambda_{ex}$) is kept at 60 nm or 15 nm, the synchronous fluorescence spectra expose the information about the microenvironment of tryptophan and tyrosine residues, respectively. Figure 3 shows the synchronous fluorescence spectra of free α-amylase and α-amylase with varying concentrations of caffeic acid (upper panel) and coumaric acid (lower panel).

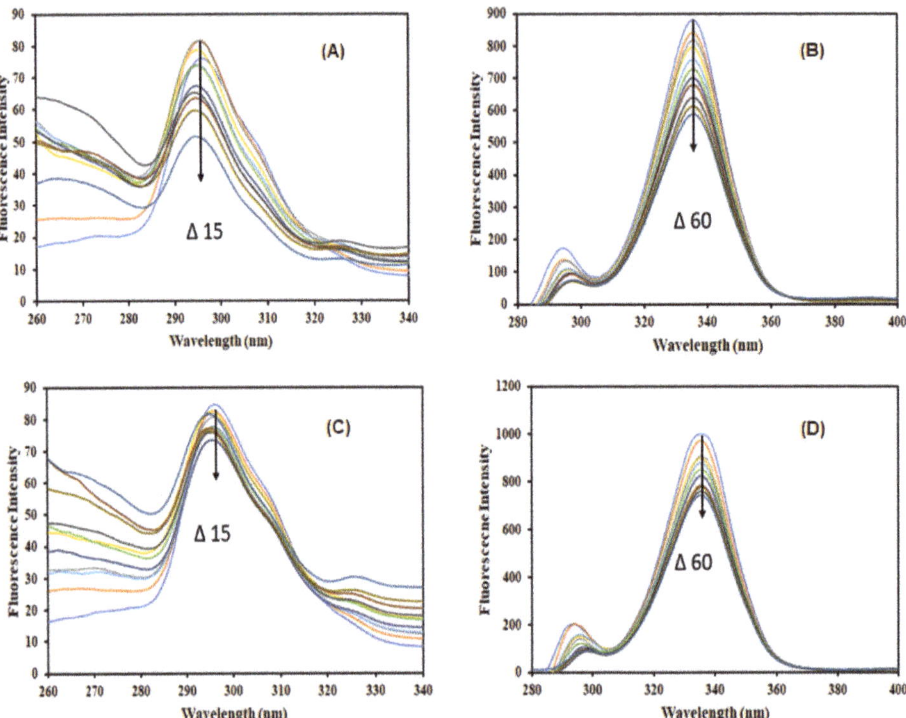

Figure 3. Synchronous fluorescence spectra for (**A**) Tyrosine residue (Δλ = 15) and (**B**) Tryptophan (Δλ = 60) of α-amylase (4 μM) in the absence and presence of caffeic acid (0–40 μM). Panel (**C,D**) are the spectra obtained under similar conditions for p-coumaric acid.

In the case of Δλ = 15 nm, a shift in the fluorescence emission maxima of α-amylase in the presence of caffeic acid (Figure 3A) and coumaric acid (Figure 3C) implies that the local environment around tyrosine residue changed significantly in the presence of both the ligands. However, for Δλ = 60 nm, no shift in the emission maxima of α-amylase is observed for both the ligands (Figure 3B,D), suggesting no change in the local environment around tryptophan residues.

3.3. Inhibition of Advanced Glycation End-Products (AGEs)

AGE formation is common at the later glycation stages in proteins, and most of the end products are fluorogenic in nature [53]. Caffeic acid and coumaric acid were studied for their inhibitory effects on AGE formation (Figure 4). HSA was incubated with MG, fluorescence was taken of native HSA, and MG + HSA was incubated. HSA incubated with MG showed high fluorescence compared to the native HSA, concluding the formation of AGEs. In the presence of varying concentrations (50–200 μM) of ligands, a dose-dependent decrease in the fluorescence emission was evident, with maximum inhibition observed for 200 μM for both the ligands (Figure 4A).

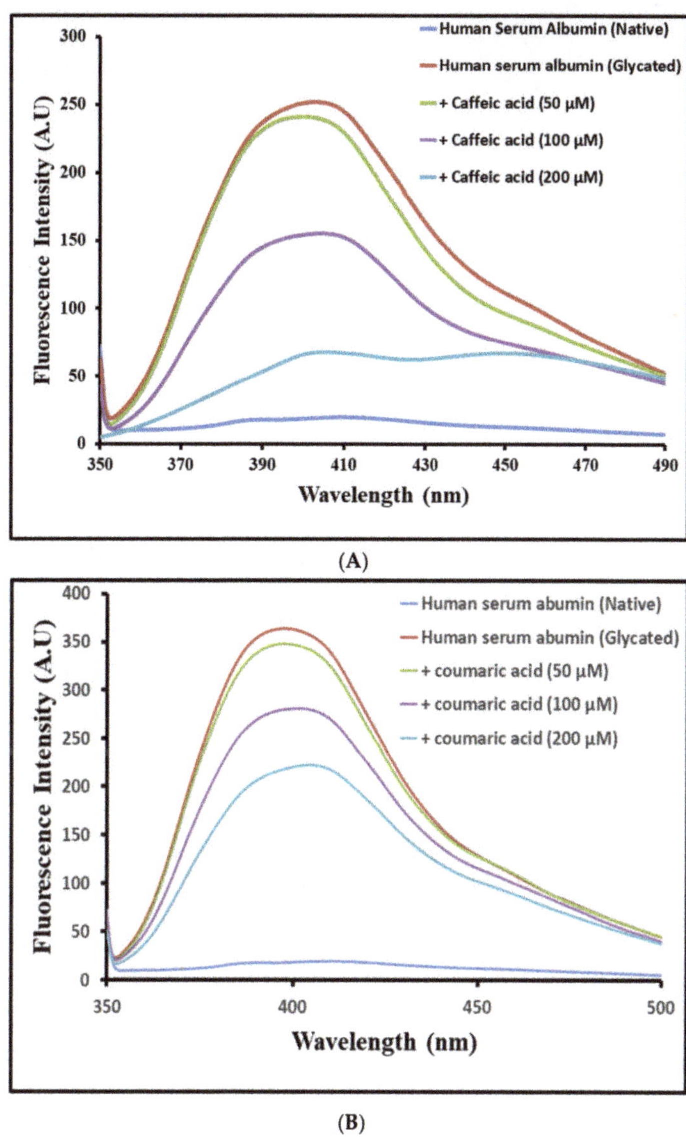

Figure 4. Inhibition of fluorescent advance glycation product (AGEs) by (**A**) caffeic acid and (**B**) p-coumaric acid.

Treatment of MG incubated HSA in the presence of 50, 100 and 200 μM ligands showed a reduction in fluorescent AGEs. Caffeic and coumaric acid both showed a significant decrease (79.2% and 43.6%) respectively at 200 μM concentration (Figure 4A,B). The antiglycative properties of caffeic acid and coumaric acid have been investigated in previous studies [53,54]. AGEs are the key players in the pathophysiology and progression of many diseases highlighting their clinical significance [30]. Strong inhibition of AGEs by caffeic acid compared to coumaric acid could be due to structural differences among them, where caffeic acid possesses 1 more OH group and might play an important role in AGE inhibition.

The results indicated that the binding of caffeic acid and coumaric acid inhibits the formation of AGEs. Thus, a conclusion was obtained that both the ligands attenuate the effect of MG by networking with HSA and reduce the fluorescence by AGEs. AGEs not only create a menace in diabetes but also contribute to other fatal diseases [29,53], signifying an urgent need to stop AGEs formation [54]. Antiglycation activities of caffeic and coumaric acid could be due to its antioxidant, ROS scavenging activity and protein-stabilizing potential. Earlier, caffeic and ferulic acid have been found to attenuate glycation and thus diabetic complications [55,56]. Additionally, the binding analysis of these phenolics with α-amylase hypothesized its inhibitory potential and thereby reduced glucose concentrations in serum.

3.4. Inhibition of Early (Amadori) Glycation Products

Fructosamine is formed by covalent attachments of sugar molecule glucose to a primary amine, followed by isomerization. The molecule undergoes Amadori rearrangement and is an indicator of early glycation products. Thus, we aimed to understand the role of caffeic acid and coumaric acid in the inhibition of glycation. Figure 5A indicates the level of fructosamine in different samples.

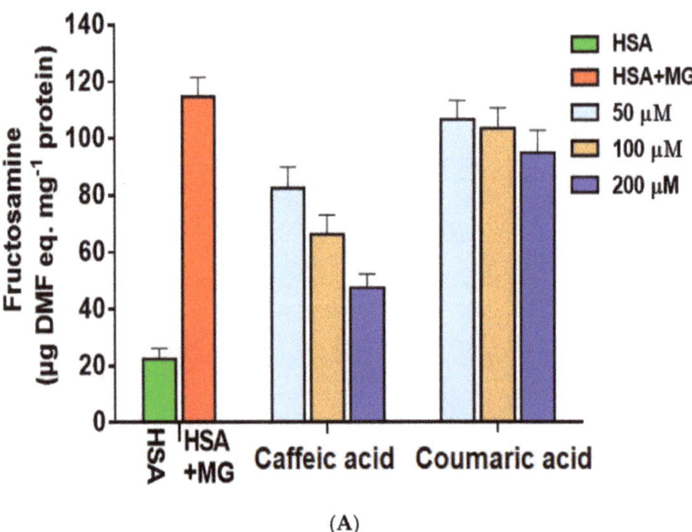

Figure 5. *Cont.*

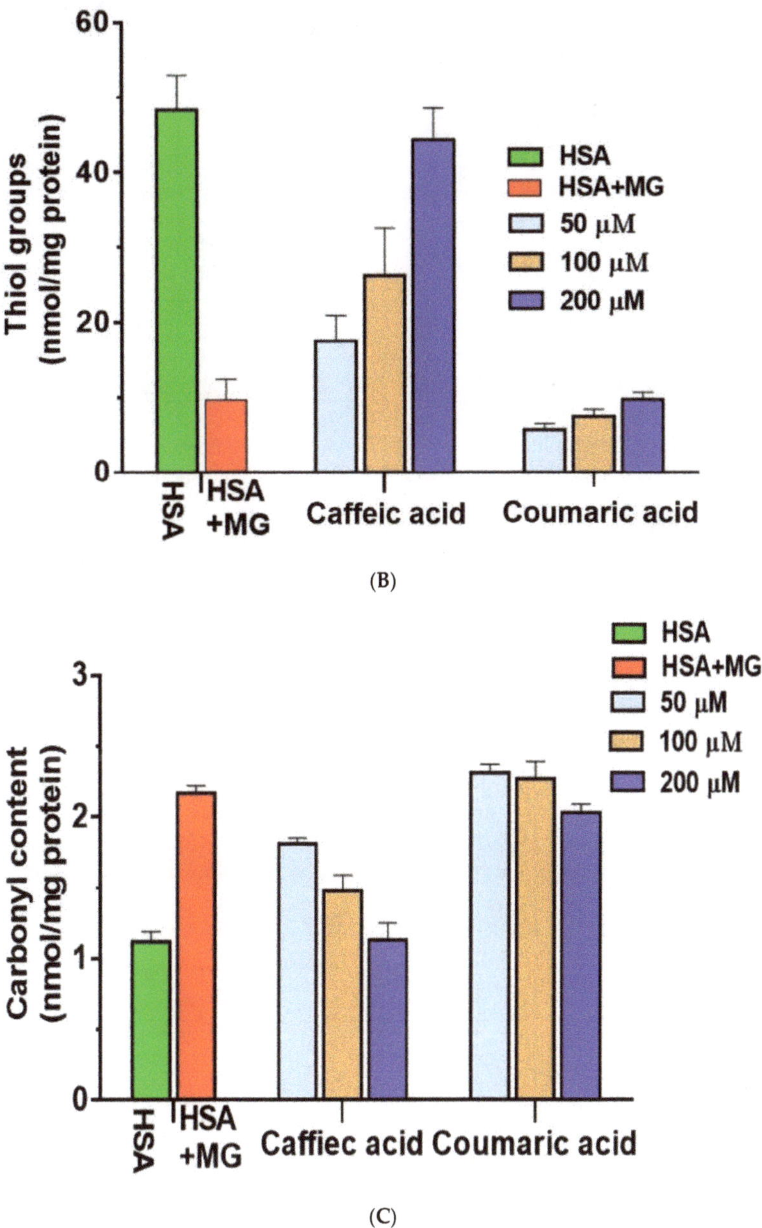

Figure 5. Estimation of antiglycation activity of caffeic and p-coumaric acid by monitoring (**A**) fructosamine content and protein oxidation by measuring (**B**) free thiol groups and (**C**) carbonyl content.

In control HSA, the fructosamine level was nearly 22.12 nmol/mg protein. However, incubation of HSA with MG showed a hike in fructosamine content to 114.63 nmol/mg protein (Table 3). Fructosamine levels showed a decline with successive increases in ligand concentration. In the presence of 200 μM caffeic acid, the fructosamine level declined to

46.13 nmol/mg, while coumaric acid declined the levels of fructosamine to 96.81 nmol/mg. The results showed that early end product formation viz fructosamine declines in the presence of both the ligands. Amadori product accumulations are associated with diabetic complications, and the selected natural polyphenols have the potential to reduce fructosamine content. The correlation shows the importance of caffeic acid and coumarin in managing complications due to diabetes.

Table 3. Effect of different concentrations of caffeic acid on alpha amylase.

Group	(nmol/mg Protein)		
	Fructosamine	Thiol Groups (SH)	Carbonyl Content
Control	22.12 ± 2.3	48.23 ± 7.8	1.12 ± 0.06
Glycated	114.63 ± 6.8	6.86 ± 0.64	2.17 ± 0.08
Caffeic acid:			
50 μM	82.41 ± 7.2	17.74 ± 3.4	1.81 ± 0.07
100 μM	67.34 ± 6.3	26.24 ± 1.8	1.48 ± 0.11
200 μM	46.63 ± 5.6	44.31 ± 4.7	1.13 ± 0.07

3.5. Free–SH Groups Content

Free sulfhydryl group alterations are an important parameter to estimate oxidative modification. A disulfide bond between the sulfhydryl groups is formed due to oxidation. The oxidative modifications in HSA resultant of glycation were correlated with that of free sulfhydryl content. Figure 5B shows the free thiol group estimated for native HSA, protein incubated with MG and different concentrations of caffeic acid and coumarin. Native HSA was used as a control and was found to have 48.23 nmol/mg protein, while MG treated protein had 6.68 nmol/mg protein. Free sulfhydryl contents relate directly to the oxidation of HSA [53]. With the increase in the concentration of both the ligands, an increase in the free thiol group of HSA was observed. The maximum value was obtained at 200 μM of caffeic acid and coumaric acid. Caffeic acid increased the content of thiol to 17.8, 26.2 and 44.3 nmol/mg for 50, 100 and 200 μM, respectively. Similarly, for the same concentration of coumaric acid, the values were 5.8, 7.2 and 9.7 nmol/mg, respectively (Tables 3 and 4).

Table 4. Effect of different concentrations of p-coumaric acid on alpha amylase.

Group	(nmol/mg Protein)		
	Fructosamine	Thiol Groups (SH)	Carbonyl Content
Control	22.12 ± 2.3	48.23 ± 7.8	1.12 ± 0.06
Glycated	114.63 ± 6.8	6.86 ± 0.64	2.17 ± 0.08
p-coumaric acid:			
50 μM	106.68 ± 6.4	5.82 ± 0.54	2.31 ± 0.03
100 μM	103.42 ± 7.3	7.16 ± 0.41	2.27 ± 0.12
200 μM	96.81 ± 6.6	9.72 ± 0.63	2.03 ± 0.06

Glycation reactions generate ROS that works against the oxidative defense mechanism of the protein group [57]. Thus, the above observations show that caffeic acid and coumarin have the potential to reduce free thiol group, resisting glycation. The results that are presented here are supported by previously published results [29,57,58].

3.6. Carbonyl Content

Glycation is a non-enzymatic reaction involving protein and sugar. It results in the formation of an unstable Schiff base, further leading to ketoamine production [33]. HSA was incubated with MG and was studied for its carbonyl content. Native HSA and MG-incubated HSA showed differences in their concentration by more than double. Native HSA was estimated 1.12 nmol/mg, while glycated HSA had 2.17 nmol/mg carbonyl content. With successive addition of caffeic acid and coumaric acid, carbonyl content decreased slightly (Figure 5C, Tables 3 and 4).

3.7. Molecular Docking and Dynamics

The interaction between pancreatic α-amylase and both caffeic acid and p-coumaric acid was performed using Autodock 4.2.6. Molecular docking has been widely utilized to study the critical residues and sites involved in protein–ligand interactions. Our in vitro results demonstrated the mode of binding between α-amylase and polyphenols (caffeic and coumaric acid), and these are further validated by employing docking studies. Figure 6A depicts the three-dimensional structure of α-amylase in cartoon form with caffeic acid shown in the catalytic pocket depicted in balls and stick model. Caffeic acid formed six hydrogen bonds (Trp 59, Gln 63, Arg 195, Arg 195, Asp 197 and Asp 197) and three hydrophobic interactions (Trp 58, Trp 59 and Tyr 62) with α-amylase (Figure 6B, Table 5) showing a binding affinity of −5.09 kcal/mol. In comparison, coumaric acid formed H-bonds with Trp 59, Gln, 63, Arg 195, Aand sp 300 (Table 6) and shared the common hydrophobic residues as of caffeic acid.

Table 5. Molecular docking parameters for caffeic acid–pancreatic α- amylase interactions obtained through (PLIP).

Hydrophobic Interactions		Hydrogen Bonds		Pi-Stacking			Docking Energy (kcal mol^{-1})
AA	Distance (Å)	AA	Distance (Å)	AA	Distance (Å)	Type	
TRP58	3.18	TRP59	1.83	TYR62	3.91	Parallel	−5.09
TRP59	3.67	GLN63	1.86				
TYR62	3.35	ARG195	2.85				
		ARG195	2.29				
		ASP197	1.85				

Table 6. Molecular docking parameters for p-coumaric acid–pancreatic α- amylase interactions obtained through PLIP.

Hydrophobic. Interactions		Hydrogen Bonds		Pi-Stacking			Docking Energy (kcal mol^{-1})
AA	Distance (Å)	AA	Distance (Å)	AA	Distance (Å)	Type	
TRP58	3.64	TRP59	2.65	TYR62	4.14	Parallel	−5.04
TRP59	3.54	GLN63	3.14				
TYR62	3.33	ARG195	3.69				
		ARG195	2.99				
		ASP197	3.15				
		HIS299	3.66				
		ASP300	2.91				

The binding affinity observed for the interaction of coumaric acid to α-amylase was −5.04 kcal/mol. It is apparent that caffeic acid forms six hydrogen bonds as compared to three hydrogen bonds for coumaric acid, revealing that the binding of caffeic acid binds to α-amylase is more significant, validating our earlier observations.

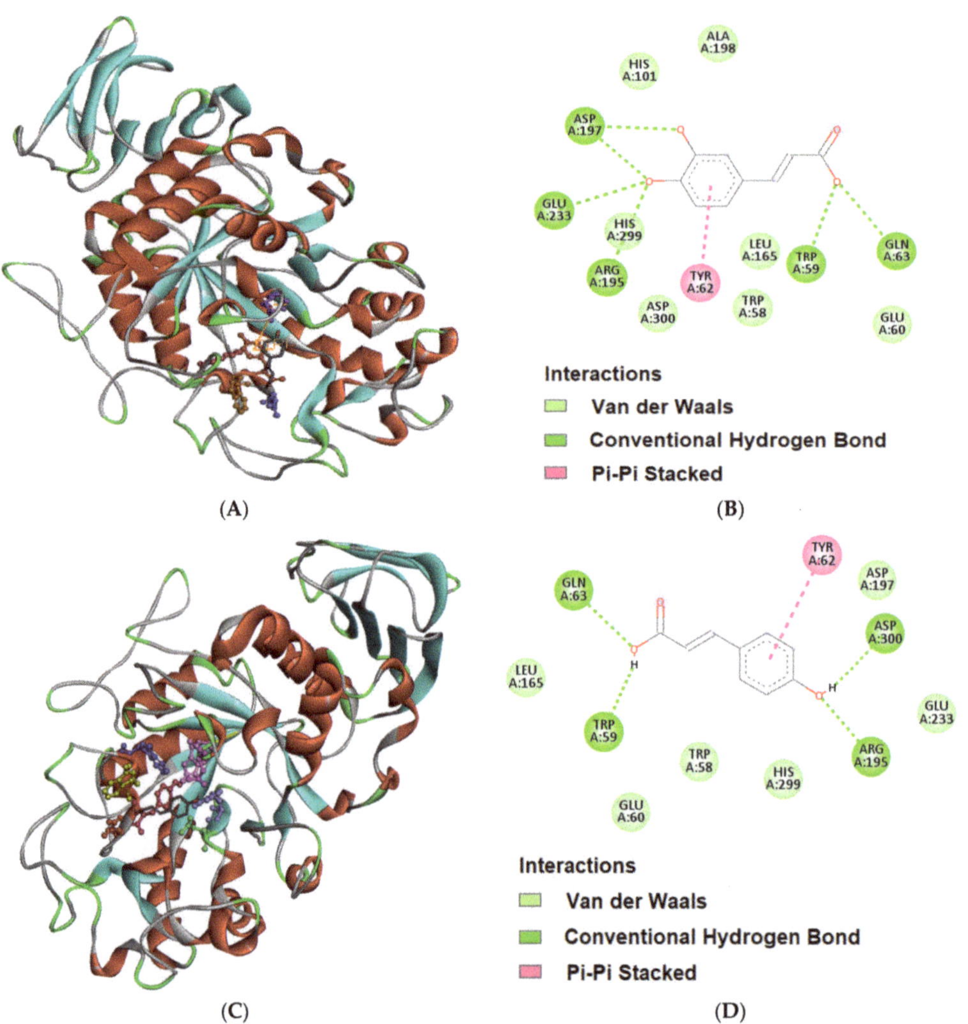

Figure 6. Molecular docking of caffeic acid with pancreatic α-amylase. (**A**) Binding of caffeic acid at the catalytic site of pancreatic α-amylase. (**B**) Amino acid residues and the types of forces in stabilizing the pancreatic α-amylase–caffeic acid complex (Discovery Studio). Similarly, for p-coumaric acid with α-amylase (Panels (**C,D**)).

4. Conclusions

Enzymes such as amylase break down polysaccharides into monomeric sugars and thereby increase glucose concentrations in the serum. Furthermore, prolonged exposure of proteins to sugars and dicarbonyl intermediates led to the formation of advanced glycation end-products (AGEs). In our study, neutraceuticals molecules such as caffeic and coumaric acid bind with α-amylase and also inhibit the AGEs formation to a different extent. Caffeic acid possesses more inhibitory activity, which could be due to its planarity and hydrogen bonding potential. Van der Waals and hydrogen bonding are the major forces between the polyphenols–protein interactions. Molecular docking along with fluorescence quenching and synchronous fluorescence displayed the ability of phenolics to form stable complex with amylase. Moreover, these phenolics decrease AGE formation by inhibiting fructosamine. Furthermore, oxidation of proteins boosted the effect of glycation; caffeic and coumaric acid attenuate it by protecting thiol and carbonyl groups. The scheme for our probable mechanism for AGE inhibition is depicted in Figure 7. More research on similar structures along with in vivo studies is warranted to design inhibitors for diabetic complications.

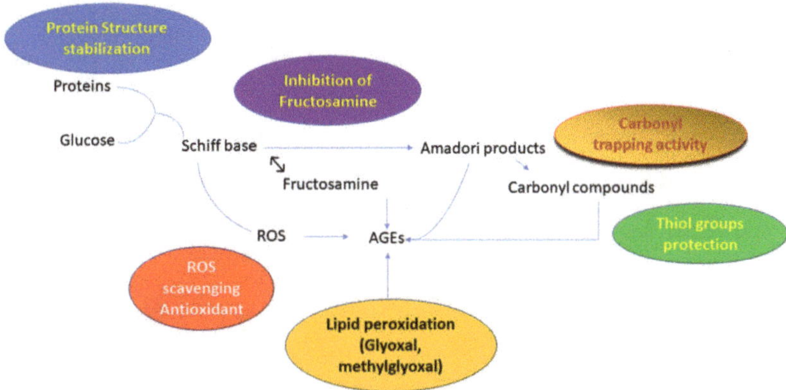

Figure 7. Mechasnidtic pathway of caffeic and coumaric acid to inhibit advanced glycation end-products (AGEs).

Author Contributions: Conceptualization, M.S.K.; validation M.S.K., M.S.A., A.M.H.A., N.A., N.O.A., A.M.A. and M.A.Z.; formal analysis, M.S.K., M.S.A., A.M.H.A., N.A., N.O.A., A.M.A. and M.A.Z.; funding acquisition writing—original draft preparation, M.S.K., M.S.A., A.M.H.A., N.A., N.O.A., A.M.A. and M.A.Z.; project administration, M.S.A.; funding acquisition, M.S.K. All authors have read and agreed to the published version of the manuscript.

Funding: MSK acknowledge the generous support from Research Supporting Project (RSP-2021/352) by King Saud University, Riyadh, Kingdom of Saudi Arabia.

Institutional Review Board Statement: Not Applicable.

Informed Consent Statement: Not Applicable.

Data Availability Statement: Data will be available on request to corresponding author.

Acknowledgments: MSK acknowledge the generous support from Research Supporting Project (RSP-2021/352) by King Saud University, Riyadh, Kingdom of Saudi Arabia.

Conflicts of Interest: The authors declare no conflict of interest.

References

1. Sapra, A.; Bhandari, P. *Diabetes Mellitus*; StatPearls Publishing: Treasure Island, FL, USA, 2019.
2. Sabharwal, R.; Mahajan, A. Diabetes mellitus, dyslipidemia: Cause for acute myocardial infarction. *JK Sci.* **2020**, *22*, 1–2.
3. Dubey, S.; Ganeshpurkar, A.; Ganeshpurkar, A.; Bansal, D.; Dubey, N. Glycolytic enzyme inhibitory and antiglycation potential of rutin. *Future J. Pharm. Sci.* **2017**, *3*, 158–162. [CrossRef]
4. Younus, H.; Anwar, S. Prevention of non-enzymatic glycosylation (glycation): Implication in the treatment of diabetic complication. *Int. J. Health Sci.* **2016**, *10*, 261. [CrossRef]
5. Zhang, Q.; Ames, J.M.; Smith, R.D.; Baynes, J.W.; Metz, T.O. A perspective on the Maillard reaction and the analysis of protein glycation by mass spectrometry: Probing the pathogenesis of chronic disease. *J. Proteome Res.* **2009**, *8*, 754–769. [CrossRef] [PubMed]
6. Neglia, C.I.; Cohen, H.J.; Garber, A.R.; Ellis, P.D.; Thorpe, S.R.; Baynes, J.W. 13C NMR investigation of nonenzymatic glucosylation of protein. Model studies using RNase A. *J. Biol. Chem.* **1983**, *258*, 14279–14283. [CrossRef]

7.   Baynes, J.; Watkins, N.; Fisher, C.; Hull, C.; Patrick, J.; Ahmed, M.; Dunn, J.; Thorpe, S. The Amadori product on protein: Structure and reactions. *Prog. Clin. Biol. Res.* **1989**, *304*, 43–67.

8.   Han, D.; Yamamoto, Y.; Munesue, S.; Motoyoshi, S.; Saito, H.; Win, M.T.T.; Watanabe, T.; Tsuneyama, K.; Yamamoto, H. Induction of receptor for advanced glycation end products by insufficient leptin action triggers pancreatic β-cell failure in type 2 diabetes. *Genes Cells* **2013**, *18*, 302–314. [CrossRef]

9.   Di Meo, S.; Reed, T.T.; Venditti, P.; Victor, V.M. Harmful and beneficial role of ROS 2017. *Oxidative Med. Cell. Longev.* **2018**, *2018*, 5943635. [CrossRef]

10.   Reddy, V.P.; Beyaz, A. Inhibitors of the Maillard reaction and AGE breakers as therapeutics for multiple diseases. *Drug Discov. Today* **2006**, *11*, 646–654. [CrossRef]

11.   Kalousova, M.; Skrha, J.; Zima, T. Advanced glycation end-products and advanced oxidation protein products in patients with diabetes mellitus. *Physiol. Res.* **2002**, *51*, 597–604.

12.   Fu, M.-X.; Requena, J.R.; Jenkins, A.J.; Lyons, T.J.; Baynes, J.W.; Thorpe, S.R. The Advanced Glycation End Product, Nε-(Carboxymethyl) lysine, Is a Product of both Lipid Peroxidation and Glycoxidation Reactions (∗). *J. Biol. Chem.* **1996**, *271*, 9982–9986. [CrossRef] [PubMed]

13.   Willemsen, S.; Hartog, J.W.; Heiner-Fokkema, M.R.; van Veldhuisen, D.J.; Voors, A.A. Advanced glycation end-products, a pathophysiological pathway in the cardiorenal syndrome. *Heart Fail. Rev.* **2012**, *17*, 221–228. [CrossRef] [PubMed]

14.   Singh, V.P.; Bali, A.; Singh, N.; Jaggi, A.S. Advanced glycation end products and diabetic complications. *Korean J. Physiol. Pharmacol.* **2014**, *18*, 1–14. [CrossRef] [PubMed]

15.   Brownlee, M. The pathological implications of protein glycation, Clinical and investigative medicine. *Med. Clin. Exp.* **1995**, *18*, 275–281.

16.   Kotowaroo, M.; Mahomoodally, M.; Gurib-Fakim, A.; Subratty, A. Screening of traditional antidiabetic medicinal plants of mauritius for possible α-amylase inhibitory effects in vitro. *Phytother. Res. Int. J. Devoted Pharmacol. Toxicol. Eval. Nat. Prod. Deriv.* **2006**, *20*, 228–231.

17.   Kim, Y.-M.; Jeong, Y.-K.; Wang, M.-H.; Lee, W.-Y.; Rhee, H.-I. Inhibitory effect of pine extract on α-glucosidase activity and postprandial hyperglycemia. *Nutrition* **2005**, *21*, 756–761. [CrossRef]

18.   Shamsi, A.; Mohammad, T.; Anwar, S.; Alajmi, M.F.; Hussain, A.; Hassan, M.I.; Ahmad, F.; Islam, A. Probing the interaction of Rivastigmine Tartrate, an important Alzheimer's drug, with serum albumin: Attempting treatment of Alzheimer's disease. *Int. J. Biol. Macromol.* **2020**, *148*, 533–542. [CrossRef]

19.   Anwar, S.; Shamsi, A.; Mohammad, T.; Islam, A.; Hassan, M.I. Targeting pyruvate dehydrogenase kinase signaling in the development of effective cancer therapy. *Biochim. Biophys. Acta Rev. Cancer* **2021**, *1876*, 188568. [CrossRef]

20.   Anwar, S.; Mohammad, T.; Shamsi, A.; Queen, A.; Parveen, S.; Luqman, S.; Hasan, G.M.; Alamry, K.A.; Azum, N.; Asiri, A.M. Discovery of Hordenine as a potential inhibitor of pyruvate dehydrogenase kinase 3: Implication in lung Cancer therapy. *Biomedicines* **2020**, *8*, 119. [CrossRef]

21.   Anwar, S.; Khan, S.; Anjum, F.; Shamsi, A.; Khan, P.; Fatima, H.; Shafie, A.; Islam, A.; Hassan, M.I. Myricetin inhibits breast and lung cancer cells proliferation via inhibiting MARK4. *J. Cell. Biochem.* **2021**, *123*, 359–374. [CrossRef]

22.   Anwar, S.; Shamsi, A.; Kar, R.K.; Queen, A.; Islam, A.; Ahmad, F.; Hassan, M.I. Structural and biochemical investigation of MARK4 inhibitory potential of cholic acid: Towards therapeutic implications in neurodegenerative diseases. *Int. J. Biol. Macromol.* **2020**, *161*, 596–604. [CrossRef] [PubMed]

23.   Anwar, S.; Khan, S.; Shamsi, A.; Anjum, F.; Shafie, A.; Islam, A.; Ahmad, F.; Hassan, M.I. Structure-based investigation of MARK4 inhibitory potential of Naringenin for therapeutic management of cancer and neurodegenerative diseases. *J. Cell. Biochem.* **2021**, *122*, 1445–1459. [CrossRef] [PubMed]

24.   Khandouzi, N.; Zahedmehr, A.; Nasrollahzadeh, J. Effect of polyphenol-rich extra-virgin olive oil on lipid profile and inflammatory biomarkers in patients undergoing coronary angiography: A randomised, controlled, clinical trial. *Int. J. Food Sci. Nutr.* **2021**, *72*, 548–558. [CrossRef] [PubMed]

25.   Rienks, J.; Barbaresko, J.; Oluwagbemigun, K.; Schmid, M.; Nöthlings, U. Polyphenol exposure and risk of type 2 diabetes: Dose-response meta-analyses and systematic review of prospective cohort studies. *Am. J. Clin. Nutr.* **2018**, *108*, 49–61. [CrossRef] [PubMed]

26.   Martin, K.R.; Appel, C.L. Polyphenols as dietary supplements: A double-edged sword. *Nutr. Diet. Suppl.* **2009**, *2*, 1–12. [CrossRef]

27.   Magnani, C.; Isaac, V.L.B.; Correa, M.A.; Salgado, H.R.N. Caffeic acid: A review of its potential use in medications and cosmetics. *Anal. Methods* **2014**, *6*, 3203–3210. [CrossRef]

28.   Grabska-Kobylecka, I.; Kaczmarek-Bak, J.; Figlus, M.; Prymont-Przyminska, A.; Zwolinska, A.; Sarniak, A.; Wlodarczyk, A.; Glabinski, A.; Nowak, D. The presence of caffeic acid in cerebrospinal fluid: Evidence that dietary polyphenols can cross the blood-brain barrier in humans. *Nutrients* **2020**, *12*, 1531. [CrossRef]

29.   Kong, C.S.; Jeong, C.H.; Choi, J.S.; Kim, K.J.; Jeong, J.W. Antiangiogenic effects of p-coumaric acid in human endothelial cells. *Phytother. Res.* **2013**, *27*, 317–323. [CrossRef]

30.   El-Seedi, H.R.; El-Said, A.M.; Khalifa, S.A.; Goransson, U.; Bohlin, L.; Borg-Karlson, A.-K.; Verpoorte, R. Biosynthesis, natural sources, dietary intake, pharmacokinetic properties, and biological activities of hydroxycinnamic acids. *J. Agric. Food Chem.* **2012**, *60*, 10877–10895. [CrossRef]

31. Pei, K.; Ou, J.; Huang, J.; Ou, S. p-Coumaric acid and its conjugates: Dietary sources, pharmacokinetic properties and biological activities. *J. Sci. Food Agric.* **2016**, *96*, 2952–2962. [CrossRef]
32. Shamsi, A.; Mohammad, T.; Khan, M.S.; Shahwan, M.; Husain, F.M.; Rehman, M.; Hassan, M.; Ahmad, F.; Islam, A. Unraveling binding mechanism of Alzheimer's drug rivastigmine tartrate with human transferrin: Molecular docking and multi-spectroscopic approach towards neurodegenerative diseases. *Biomolecules* **2019**, *9*, 495. [CrossRef] [PubMed]
33. Shamsi, A.; Ahmed, A.; Khan, M.S.; Husain, F.M.; Bano, B. Rosmarinic acid restrains protein glycation and aggregation in human serum albumin: Multi spectroscopic and microscopic insight-possible therapeutics targeting diseases. *Int. J. Biol. Macromol.* **2020**, *161*, 187–193. [CrossRef] [PubMed]
34. Khan, M.S.; Rehman, M.T.; Ismael, M.A.; AlAjmi, M.F.; Alruwaished, G.I.; Alokail, M.S.; Khan, M.R. Bioflavonoid (Hesperidin) Restrains Protein Oxidation and Advanced Glycation End Product Formation by Targeting AGEs and Glycolytic Enzymes. *Cell Biochem. Biophys.* **2021**, *79*, 833–844. [CrossRef] [PubMed]
35. Shamsi, A.; Ahmed, A.; Khan, M.S.; al Shahwan, M.; Husain, F.M.; Bano, B. Understanding the binding between Rosmarinic acid and serum albumin: In vitro and in silico insight. *J. Mol. Liq.* **2020**, *311*, 113348. [CrossRef]
36. Ahmed, A.; Shamsi, A.; Khan, M.S.; Husain, F.M.; Bano, B. Methylglyoxal induced glycation and aggregation of human serum albumin: Biochemical and biophysical approach. *Int. J. Biol. Macromol.* **2018**, *113*, 269–276. [CrossRef] [PubMed]
37. Shamsi, A.; Anwar, S.; Shahbaaz, M.; Mohammad, T.; Alajmi, M.F.; Hussain, A.; Hassan, I.; Ahmad, F.; Islam, A. Evaluation of Binding of Rosmarinic Acid with Human Transferrin and Its Impact on the Protein Structure: Targeting Polyphenolic Acid-Induced Protection of Neurodegenerative Disorders. *Oxidative Med. Cell. Longev.* **2020**, *2020*, 1245875. [CrossRef]
38. Khan, S.N.; Islam, B.; Yennamalli, R.; Sultan, A.; Subbarao, N.; Khan, A.U. Interaction of mitoxantrone with human serum albumin: Spectroscopic and molecular modeling studies. *Eur. J. Pharm. Sci.* **2008**, *35*, 371–382. [CrossRef]
39. Pacheco, M.E.; Bruzzone, L. Synchronous fluorescence spectrometry: Conformational investigation or inner filter effect? *J. Lumin.* **2013**, *137*, 138–142. [CrossRef]
40. Shamsi, A.; Shahwan, M.; Husain, F.M.; Khan, M.S. Characterization of methylglyoxal induced advanced glycation end products and aggregates of human transferrin: Biophysical and microscopic insight. *Int. J. Biol. Macromol.* **2019**, *138*, 718–724. [CrossRef]
41. Giannoukakis, N. Drug evaluation: Ranirestat—An aldose reductase inhibitor for the potential treatment of diabetic complications. *Curr. Opin. Investig. Drugs* **2006**, *7*, 916–923.
42. Khan, M.S.; Qais, F.A.; Rehman, M.T.; Ismail, M.H.; Alokail, M.S.; Altwaijry, N.; Alafaleq, N.O.; AlAjmi, M.F.; Salem, N.; Alqhatani, R. Mechanistic inhibition of non-enzymatic glycation and aldose reductase activity by naringenin: Binding, enzyme kinetics and molecular docking analysis. *Int. J. Biol. Macromol.* **2020**, *159*, 87–97. [CrossRef]
43. Johnson, R.N.; Metcalf, P.A.; Baker, J.R. Fructosamine: A new approach to the estimation of serum glycosylprotein. An index of diabetic control. *Clin. Chim. Acta* **1983**, *127*, 87–95. [CrossRef]
44. Ellman, G.L. A colorimetric method for determining low concentrations of mercaptans. *Arch. Biochem. Biophys.* **1958**, *74*, 443–450. [CrossRef]
45. Brayer, G.D.; Luo, Y.; Withers, S.G. The structure of human pancreatic α-amylase at 1.8 Å resolution and comparisons with related enzymes. *Protein Sci.* **1995**, *4*, 1730–1742. [CrossRef] [PubMed]
46. Zhang, H.; Wang, Y.; Fei, Z.; Wu, L.; Zhou, Q. Characterization of the interaction between Fe (III)-2, 9, 16, 23-tetracarboxyphthalocyanine and blood proteins. *Dye. Pigment.* **2008**, *78*, 239–247. [CrossRef]
47. Wang, Q.; Huang, C.-R.; Jiang, M.; Zhu, Y.-Y.; Wang, J.; Chen, J.; Shi, J.-H. Binding interaction of atorvastatin with bovine serum albumin: Spectroscopic methods and molecular docking. *Spectrochim. Acta Part A Mol. Biomol. Spectrosc.* **2016**, *156*, 155–163. [CrossRef]
48. Zhang, Y.-F.; Zhou, K.-L.; Lou, Y.-Y.; Pan, D.-Q.; Shi, J.-H. Investigation of the binding interaction between estazolam and bovine serum albumin: Multi-spectroscopic methods and molecular docking technique. *J. Biomol. Struct. Dyn.* **2017**, *35*, 3605–3614. [CrossRef] [PubMed]
49. Wang, B.-L.; Pan, D.-Q.; Zhou, K.-L.; Lou, Y.-Y.; Shi, J.-H. Multi-spectroscopic approaches and molecular simulation research of the intermolecular interaction between the angiotensin-converting enzyme inhibitor (ACE inhibitor) benazepril and bovine serum albumin (BSA). *Spectrochim. Acta Part A Mol. Biomol. Spectrosc.* **2019**, *212*, 15–24. [CrossRef]
50. Shamsi, A.; Mohammad, T.; Anwar, S.; Nasreen, K.; Hassan, M.I.; Ahmad, F.; Islam, A. Insight into the binding of PEG-400 with eye protein alpha-crystallin: Multi spectroscopic and computational approach: Possible therapeutics targeting eye diseases. *J. Biomol. Struct. Dyn.* **2020**, 1–11. [CrossRef]
51. Ross, P.D.; Subramanian, S. Thermodynamics of protein association reactions: Forces contributing to stability. *Biochemistry* **1981**, *20*, 3096–3102. [CrossRef]
52. Ahmed, A.; Shamsi, A.; Khan, M.S.; Husain, F.M.; Bano, B. Probing the interaction of human serum albumin with iprodione, a fungicide: Spectroscopic and molecular docking insight. *J. Biomol. Struct. Dyn.* **2019**, *37*, 857–862. [CrossRef] [PubMed]
53. Chao, C.Y.; Mong, M.C.; Chan, K.C.; Yin, M.C. Anti-glycative and anti-inflammatory effects of caffeic acid and ellagic acid in kidney of diabetic mice. *Mol. Nutr. Food Res.* **2010**, *54*, 388–395. [CrossRef] [PubMed]
54. Moselhy, S.S.; Razvi, S.S.; Lshibili, F.A.A.; Kuerban, A.; Hasan, M.N.; Balamash, K.S.; Huwait, E.A.; Abdulaal, W.H.; Al-Ghamdi, M.A.; Kumosani, T.A. m-Coumaric acid attenuates non-catalytic protein glycosylation in the retinas of diabetic rats. *J. Pestic. Sci.* **2018**, *43*, 180–185. [CrossRef] [PubMed]

55. Shiozawa, R.; Inoue, Y.; Murata, I.; Kanamoto, I. Effect of antioxidant activity of caffeic acid with cyclodextrins using ground mixture method. *Asian J. Pharm. Sci.* **2018**, *13*, 24–33. [CrossRef] [PubMed]
56. Liu, J.; He, Y.; Wang, S.; He, Y.; Wang, W.; Li, Q.; Cao, X. Ferulic acid inhibits advanced glycation end products (AGEs) formation and mitigates the AGEsinduced inflammatory response in HUVEC cells. *J. Funct. Foods* **2018**, *48*, 19–26. [CrossRef]
57. Wu, C.-H.; Huang, H.-W.; Lin, J.-A.; Huang, S.-M.; Yen, G.C. The proglycation effect of caffeic acid leads to the elevation of oxidative stress and inflammation in monocytes, macrophages and vascular endothelial cells. *J. Nutr. Biochem.* **2011**, *22*, 585–594. [CrossRef]
58. Ying, X.; Meng, Z.; Weiyu, W.; Yin, H.; Jianli, L. Caffeic Acid Inhibits the Formation of Advanced Glycation End Products (AGEs) and Mitigates the AGEs-Induced Oxidative Stress and Inflammation Reaction in Human Umbilical Vein Endothelial Cells (HUVECs). *Chem. Biodivers.* **2019**, *16*, e1900174. [CrossRef]

Article

Screening a Panel of Topical Ophthalmic Medications against MMP-2 and MMP-9 to Investigate Their Potential in Keratoconus Management

Amany Belal [1,*], Mohamed A. Elanany [2], Eman Y. Santali [1], Ahmed A. Al-Karmalawy [3], Moustafa O. Aboelez [4], Ali H. Amin [5,6], Magda H. Abdellattif [7], Ahmed B. M. Mehany [8] and Hazem Elkady [9]

1. Department of Pharmaceutical Chemistry, College of Pharmacy, Taif University, P.O. Box 11099, Taif 21944, Saudi Arabia; eysantali@tu.edu.sa
2. School of Pharmacy and Pharmaceutical Industries, Badr University in Cairo (BUC), Cairo 11884, Egypt; mohamed.a.elanany@hotmail.com
3. Department of Pharmaceutical Medicinal Chemistry, Faculty of Pharmacy, Horus University-Egypt, New Damietta 34518, Egypt; akarmalawy@horus.edu.eg
4. Department of Pharmaceutical Chemistry, Faculty of Pharmacy, Sohag University, Sohag 82524, Egypt; moustafaaboelez@pharm.sohag.edu.eg
5. Deanship of Scientific Research, Umm Al-Qura University, Makkah 21955, Saudi Arabia; ahamin@uqu.edu.sa
6. Zoology Department, Faculty of Science, Mansoura University, Mansoura 35516, Egypt
7. Department of Chemistry, College of Sciences, Taif University, P.O. Box 11099, Taif 21944, Saudi Arabia; m.hasan@tu.edu.sa
8. Zoology Department, Faculty of Science (Boys), Al-Azhar University, Cairo 11884, Egypt; abelal_81@azhar.edu.eg
9. Pharmaceutical Medicinal Chemistry & Drug Design Department, Faculty of Pharmacy (Boys), Al-Azhar University, Cairo 11884, Egypt; hazemelkady@azhar.edu.eg
* Correspondence: a.belal@tu.edu.sa or abilalmoh1@yahoo.com or amany.mehani@pharm.bsu.edu.eg

Citation: Belal, A.; Elanany, M.A.; Santali, E.Y.; Al-Karmalawy, A.A.; Aboelez, M.O.; Amin, A.H.; Abdellattif, M.H.; Mehany, A.B.M.; Elkady, H. Screening a Panel of Topical Ophthalmic Medications against MMP-2 and MMP-9 to Investigate Their Potential in Keratoconus Management. *Molecules* **2022**, *27*, 3584. https://doi.org/10.3390/molecules27113584

Academic Editors: Tanveer A. Wani, Seema Zargar, Afzal Hussain and Anna Maria Almerico

Received: 19 April 2022
Accepted: 31 May 2022
Published: 2 June 2022

Publisher's Note: MDPI stays neutral with regard to jurisdictional claims in published maps and institutional affiliations.

Abstract: Keratoconus (KC) is a serious disease that can affect people of any race or nationality, although the exact etiology and pathogenic mechanism are still unknown. In this study, thirty-two FDA-approved ophthalmic drugs were exposed to virtual screening using docking studies against both the MMP-2 and MMP-9 proteins to find the most promising inhibitors as a proposed computational mechanism to treat keratoconus. Matrix metalloproteinases (MMPs) are zinc-dependent proteases, and MMP inhibitors (MMPIs) are usually designed to interact with zinc ion in the catalytic (CAT) domain, thus interfering with enzymatic activity. In our research work, the FDA-approved ophthalmic medications will be investigated as MMPIs, to explore if they can be repurposed for KC treatment. The obtained findings of the docking study suggest that atenolol and ampicillin are able to accommodate into the active sites of MMP-2 and MMP-9. Additionally, both exhibited binding modes similar to inhibitors used as references, with an ability to bind to the zinc of the CAT. Molecular dynamic simulations and the MM-GBSA binding free-energy calculations revealed their stable binding over the course of 50 ns. An additional pharmacophoric study was carried out on MMP-9 (PDB ID: 1GKC) using the co-crystallized ligand as a reference for the future design and screening of the MMP-9 inhibitors. These promising results open the door to further biological research to confirm such theoretical results.

Keywords: keratoconus; MMP-2; MMP-9; molecular docking; molecular dynamics; MM-GBSA calculations; pharmacophore mapping

1. Introduction

Keratoconus (KC) is the most common primary cause of corneal ectasia. It commonly strikes in one's second decade of life, affecting people of all races and nationalities. In the general population, the prevalence is estimated to be 54 per 100,000 [1]. Although the exact etiology and pathogenic mechanism are unknown, environmental and genetic variables are considered to play a role in the disease's progression [2]. Hay fever and allergies are linked to an increased risk, whereas diabetes is thought to be protective [3]. Keratoconus is caused by a combination of genes, with a relatively high prevalence of positive family history. Even though both genders are affected, men appear to be more frequently involved [4].

The disease's stage and progression determine the treatment of keratoconus. Spectacles can give an acceptable vision to people in the early stages of their condition, and they are beneficial for those with a visual acuity of 20/40 or greater. On the other hand, spectacles cannot rectify irregular astigmatism, and in such circumstances, hard contact lenses can improve the patient's vision [4,5]. Penetrating or deep anterior lamellar keratoplasty has been the cornerstone of treatment for advanced KC [6]. Collagen cross-linking (CXL) is a new technology recently introduced. Compared to those that were not treated, eyes treated with CXL were less likely to have difficulties with bulging progression [7].

Many investigations have found higher levels of collagenolytic and gelatinolytic activities in laboratory cultures of KC. Collagenases and gelatinases are members of the matrix metalloproteinases (MMPs) family of zinc-dependent proteins [8]. Compared to tears from controls, tears from persons with keratoconus had 1.9 times greater levels of proteolytic activity and overexpression of various MMPs and cytokines [9].

Selective MMP inhibition is an essential objective in medicinal chemistry research [10]. MMP-2 and MMP-9 play important roles in cancer, heart disease, and inflammatory etiology. Many orally accessible broad-spectrum MMP inhibitors (MMPIs) have been discovered in recent years. MMPs typically consist of a pro-peptide sequence, a catalytic (CAT) domain, a hinge region or linker peptide, and a hemopexin domain. MMPs have two zinc ions, one structural zinc and the other in the CAT domain. The early design of the MMPIs relied on the ability of compounds to mimic the amide nature of collagen in addition to having a group that can interact with zinc [11]. In most cases, the hydroxamate group was used for this purpose as in batimastat **I** and marimastat **II** [12]. Other groups were also used for that purpose, such as the carboxylic group of tanomastat **III** and the mercapto group of rebimastat **IV** (Figure 1) [11]. Furthermore, questions have been raised about the real therapeutic efficacy of this family of MMP inhibitors and their considerable toxicity [13]. As a result, researchers are paying more attention to finding novel zinc binding groups (ZBGs) that could be a viable replacement to the hydroxamate function [14–20].

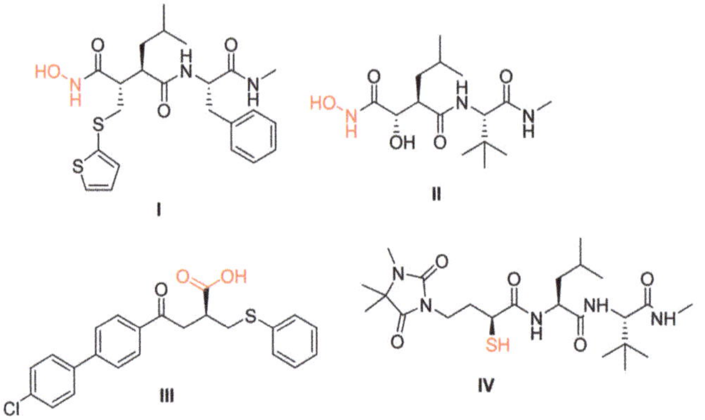

Figure 1. Reported molecules having a group that can interact with zinc in MMPs.

The primary strategy of our design implemented the repurposing of various FDA-approved ophthalmic medications for targeting MMP-2 and MMP-9. The first criterium is the market availability of these drugs as ophthalmic systems, which enabled us to shift our whole focus on the pharmacodynamic potentials rather than pharmacokinetics and drug delivery factors. The second criterium is the presence of groups with a high potential of interaction with zinc, such as carboxylic, mercapto, and hydroxyl groups. For these reasons, a group of thirty-two FDA-approved drugs were chosen (Figure 2) [21–23]. The selection involved a variety of drugs for different conditions, such as glaucoma (acetazolamide), antivirals (acyclovir and ganciclovir), antibacterials (ampicillin and aztreonam), and analgesics (diclofenac and ketorolac). The drugs were subjected to virtual screening using docking studies against both MMP-2 and MMP-9 to reach a promising candidate against these proteins. The results indicate that some drugs may have potential activities against these proteins, opening the field to further biological studies.

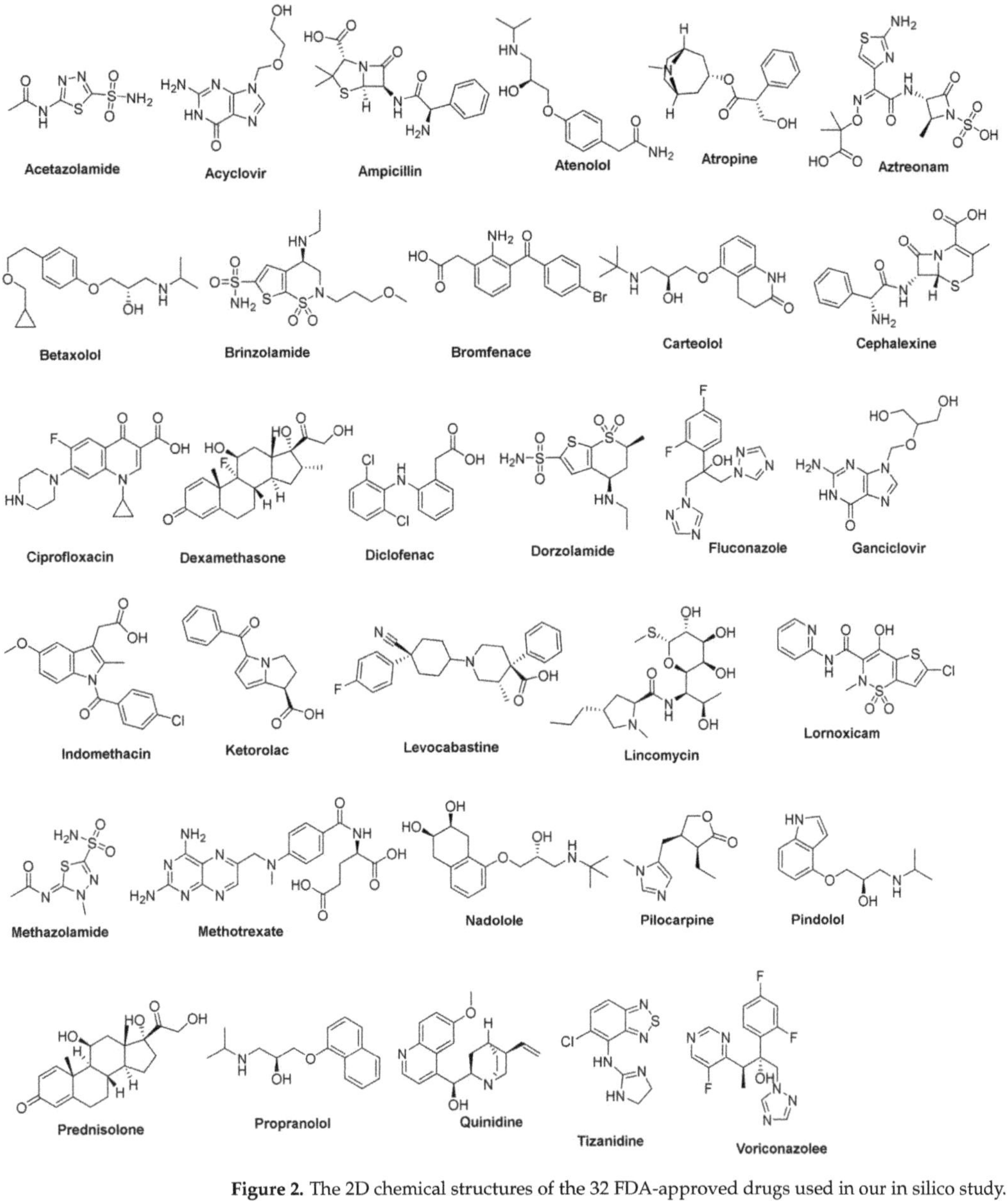

Figure 2. The 2D chemical structures of the 32 FDA-approved drugs used in our in silico study.

2. Results and Discussion

2.1. Docking Studies

A molecular operating environment (MOE) program was used in the current docking study. The validation of the docking accurately reproduced the binding conformation of the co-crystallized ligands with MMP proteins. The RMSD values were calculated between the co-crystallized poses and the docked poses of the same ligands in MMP-2 (PDB ID: 1HOV) and MMP-9 (PDB ID: 1GKC). Minor deviations of 1.30 and 0.75 A^0, for MMP-2 and MMP-9, were observed, respectively (Figures 3 and 4). Such results indicated the validity of the docking studies.

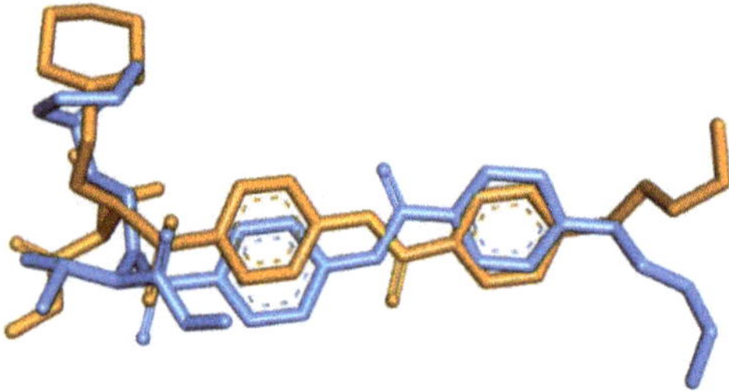

Figure 3. Overlay of the co-crystallized pose (brown) and the re-docked pose (blue) of **I52** inside MMP-2 (PDB ID: 1HOV) during validation (RMSD = 1.30 A^0).

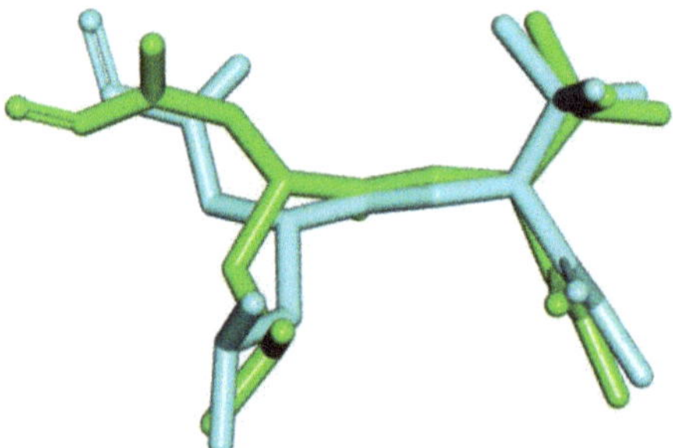

Figure 4. Overlay of the co-crystallized pose (turquoise) and the re-docked pose (green) of **NFH** inside MMP-9 (PDB ID: 1GKC) during validation (RMSD = 0.75 A^0).

2.1.1. Docking of the Target Compounds into MMP-2 Catalytic Domain

All the selected compounds showed favorable binding, demonstrating ΔG (binding free energies) values in negative $Kcal.mol^{-1}$, as shown in Table 1. The most promising candidates were found to be lincomycin, atenolol, and ciprofloxacin, which were able to accommodate into the MMP-2 active site with the highest binding energy score (ΔG = −29.06, −28.20, and −27.87 kcal/mol, respectively) and bind the active site essential residues via several hydrogen bonding, electrostatic, and hydrophobic interactions. An analysis of the binding modes of the co-crystalized ligand (**I52**) and our top hits was then performed for a comparative study of how well our compounds conform to the intended design.

The 2D and 3D interactions of **I52** (Figure 5) revealed that the pentylbenzamide moiety formed three hydrophobic interactions with Leu137, Phe148, and Leu150, besides two hydrogen bonds with Ile141 and Thr143. Moreover, the sulfamoylphenyl moiety formed three hydrogen bonds with Leu 82, Leu83, and Ala84 and three hydrophobic interactions with Leu83 and His120, and the zinc ion. The morpholine ring interacted with His130 via a hydrogen bond and the isopropyl moiety showed a hydrophobic interaction with His85. Finally, the ZBG (hydroxamate) interacted with the zinc group as well as Glu121 via a hydrophobic interaction.

Table 1. The calculated ΔG (binding free energies) of the tested drugs against MMP-2 (PDB ID 1HOV).

	Compound	ΔG (kcal.mol^{-1})		Compound	ΔG (kcal.mol^{-1})
1	Acetazolamide	−11.39	17	Ganciclovir	−15.93
2	Acyclovir	−15.51	18	Indomethacin	−21.81
3	Ampicillin	−19.31	19	Ketorolac	−18.35
4	Atenolol	−28.20	20	Levocabastine	−22.00
5	Atropine	−15.61	21	Lincomycin	−29.06
6	Aztreonam	−19.01	22	Lornoxicam	−18.51
7	Betaxolol	−26.50	23	Methazolamide	−12.86
8	Brinzolamide	−20.96	24	Methotrexate	−26.00
9	Bromfenac	−18.04	25	Nadolole	−25.86
10	Carteolol	−24.09	26	Pilocarpine	−17.28
11	Cephalexine	−21.07	27	Pindolol	−21.05
12	Ciprofloxacin	−27.87	28	Prednisolone	−19.66
13	Dexamethasone	−20.70	29	Propranolol	−23.93
14	Diclofenac	−17.33	30	Quinidine	−20.97
15	Dorzolamide	−19.06	31	Tizanidine	−15.93
16	Fluconazole	−17.47	32	Voriconazolee	−18.24

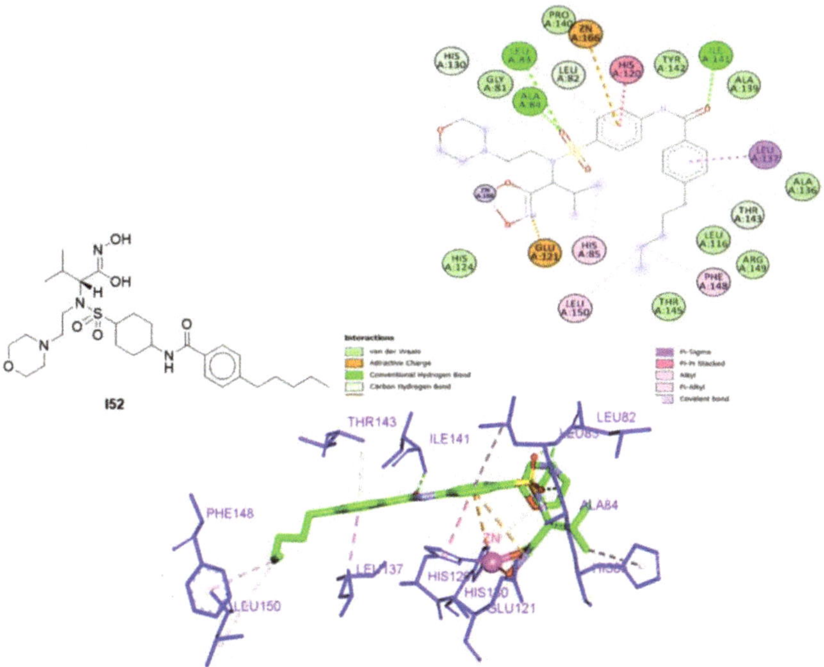

Figure 5. The 2D and 3D interactions of the co-crystallized ligand (**I52**) with amino acid residues of the catalytic domain of MMP-2 (PDB ID: 1HOV).

As described in Figure 6, the docking pose of Lincomycin into the MMP-2 active site showed that the *2H*-pyran arm was engaged in six hydrophobic interactions with Leu116, His120, Leu137, and Tyr142, in addition to two hydrogen bonds with Ala139 and Ile141. The sugar moiety formed three hydrogen bonds with Leu83, Ala84, and Glu121. Finally, the methyl thio group formed two hydrophobic bonds with His130. Although lincomycin showed the highest binding score, it lacked the ability to interact with the zinc of the CAT. As a result, it was excluded from subsequent studies.

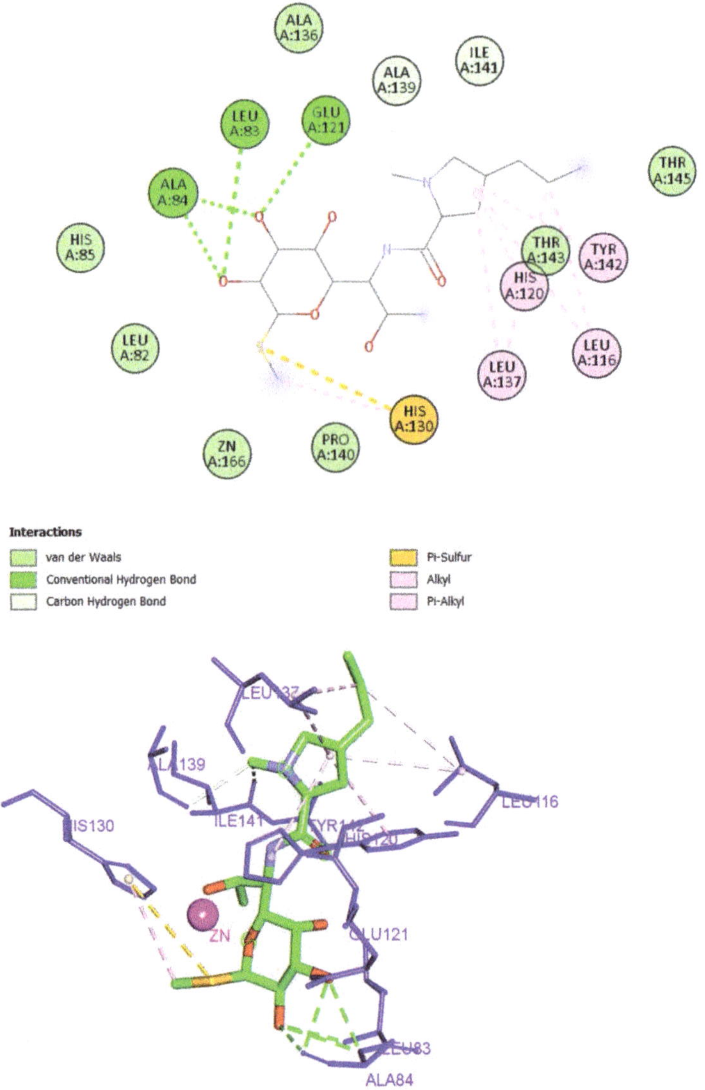

Figure 6. The 2D and 3D interactions of lincomycin with amino acid residues in the catalytic domain of MMP-2 (PDB ID: 1HOV) (hydrogen bonds = green dashed lines, electrostatic interactions = orange dashed lines, pi-pi interactions = deep pink dashed lines, and pi-alkyl interactions = light pink dashed lines).

Although atenolol interacted with the active site with a relatively lower binding, it did also bind successfully to zinc through a hydrophobic pi-cation interaction (Figure 7). Additionally, it formed five hydrogen bonds with Leu83, Ala84, Glu121, Thr142, and Thr143 through its acetamide and hydroxy moieties. Three hydrophobic interactions were observed with His120, Leu137, and Arg149.

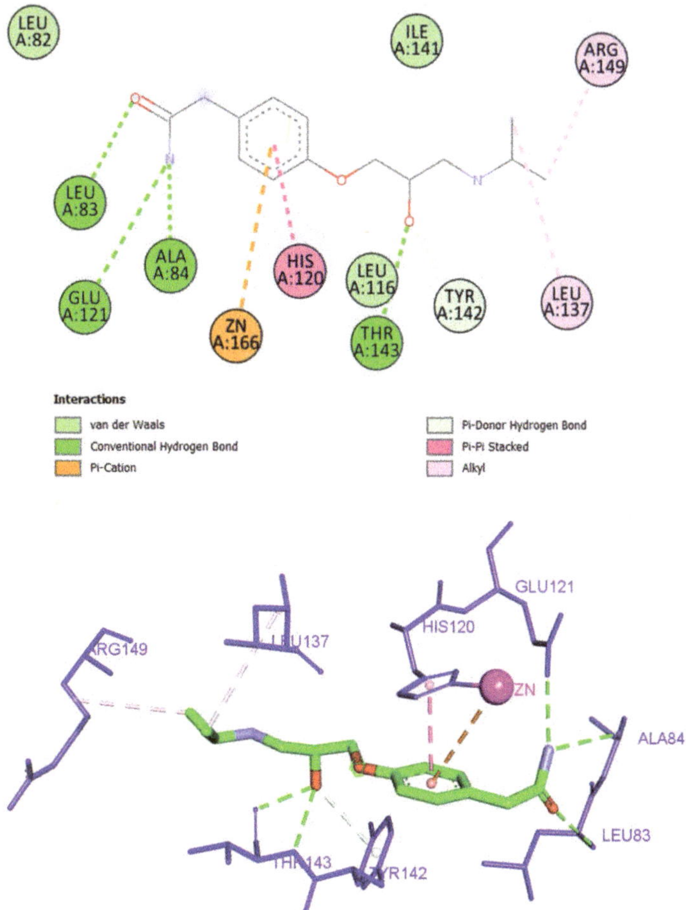

Figure 7. The 2D and 3D interactions of atenolol with amino acid residues in the catalytic domain of MMP-2 (PDB ID: 1HOV).

Ciprofloxacin (a fluoroquinolone drug) occupied the MMP-2 active site via the formation of six hydrogen bonds and eight hydrophobic interactions. The fluoroquinolone nucleus formed three hydrogen bonds with Leu83 and Ala84, as well as six hydrophobic interactions with zinc, Leu83, and His120. The piperazine arm was incorporated in three hydrogen bonds with Leu137, Ala139, and Thr143, as presented in Figure 8.

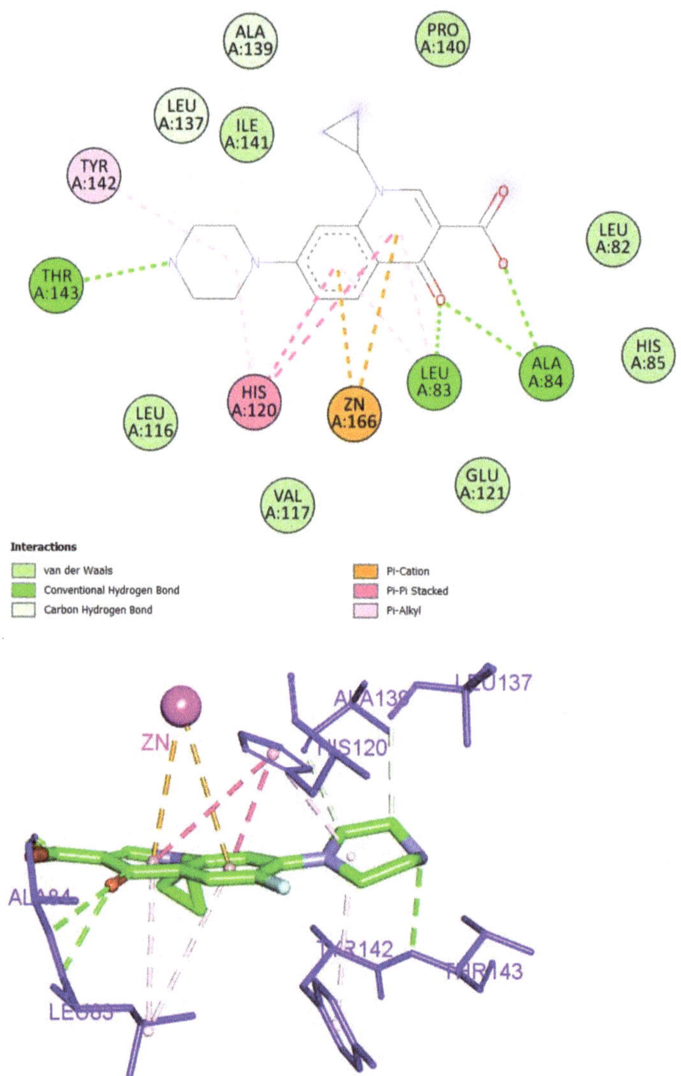

Figure 8. The 2D and 3D interactions of ciprofloxacin with amino acid residues in the catalytic domain of MMP-2 (PDB ID: 1HOV).

2.1.2. Docking of the Target Compounds into MMP-9 Catalytic Domain

The tested compounds revealed to be able to bind into MMP-9 and showed negative ΔG (Kcal.mol^{-1}) scores, as shown in Table 2. It was found that ampicillin, aztreonam, and ganciclovir, the most promising candidates, achieved the highest energy score and accommodated into the MMP-9 active site ($\Delta G = -30.81$, -29.97, and -28.89 kcal/mol, respectively).

By examining the binding interactions of the co-crystalized ligand (**NFH**) to the active site of MMP-9, it showed six hydrogen bonds with Gly186, Leu188, Ala189, Tyr393, and Tyr423, in addition to two hydrophobic interactions with Leu188. The hydroxamate group interacted as expected with the zinc of the CAT, as presented in Figure 9.

Table 2. The calculated ΔG (binding free energies) of the tested drugs against the catalytic domain of MMP-9 (PDB ID: 1GKC).

Serial	Compound	ΔG (kcal.mol^{-1})	Serial	Compound	ΔG (kcal.mol^{-1})
1	Acetazolamide	−15.34	17	Ganciclovir	−28.89
2	Acyclovir	−20.59	18	Indomethacin	−26.12
3	Ampicillin	−30.81	19	Ketorolac	−23.59
4	Atenolol	−27.60	20	Levocabastine	−24.62
5	Atropine	−26.99	21	Lincomycin	−26.74
6	Aztreonam	−29.97	22	Lornoxicam	−21.02
7	Betaxolol	−25.97	23	Methazolamide	−15.99
8	Brinzolamide	−26.07	24	Methotrexate	−27.70
9	Bromfenac	−22.24	25	Nadolole	−25.62
10	Carteolol	−25.96	26	Pilocarpine	−23.69
11	Cephalexine	−24.47	27	Pindolol	−23.56
12	Ciprofloxacin	−25.13	28	Prednisolone	−29.51
13	Dexamethasone	−23.65	29	Propranolol	−24.36
14	Diclofenac	−20.26	30	Quinidine	−28.53
15	Dorzolamide	−20.12	31	Tizanidine	−18.29
16	Fluconazole	−22.03	32	Voriconazolee	−23.46

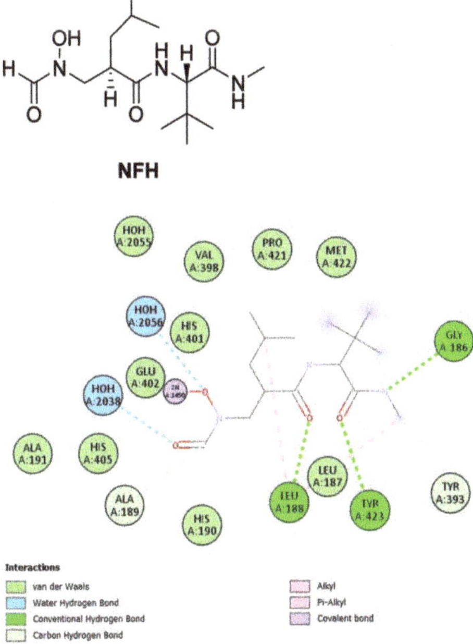

Figure 9. *Cont.*

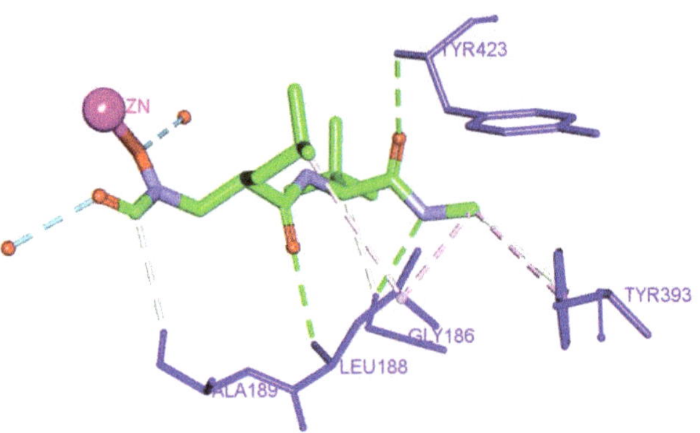

Figure 9. The 2D and 3D interactions of the co-crystallized ligand (**NFH**) with amino acid residues of the catalytic domain of MMP-9 (PDB ID: 1GKC).

This high binding score of ampicillin is attributed to the formation of six hydrogen bonds, five hydrophobic interactions, and zinc binding as well. The 3,3-dimethyl moiety of ampicillin was involved in three hydrophobic bonds with Leu187, His411, and Pro421, while the carboxylic acid and 7-oxo groups formed hydrogen bonds with Leu187, Leu188, and Tyr423. The phenyl acetamide moiety interacted with the active site by three hydrogen bonds with Leu397 and Val398, as well as two hydrophobic interactions with Leu418 and Arg424. Finally, the sulfur atom of the thiazolidine ring was able to interact with the zinc ion of the CAT through a metal–acceptor bond (Figure 10).

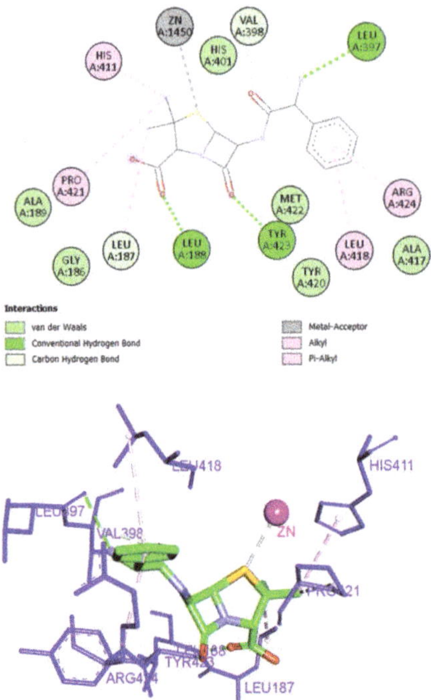

Figure 10. The 2D and 3D interactions of ampicillin with amino acid residues of the catalytic domain of MMP-9 (PDB ID: 1GKC).

On the other hand, aztreonam was unable to bind to the zinc ion of the CAT. An investigation of the top docking pose of aztreonam showed that it interacted with the MMP-9 active site by forming five hydrogen bond interactions (Leu188, Tyr420, Met422, and Tyr423) and nine hydrophobic interactions (Val398, His401, His405, His411, Met422, and Tyr423) (Figure 11).

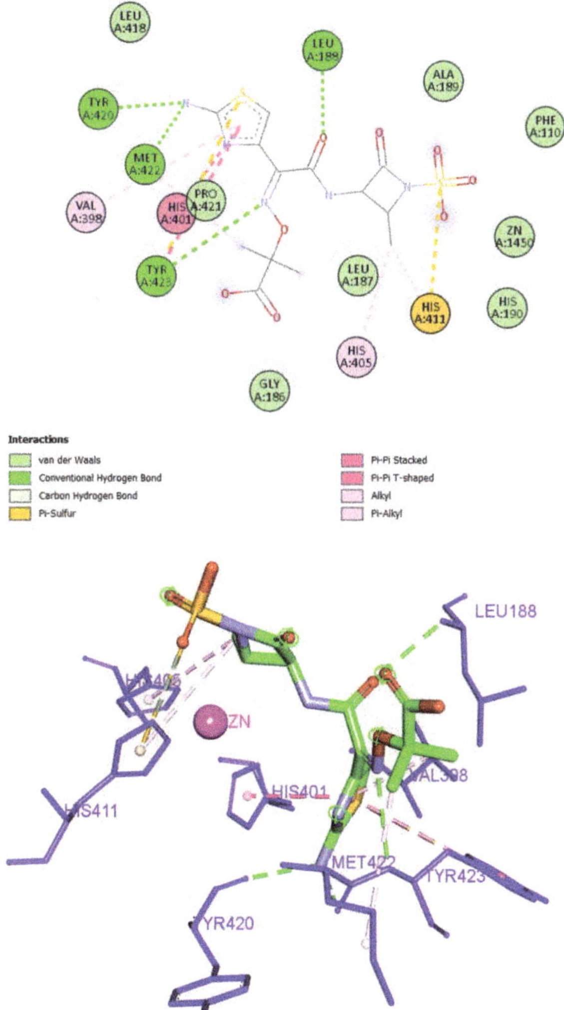

Figure 11. The 2D and 3D interactions of Aztreonam with amino acid residues of the catalytic domain of MMP-9 (PDB ID: 1GKC).

Finally, the binding of ganciclovir was through four hydrogen bonds with Leu188, Glu402, and Pro421. Four hydrophobic interactions were also detected (Val398, His401, and Tyr423). It was also able to bind with zinc ion through a metal–acceptor interaction (Figure 12).

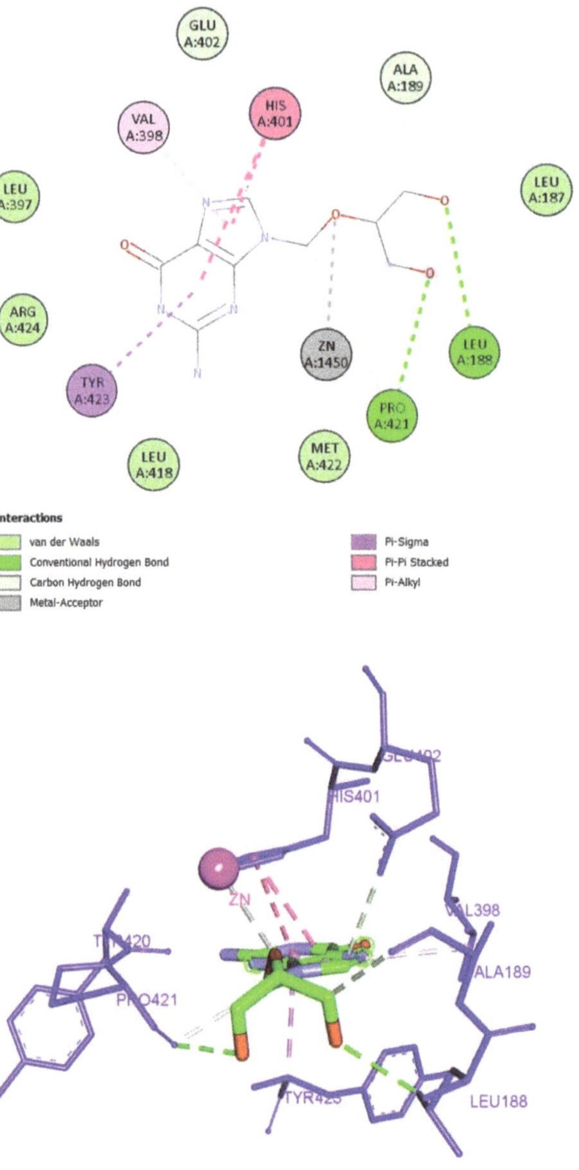

Figure 12. The 2D and 3D interactions of Ganciclovir with amino acid residues of the catalytic domain of MMP-9 (PDB ID: 1GKC).

Because atenolol and ampicillin were the most promising compounds that achieved high docking scores and similar binding modes to co-crystallized ligands with the ability to interact with zinc ions in both enzymes, they were both promoted for further analysis through molecular dynamics.

2.2. Molecular Dynamics and Molecular Mechanics–Generalized Born Surface Area (MM-GBSA) Calculations

The potential of atenolol and ampicillin to bind to MMP-2 and MMP-9, respectively, was further investigated through molecular dynamics. This allows the extensive analysis of the binding modes under realistic physiological conditions. Both protein files with the corresponding drugs were processed by the Schrodinger Maestro suite and simulated for 50 ns. The RMSD of the protein residues (Figure 13) in both complexes showed uniform values around 3.7 and 1.75 A^0 deviation for MMP-2 and MMP-9, respectively. Furthermore, both atenolol and ampicillin exhibited stable conformations with an RMSD (Supplementary Figures S1 and S2) of around 1.2 and 1.8 A^0, respectively.

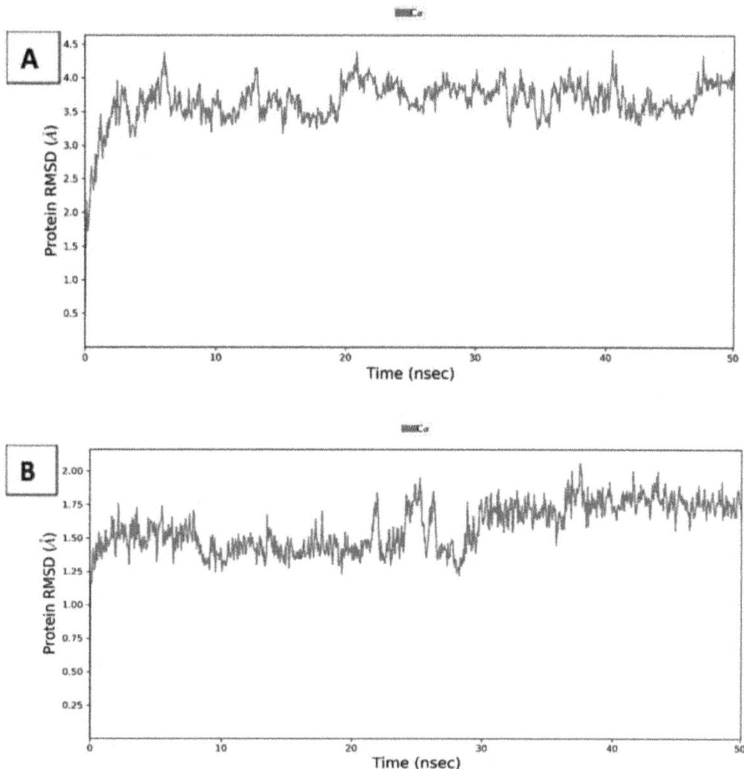

Figure 13. Root mean square deviation (RMSD) of *C*-alpha of MMP-2 (**A**) and MMP-9 (**B**) complexes with atenolol and ampicillin through 50 ns simulations.

Additionally, the flexibility of the conformers was assessed through the calculation of the root mean square fluctuations (RMSF) of the residues of the proteins and ligand atoms across the simulation time. Consistent with the calculated RMSD, the protein residues showed a low degree of fluctuations, especially with the ones in contact with the ligands as shown in Figure 14. The RMSF of the protein residues showed around 3.5 A^0 fluctuations of the residues exposed to drugs, while the fluctuations were lower in the case of both ligands, especially ampicillin which showed around $1A^0$ only (Supplementary Figures S3 and S4). These uniform values obtained points to the relative stability of both proteins and drugs conformations for the entire simulation duration.

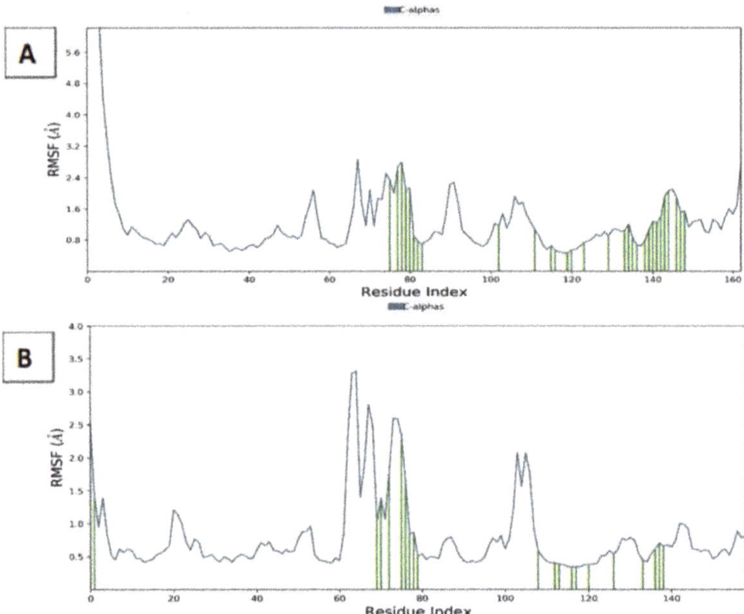

Figure 14. Root mean square fluctuation (RMSF) of *C*-alpha of MMP-2 (**A**) and MMP-9 (**B**) complexes with atenolol and ampicillin through 50 ns simulations. (Ligand contacts are marked green.).

For a better understanding of the binding modes, a further analysis of the interactions throughout the whole 50 ns simulation time was performed (Figure 15). For MMP-9, ampicillin successfully formed metal coordination with the zinc ion constantly through its amide carbonyl group. On the other hand, the interaction of atenolol with the zinc ion of MMP-2 was not observed; however, it interacted extensively with the binding site residues as well compensating this inability as shown in Figure 16.

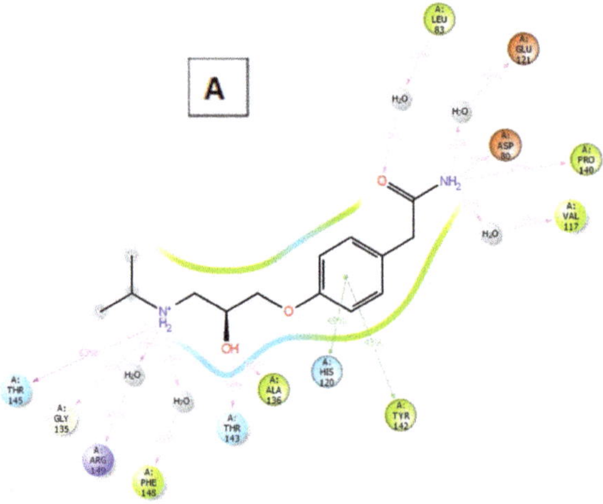

Figure 15. *Cont.*

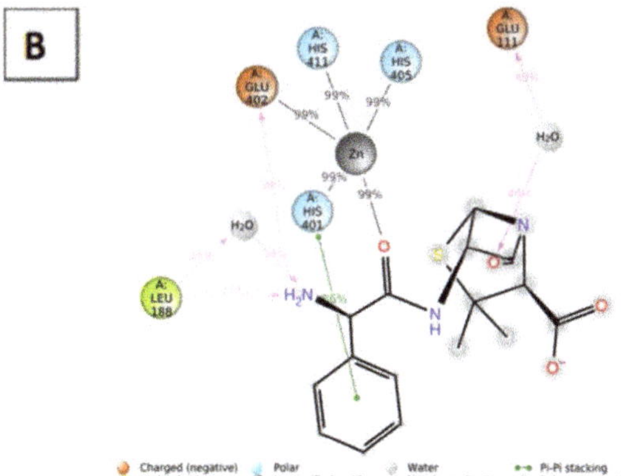

Figure 15. The 2D ligand-protein contact summary of atenolol-MMP-2 (**A**) and ampicillin-MMP-9 (**B**) complexes through 50 ns simulations.

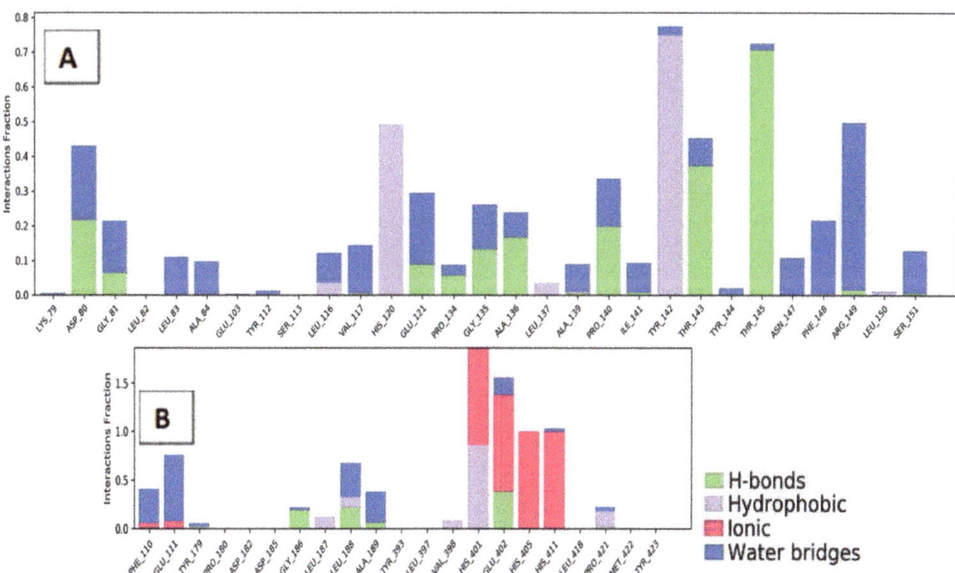

Figure 16. Ligand-protein contact histogram of atenolol-MMP-2 (**A**) and ampicillin-MMP-9 (**B**) complexes through 50 ns simulations.

One of the most commonly used methods for calculating the binding free energy is molecular mechanics–generalized born surface area (MM-GBSA). The lower a ligand-protein complex's projected binding free energy is, the more stable the complex is expected to be, and the higher the ligand's activity and potency (Table 3). Both complexes showed stable binding throughout the dynamic simulation.

Table 3. The MM-GBSA binding free energies (Kcal.mol^{-1}) of MMP-2/atenolol and MMP-9/ampicillin complexes.

	MMP-2/Atenolol		MMP-9/Ampicillin	
	Start	**End**	**Start**	**End**
dG Binding	−38.07	−47.75	−9.35	−26.96
dG binding Coulomb	−22.89	−51.36	39.63	11.89
dG Binding (NS)	−49.77	−52.57	−19.88	−28.70
dG binding (NS) Coulomb	−22.14	−52.42	39.35	13.00

2.3. Pharmacophore Study

The prospect of repositioning ampicillin for KC treatment may seem beneficial especially for complicated cases suffering from secondary bacterial infections. However, its unattended use for prolonged periods of time increases the risk of bacterial resistance. For this purpose, we extended our study on the MMP-9 inhibitors to propose pharmacophoric features with a high potential of a MMP-9 inhibitory goal for future use.

The co-crystallized ligand of MMP-9 (PDB ID: 1GKC) was used to generate the pharmacophore model using the Discovery studio software. In this test, the protocol of receptor-ligand pharmacophore generation was applied. In this protocol, the software identifies the essential features of **NFH** (co-crystalized ligand) during its interaction with the receptor. The library was then screened, and fit value data were calculated (Supplementary Table S1). Ampicillin, aztreonam, cephalexin, and lincomycin achieved the highest fit value, even comparable to **NFH,** as shown in Table 4.

Table 4. Fit value of the top four compounds and the co-crystallized ligand of MMP-9 (PDB ID: 1GKC).

Compound	Mapped Features	Fit Value
NFH	HBD1, HBD2, HBA1, HBA2, H1, H2	4.74
Ampicillin	HBD1, HBD2, HBA1, HBA2, H1, H2	3.76
Aztreonam	HBD1, HBD2, HBA1, HBA2, H1, H2	4.13
Cephalexine	HBD1, HBD2, HBA1, HBA2, H1, H2	4.06
Lincomycin	HBD1, HBD2, HBA1, HBA2, H1, H2	3.76

The formed pharmacophore model consisted of six features: two H-bond donors (HBD), two H-bond acceptors (HBA), and two hydrophobic centers (H). The features were distributed in a pyramidal shape with each feature occupying a 1.6 A^0 radius (Figure 17). In accordance with the data acquired so far, the overlay of each compound on the pharmacophore (Figure 18) demonstrates the ability of the compounds to span across the entire pharmacophoric features in a similar fashion to the co-crystallized inhibitor **NFH.**

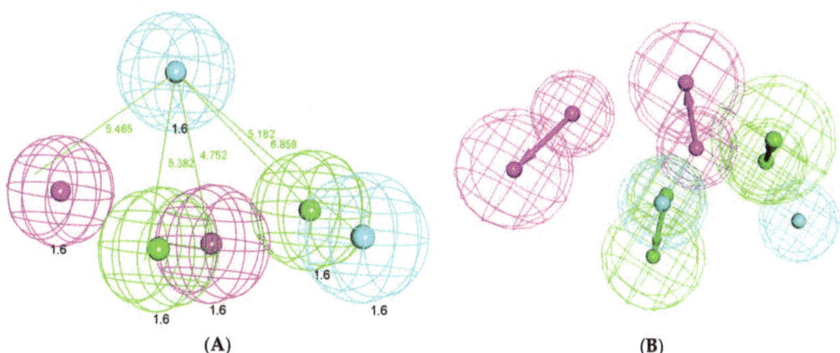

Figure 17. (**A**) The generated 3D pharmacophore geometry with six features: two hydrogen bond donors (pink color) and two hydrogen bond acceptors (green), and two hydrophobic centers (blue). (**B**) The 3D-pharmacophore with vector direction of each feature.

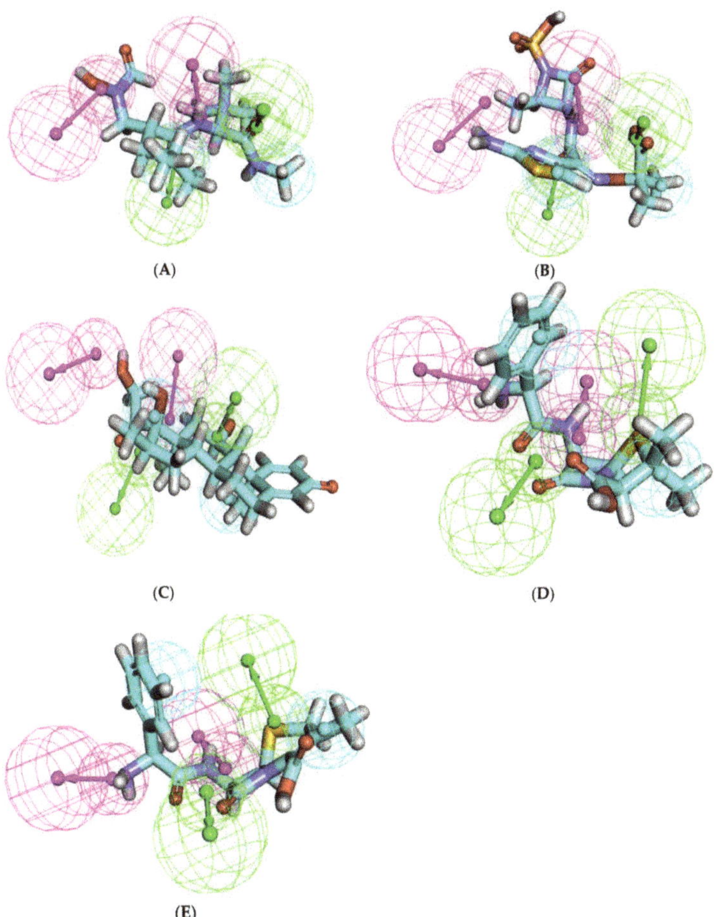

Figure 18. Mapping of the tested compounds on the generated pharmacophore: (**A**) **NFH** on the generated pharmacophore (fit value = 4.74), (**B**) ampicillin (fit value = 3.76), (**C**) aztreonam (fit value = 4.13), (**D**) cephalexin (fit value = 4.06), and (**E**) lincomycin (Fit value = 3.76).

3. Conclusions

In an attempt to fasten the drug discovery process, a repositioning approach was enforced to discover potential medications for KC in a cost/efficient manner. Because high levels of collagenolytic and gelatinolytic actions were observed, collagenase and gelatinases became a promising target for tackling KC. Among those proteolytic enzymes are MMP-2 and MMP-9. Both are metalloenzymes that are characterized by the presence of zinc ion in the CAT. Thus, inhibition techniques focused on drugs that can interact with zinc ions through their ZBGs. As a result, thirty-two FDA-approved drugs were subjected to virtual screening through docking against MMP-2 and MMP-9 proteins to identify the most promising inhibitors as a proposed computational mechanism to treat KC. The docking results showed the ability of atenolol and ampicillin to accommodate well into the active sites of MMP-2 and MMP-9, respectively. Additionally, both exhibited similar binding modes as **I52** and **NHF** (co-crystallized inhibitors of MMP-2 and MMP-9, respectively) and interacted with the zinc ion of the CAT successfully. Subsequent molecular dynamic simulations and MM-GBSA calculations point to the stability of the binding of both drugs to the respective enzyme, thus adding to the potential of both compounds in KC management. The dual potential properties of ampicillin for the treatment of KC, especially with bacterial infections, pushed for the design of alternatives that could be used for prolonged treatment times without risk of bacterial resistance. An additional pharmacophore study was carried out using the co-crystallized ligand of MMP-9 (PDB ID: 1GKC) as a reference molecule for future designs. These encouraging findings pave the way for additional clinical investigations to confirm such theoretical findings.

4. Experimental

4.1. Literature Search and Library Generation

The designed compounds' structures were retrieved online (Pubchem; https://pubchem.ncbi.nlm.nih.gov/) (accessed on 1 April 2022) and sketched using ChemBioDraw Ultra 14.0 and saved in MDL-SD file format.

4.2. Docking Studies

The crystal structures of MMP-2 and MMP-9 (PDB ID: 1HOV [16] and PDB ID: 1GKC [17], respectively) were downloaded from the Protein Data Bank (http://www.rcsb.org/pdb) (accessed on 1 April 2022). Molecular operating environment was used for docking. At first, the protein files were prepped using built-in "Quickprep" function. Initial validation was performed through docking of each co-crystallized ligand to its protein file, followed by the calculation of root mean square deviation (RMSD) between the docked pose and the co-crystallized one. After successful validation, the library of compounds was imported and prepped into MOE database file, that was then docked using "Induced fit" protocol. The interactions were then viewed using Discovery Studio Visualizer 2021 [24–32].

4.3. Molecular Dynamics and Molecular Mechanics–Generalized Born Surface Area (MM-GBSA) Calculations

Schrödinger Desmond [18] package was used for molecular dynamics simulations using "OPLS4" forcefield as described in past research. The MM-GBSA technique was used to compute the binding free energy of the protein-ligand complexes studied, which integrated molecular mechanics (MM) force fields with a generalized born and surface area continuum solvation solvent model using Schrodinger Prime package [23,33–36].

4.4. Pharmacophore Studies

The pharmacophore model was carried out using Discovery Studio 4.0 software. The protocol of receptor-ligand pharmacophore generation was applied. This protocol used the co-crystallized ligand of MMP-9 (PDB ID: 1GKC) as a reference molecule. The tested compounds were used as a training set. In this protocol, we used the following features in pharmacophore generation: (i) hydrogen bond donor (HBD), (ii) hydrogen bond acceptor (HBA), (iii) hydrophobic aliphatic (HA), (iv) hydrophobic aromatic (HAr), and ring aromatic (RA). Then, the ligand pharmacophore mapping protocol was used in the virtual screening process. The most predictive model was used as 3D queries to identify compounds with high fit values [36–39].

Supplementary Materials: The following supporting information can be downloaded at: https://www.mdpi.com/article/10.3390/molecules27113584/s1, Figure S1. Atenolol properties throughout 50 ns simulation in complex with MMP-2; Figure S2. Ampicillin properties throughout 50 ns simulation in complex with MMP-9; Figure S3. RMSF of atenolol throughout 50 ns simulation in complex with MMP-2; Figure S4. RMSF of ampicillin throughout 50 ns simulation in complex with MMP-9; Table S1. Fit values of the full library on MMP-9 pharmacophore.

Author Contributions: A.B.: Conceptualization, Investigation, Visualization, Writing—original draft, Writing—review & editing; M.A.E.: Methodology, Software, Validation and Writing—review & editing; E.Y.S.: Validation, Formal analysis, Funding acquisition, Writing—review & editing; A.A.A.-K.: Validation, Formal analysis, Funding acquisition, Writing—review & editing; M.O.A.: Data curation, Validation and Writing—review & editing; A.H.A.: Validation, Formal analysis, Funding acquisition, Writing—review & editing; M.H.A.: Validation, methodology, Writing—review & editing; A.B.M.M.: Conceptualization, Validation, Investigation, Writing—review & editing; H.E.: Methodology, Software, Validation and Writing—original draft. All authors have read and agreed to the published version of the manuscript.

Funding: Taif University Researchers Supporting, Project number (TURSP-2020/330), Taif University, Taif, Saudi Arabia, and Deanship of Scientific Research at Umm Al-Qura University, Grant Code: (22UQU4331100DSR13).

Institutional Review Board Statement: Not applicable.

Informed Consent Statement: Not applicable.

Data Availability Statement: Supplementary Materials are provided.

Acknowledgments: The authors would like to introduce their appreciation and thanks to the Taif University Researchers Supporting, Project number (TURSP-2020/330), Taif University, Taif, Saudi Arabia. The authors would like to thank the Deanship of Scientific Research at Umm Al-Qura University for supporting this work by Grant Code: (22UQU4331100DSR13).

Conflicts of Interest: The authors declare no conflict of interest.

References

1. Romero-Jiménez, M.; Santodomingo-Rubido, J.; Wolffsohn, J.S. Keratoconus: A review. *Contact Lens Anterior Eye* **2010**, *33*, 157–166. [CrossRef] [PubMed]
2. Davidson, A.E.; Hayes, S.; Hardcastle, A.J.; Tuft, S.J. The pathogenesis of keratoconus. *Eye* **2014**, *28*, 189–195. [CrossRef] [PubMed]
3. Lucas, S.E.M.; Burdon, K.P. Genetic and Environmental Risk Factors for Keratoconus. *Annu. Rev. Vis. Sci.* **2020**, *6*, 25–46. [CrossRef] [PubMed]
4. Mohammadpour, M.; Heidari, Z.; Hashemi, H. Updates on Managements for Keratoconus. *J. Curr. Ophthalmol.* **2018**, *30*, 110–124. [CrossRef] [PubMed]
5. Smiddy, W.E.; Hamburg, T.R.; Kracher, G.P.; Stark, W.J. Keratoconus. *Ophthalmology* **1988**, *95*, 487–492. [CrossRef]
6. Parker, J.S.; van Dijk, K.; Melles, G.R.J. Treatment options for advanced keratoconus: A review. *Surv. Ophthalmol.* **2015**, *60*, 459–480. [CrossRef] [PubMed]
7. Sykakis, E.; Karim, R.; Evans, J.R.; Bunce, C.; Amissah-Arthur, K.N.; Patwary, S.; McDonnell, P.J.; Hamada, S. Corneal collagen cross-linking for treating keratoconus. *Cochrane Database Syst. Rev.* **2015**, *24*, CD010621. [CrossRef]
8. Balasubramanian, S.A.; Pye, D.C.; Willcox, M.D.P. Are Proteinases the Reason for Keratoconus? *Curr. Eye Res.* **2010**, *35*, 185–191. [CrossRef]
9. Balasubramanian, S.A.; Mohan, S.; Pye, D.C.; Willcox, M.D.P. Proteases, proteolysis and inflammatory molecules in the tears of people with keratoconus. *Acta Ophthalmol.* **2012**, *90*, e303–e309. [CrossRef]
10. Dormán, G.; Cseh, S.; Hajdú, I.; Barna, L.; Kónya, D.; Kupai, K.; Kovács, L.; Ferdinandy, P. Matrix Metalloproteinase Inhibitors. *Drugs* **2010**, *70*, 949–964. [CrossRef]
11. Das, S.; Amin, S.A.; Jha, T. Inhibitors of gelatinases (MMP-2 and MMP-9) for the management of hematological malignancies. *Eur. J. Med. Chem.* **2021**, *223*, 113623. [CrossRef] [PubMed]

12. Rasmussen, H.S.; McCann, P.P. Matrix Metalloproteinase Inhibition as a Novel Anticancer Strategy: A Review with Special Focus on Batimastat and Marimastat. *Pharmacol. Ther.* **1997**, *75*, 69–75. [CrossRef]

13. Rao, B. Recent Developments in the Design of Specific Matrix Metalloproteinase Inhibitors aided by Structural and Computational Studies. *Curr. Pharm. Des.* **2005**, *11*, 295–322. [CrossRef] [PubMed]

14. Breuer, E.; Frant, J.; Reich, R. Recent non-hydroxamate matrix metalloproteinase inhibitors. *Expert Opin. Ther. Pat.* **2005**, *15*, 253–269. [CrossRef]

15. Nicolotti, O.; Catto, M.; Giangreco, I.; Barletta, M.; Leonetti, F.; Stefanachi, A.; Pisani, L.; Cellamare, S.; Tortorella, P.; Loiodice, F.; et al. Design, synthesis and biological evaluation of 5-hydroxy, 5-substituted-pyrimidine-2,4,6-triones as potent inhibitors of gelatinases MMP-2 and MMP-9. *Eur. J. Med. Chem.* **2012**, *58*, 368–376. [CrossRef] [PubMed]

16. Mehany, A.B.; Belal, A.; Mohamed, A.F.; Shaaban, S.; Abdelhamid, G. Apoptotic and anti-angiogenic effects of propolis against human bladder cancer: Molecular docking and in vitro screening. *Biomarkers* **2022**, *27*, 138–150. [CrossRef]

17. Belal, A.; Elanany, M.A.; Raafat, M.; Hamza, H.T.; Mehany, A.B.M. Calendula officinalis Phytochemicals for the Treatment of Wounds Through Matrix Metalloproteinases-8 and 9 (MMP-8 and MMP-9): In Silico Approach. *Nat. Prod. Commun.* **2022**, *17*. [CrossRef]

18. Ghattas, A.-E.-B.A.; Khodairy, A.; Moustafa, H.M.; Hussein, B.R.; Farghaly, M.M.; Aboelez, M.O. Synthesis, in vitro Antibacterial and in vivo Anti-Inflammatory Activity of Some New Pyridines. *Pharm. Chem. J.* **2017**, *51*, 652–660. [CrossRef]

19. Liu, H.; Zhu, W.; Wu, Y.; Jiang, C.; Huo, L.; Belal, A. COVID-19 Pandemic Between Severity Facts and Prophylaxis. *Nat. Prod. Commun.* **2021**, *16*. [CrossRef]

20. AbdRabou, M.A.; Mehany, A.; Farrag, I.M.; Belal, A.; Abdelzaher, O.F.; El-Sharkawy, A.; El-Azez, A.; Asmaa, M.; EL-Sharkawy, S.M.; Al Badawi, M.H. Therapeutic Effect of Murine Bone Marrow-Derived Mesenchymal Stromal/Stem Cells and Human Placental Extract on Testicular Toxicity Resulting from Doxorubicin in Rats. *BioMed Res. Int.* **2021**, *2021*, 9979670. [CrossRef]

21. Feng, Y.; Likos, J.J.; Zhu, L.; Woodward, H.; Munie, G.; McDonald, J.J.; Stevens, A.M.; Howard, C.P.; De Crescenzo, G.A.; Welsch, D.; et al. Solution structure and backbone dynamics of the catalytic domain of matrix metalloproteinase-2 complexed with a hydroxamic acid inhibitor. *Biochim. Biophys. Acta-Proteins Proteom.* **2002**, *1598*, 10–23. [CrossRef]

22. Rowsell, S.; Hawtin, P.; Minshull, C.A.; Jepson, H.; Brockbank, S.M.V.; Barratt, D.G.; Slater, A.M.; McPheat, W.L.; Waterson, D.; Henney, A.M.; et al. Crystal Structure of Human MMP9 in Complex with a Reverse Hydroxamate Inhibitor. *J. Mol. Biol.* **2002**, *319*, 173–181. [CrossRef]

23. Schrödinger, L.L.C. *Schrödinger Release 2021-3: Maestro*; Schrödinger LLC: New York, NY, USA, 2021.

24. Nada, H.; Lee, K.; Gotina, L.; Pae, A.N.; Elkamhawy, A. Identification of novel discoidin domain receptor 1 (DDR1) inhibitors using E-pharmacophore modeling, structure-based virtual screening, molecular dynamics simulation and MM-GBSA approaches. *Comput. Biol. Med.* **2022**, *142*, 105217. [CrossRef]

25. Elsayed, A.; Belal, A. Formulation, characterization and in-vitro evaluation of solid lipid nanoparticles for the delivery of a new anticancer agent, 1H-pyrazolo [3, 4-d] pyrimidine derivative. *Trop. J. Pharm. Res.* **2021**, *20*, 885–891. [CrossRef]

26. Shoman, M.E.; Aboelez, M.O.; Shaykhon, M.S.; Ahmed, S.A.; Abuo-Rahma, G.E.-D.A.; Elhady, O.M. New nicotinic acid-based 3, 5-diphenylpyrazoles: Design, synthesis and antihyperlipidemic activity with potential NPC1L1 inhibitory activity. *Mol. Divers.* **2021**, *25*, 673–686. [CrossRef]

27. Belal, A. 3D-Pharmacophore Modeling, Molecular Docking, and Virtual Screening for Discovery of Novel CDK4/6 Selective Inhibitors. *Russ. J. Bioorganic Chem.* **2021**, *47*, 317–333. [CrossRef]

28. Zhaorigetu, I.M.F.; Belal, A.; Al Badawi, M.H.; Abdelhady, A.A.; Abou Galala, F.M.; El-Sharkawy, A.; El-Dahshan, A.A.; Mehany, A.B. Antiproliferative, Apoptotic Effects and Suppression of Oxidative Stress of Quercetin against Induced Toxicity in Lung Cancer Cells of Rats: In vitro and In vivo Study. *J. Cancer* **2021**, *12*, 5249. [CrossRef]

29. Khodairy, A.; Ali, A.M.; Aboelez, M.O.; El-Wassimy, M. One-Pot Multicomponent Synthesis of Novel 2-Tosyloxyphenylpyrans under Green and Conventional Condition with Anti-inflammatory Activity. *J. Heterocycl. Chem.* **2017**, *54*, 1442–1449. [CrossRef]

30. Eldeeb, E.; Belal, A. Two promising herbs that may help in delaying corona virus progression. *Int. J. Trend. Sci. Res. Dev.* **2020**, *4*, 764–766.

31. Belal, A. Pyrrolizines as Potential Anticancer Agents: Design, Synthesis, Caspase-3 activation and Micronucleus (MN) Induction. *Anti-Cancer Agents Med. Chem.* **2018**, *18*, 2124–2130. [CrossRef]

32. Elsayed, M.; Aboelez, M.O.; Elsadek, B.E.; Sarhan, H.A.; Khaled, K.A.; Belal, A.; Khames, A.; Hassan, Y.A.; Abdel-Rheem, A.A.; Elkaeed, E.B. Tolmetin Sodium Fast Dissolving Tablets for Rheumatoid Arthritis Treatment: Preparation and Optimization Using Box-Behnken Design and Response Surface Methodology. *Pharmaceutics* **2022**, *14*, 880. [CrossRef] [PubMed]

33. Ibrahim, M.K.; Eissa, I.H.; Abdallah, A.E.; Metwaly, A.M.; Radwan, M.M.; ElSohly, M.A. Design, synthesis, molecular modeling and anti-hyperglycemic evaluation of novel quinoxaline derivatives as potential PPARγ and SUR agonists. *Bioorg. Med. Chem.* **2017**, *25*, 1496–1513. [CrossRef] [PubMed]

34. Ibrahim, M.K.; Eissa, I.H.; Alesawy, M.S.; Metwaly, A.M.; Radwan, M.M.; ElSohly, M.A. Design, synthesis, molecular modeling and anti-hyperglycemic evaluation of quinazolin-4(3H)-one derivatives as potential PPARγ and SUR agonists. *Bioorg. Med. Chem.* **2017**, *25*, 4723–4744. [CrossRef] [PubMed]

35. El-Zahabi, M.A.; Elbendary, E.R.; Bamanie, F.H.; Radwan, M.F.; Ghareib, S.A.; Eissa, I.H. Design, synthesis, molecular modeling and anti-hyperglycemic evaluation of phthalimide-sulfonylurea hybrids as PPARγ and SUR agonists. *Bioorg. Chem.* **2019**, *91*, 103115. [CrossRef]

36. Kamel, M.S.; Belal, A.; Aboelez, M.O.; Shokr, E.K.; Abdel-Ghany, H.; Mansour, H.S.; Shawky, A.M.; El-Remaily, M.A.E.A.A.A. Microwave-Assisted Synthesis, Biological Activity Evaluation, Molecular Docking, and ADMET Studies of Some Novel Pyrrolo [2, 3-b] Pyrrole Derivatives. *Molecules* **2022**, *27*, 2061. [CrossRef]
37. Moustafa, A.H.; Ahmed, W.W.; Awad, M.F.; Aboelez, M.O.; Khodairy, A.; Amer, A.A. Eco-friendly and regiospecific intramolecular cyclization reactions of cyano and carbonyl groups in N, N-disubstituted cyanamide. *Mol. Divers.* **2022**, 1–11. [CrossRef]
38. Abdelkreem, E.; Mahmoud, S.M.; Aboelez, M.O.; Abd El Aal, M. Nebulized magnesium sulfate for treatment of persistent pulmonary hypertension of newborn: A pilot randomized controlled trial. *Indian J. Pediatrics* **2021**, *88*, 771–777. [CrossRef]
39. Shokr, E.K.; Kamel, M.S.; Abdel-Ghany, H.; El-Remaily, M.A.E.A.A.A. Optical characterization and effects of iodine vapor & gaseous HCl adsorption investigation of novel synthesized organic dye based on thieno [2, 3-b] thiophene. *Optik* **2021**, *243*, 167385.

molecules

Article

Increasing the Efficacy of Seproxetine as an Antidepressant Using Charge–Transfer Complexes

Walaa F. Alsanie [1,2], Abdulhakeem S. Alamri [1,2], Hussain Alyami [3], Majid Alhomrani [1,2], Sonam Shakya [4], Hamza Habeeballah [5], Heba A. Alkhatabi [6,7,8], Raed I. Felimban [6,9], Ahmed S. Alzahrani [2], Abdulhameed Abdullah Alhabeeb [10], Bassem M. Raafat [11], Moamen S. Refat [12,*] and Ahmed Gaber [2,13,*]

[1] Department of Clinical Laboratories Sciences, The Faculty of Applied Medical Sciences, Taif University, Taif 21944, Saudi Arabia; w.alsanie@tu.edu.sa (W.F.A.); a.alamri@tu.edu.sa (A.S.A.); m.alhomrani@tu.edu.sa (M.A.)

[2] Centre of Biomedical Sciences Research (CBSR), Deanship of Scientific Research, Taif University, Taif 21944, Saudi Arabia; a.s.zahrani@tu.edu.sa

[3] College of Medicine, Taif University, Taif 21944, Saudi Arabia; hmyami@tu.edu.sa

[4] Department of Chemistry, Faculty of Science, Aligarh Muslim University, Aligarh 202002, India; sonamshakya08@gmail.com

[5] Department of Medical Laboratory Technology, Faculty of Applied Medical Sciences in Rabigh, King Abdulaziz University, Jeddah 21589, Saudi Arabia; hhabeeballah@kau.edu.sa

[6] Department of Medical Laboratory Sciences, Faculty of Applied Medical Sciences, King Abdulaziz University, Jeddah 21589, Saudi Arabia; halkhattabi@kau.edu.sa (H.A.A.); faraed@kau.edu.sa (R.I.F.)

[7] Center of Excellence in Genomic Medicine Research (CEGMR), King Abdulaziz University, Jeddah 21589, Saudi Arabia

[8] King Fahd Medical Research Centre, Hematology Research Unit, King Abdulaziz University, Jeddah 21589, Saudi Arabia

[9] Center of Innovation in Personalized Medicine (CIPM), 3D Bioprinting Unit, King Abdulaziz University, Jeddah 21589, Saudi Arabia

[10] National Centre for Mental Health Promotion, Riyadh 11525, Saudi Arabia; aalhabeeb@ncmh.org.sa

[11] Department of Radiological Sciences, College of Applied Medical Sciences, Taif University, Taif 21944, Saudi Arabia; bassemraafat@tu.edu.sa

[12] Department of Chemistry, College of Science, Taif University, Taif 21944, Saudi Arabia

[13] Department of Biology, College of Science, Taif University, Taif 21944, Saudi Arabia

* Correspondence: moamen@tu.edu.sa (M.S.R.); a.gaber@tu.edu.sa (A.G.)

Citation: Alsanie, W.F.; Alamri, A.S.; Alyami, H.; Alhomrani, M.; Shakya, S.; Habeeballah, H.; Alkhatabi, H.A.; Felimban, R.I.; Alzahrani, A.S.; Alhabeeb, A.A.; et al. Increasing the Efficacy of Seproxetine as an Antidepressant Using Charge–Transfer Complexes. *Molecules* **2022**, 27, 3290. https://doi.org/10.3390/molecules27103290

Academic Editors: Tanveer A. Wani, Seema Zargar and Afzal Hussain

Received: 27 April 2022
Accepted: 19 May 2022
Published: 20 May 2022

Publisher's Note: MDPI stays neutral with regard to jurisdictional claims in published maps and institutional affiliations.

Abstract: The charge transfer interactions between the seproxetine (SRX) donor and π-electron acceptors [picric acid (PA), dinitrobenzene (DNB), p-nitrobenzoic acid (p-NBA), 2,6-dichloroquinone-4-chloroimide (DCQ), 2,6-dibromoquinone-4-chloroimide (DBQ), and 7,7',8,8'-tetracyanoquinodi methane (TCNQ)] were studied in a liquid medium, and the solid form was isolated and characterized. The spectrophotometric analysis confirmed that the charge–transfer interactions between the electrons of the donor and acceptors were 1:1 (SRX: π-acceptor). To study the comparative interactions between SRX and the other π-electron acceptors, molecular docking calculations were performed between SRX and the charge transfer (CT) complexes against three receptors (serotonin, dopamine, and TrkB kinase receptor). According to molecular docking, the CT complex [(SRX)(TCNQ)] binds with all three receptors more efficiently than SRX alone, and [(SRX)(TCNQ)]-dopamine (CTcD) has the highest binding energy value. The results of AutoDock Vina revealed that the molecular dynamics simulation of the 100 ns run revealed that both the SRX-dopamine and CTcD complexes had a stable conformation; however, the CTcD complex was more stable. The optimized structure of the CT complexes was obtained using density functional theory (B-3LYP/6-311G++) and was compared.

Keywords: seproxetine; antidepressant; charge transfer; π-acceptors; DFT

1. Introduction

Depression is the most common mental illness, affecting roughly 322 million people worldwide [1]. Depression is the main cause of disability and the fourth major contributor to the global illness burden [2]. Antidepressants are the third most commonly sold class of therapeutic drugs worldwide [3]. The majority of these treatments are based on chemicals that target the serotonin (5-hydroxytryptamine (5-HT): a group of G protein-coupled receptor and ligand-gated ion channels found in the central and peripheral nervous systems) transporter, a single protein in the brain. Selected serotonin reuptake inhibitors (SSRIs), which block 5-HT reuptake, account for around 80% of all antidepressants on the market [3]. Other antidepressants, such as serotonin and noradrenaline reuptake inhibitors, as well as traditional tricyclic antidepressants (e.g., amitryptyline, clomipramine, imipramine), prevent noradrenaline reuptake. Indeed, compared to tricyclic medicines, the success of selective serotonin reuptake inhibitors is mostly due to their safety, tolerability, and lack of severe side effects, which enhances patient compliance and quality of life [3].

Although seproxetine (SRX, also known as S-norfluoxetine) is classified as a selective serotonin reuptake inhibitor, its inhibitory action extends beyond serotonin transporters to dopamine transporters (DAT) and 5-HT2A/2C receptors [4]. It is the active N-demethylate metabolite of the commonly prescribed antidepressant fluoxetine and is deemed more potent than the parental compound itself [5]. The 5-HT(2A) and 5-HT(2C) receptors belong to the G-protein-coupled receptor (GPCR) superfamily. GPCRs interact with G-proteins to transmit extracellular signals to the inside of cells. The 5-HT(2A) and 5-HT(2C) receptors are involved in the effects of a wide range of drugs on anxiety, sleep patterns, depression, hallucinations, schizophrenia, dysthymia, eating behavior, and neuro-endocrine processes [6].

As SRX was found to be a 20 times more potent serotonin inhibitor than its sister enantiomer R-norfluoxetine, significant research efforts were focused on this drug in the 1990s [7]. However, serious cardiac side effects, such as QT prolongation (a measure of delayed ventricular repolarisation), halted further development [4,8]. The potency of SRX as a serotonin inhibitor should not be ignored, and an effort must be taken to chemically modify (charge–transfer complexation) SRX for a better serotonin inhibitor while suppressing the drawback.

Charge–transfer (CT) complexation, or electron–donor transfer, is a crucial aspect of biochemical and biological processes such as drug design, enzyme catalysis, and ion sensing [9]. The pharmacodynamics and thermodynamics of therapeutic substances and biological processes in the human body are studied using charge–transfer complexation interactions [10–14]. In biological systems, charge–transfer complexes may play a crucial function. Extensive research has been carried out on charge–transfer interactions between inorganic anions, particularly the iodide ion and pyridinium, and substituted pyridinium cations, to determine the sensitivity of their charge–transfer absorption to the solvent environment, as well as the potential role of structures of this type in enzymatic oxidation-reduction processes [15]. As the charge–transfer complexes are a simpler, cheaper, and more efficient tool of analysis than the other methods mentioned in the literature, charge–transfer interactions are an important subject employed in the determination of medicines in pharmaceutical and pure forms [16].

Many reports stated the interactions, in solution, between flavin mononucleotide, flavin adenine dinucleotide, or riboflavin and a variety of donors, including hydrocarbons [17], indoles [18], NADH [19], NADPH [19], purines and pyrimidines, as well as other compounds with no obvious donor properties. There is little doubt that complete electron transfer happens in several of these systems to generate the flavin semiquinone [20]. The new broad absorption band reported for mixes of the reduced form of flavin mononucleotide (FMNH2) and (FMN) was attributed to the creation of charge–transfer complexes [21]. 2-methyl-1,4-naphthoquinone, also known as vitamin K3, used as a synthetic substitute for K1, o-quinone adrenochrome, and many other biologically important quinones have substantial electron donor complexing capacity [22].

Tryptophan appears to be unique among amino acids in its capacity to generate charge transfer complexes due to the strong donor characteristics of the indole ring. However, another study has shown that a pyridinium model compound of NAD+ may form complexes with tyrosine and phenylalanine [23]. Spectral evidence was also found to produce charge–transfer complexes between NAD+ and model pyridinium compounds with chymotrypsinogen, a tryptophan-rich protein [24].

Molecular docking (MD) is a computer method for efficiently predicting the non-covalent binding of macromolecules (receptors) and small molecules (acceptors) based on their unbound structures, structures generated through MD simulations, homology modeling, and other methods. The prediction of small molecule binding to proteins is of particular practical significance since it is used to screen virtual libraries of drug-like compounds for leads for further drug development. As a result, MD has become an important method in drug development.

Here, we used the Autodock Vina program to investigate the interactions between the ligand (SRX and synthesized CT complexes) and receptors (serotonin, dopamine, and TrkB kinase receptors). In the 1970s and 1980s periods, selective serotonin reuptake inhibitors (SSRIs) were developed, which are as effective antidepressants as tricyclics but do not have as many side effects as other antidepressant drugs. Binding energy, along with hydrophobic properties, ionizability, aromatic, and hydrogen bond surfaces, were also investigated. The molecular dynamic simulation was achieved at 300 K for 100 ns. The dynamic properties of the complexes were compared in many characterizations such as residue flexibility, structural solidity, solvent-accessible surface area, and other measurements. DFT using the B-3LYP/6-311G++ (basis set) level of theory was employed to obtain an optimized geometry of the CT complex- [(SRX)(PA}], [(SRX)(DNB)], [(SRX)(p-NBA)], [(SRX)(DCQ)], [(SRX)(DBQ)], and [(SRX)(TCNQ)] with minimal energy. Different parameters of the complexes were obtained and compared.

2. Materials and Methods

2.1. Synthesis of [(SRX)(π-Acceptor)] Charge–Transfer Complexes

The charge–transfer complexes [(SRX)(π-acceptor)] where π-acceptor are PA, DNB, *p*-NBA, DCQ, DBQ, and TCNQ (Figure 1) were synthesized as 1:1 by the reaction of SRX donor in a solution (25 mL) of each acceptor [25].

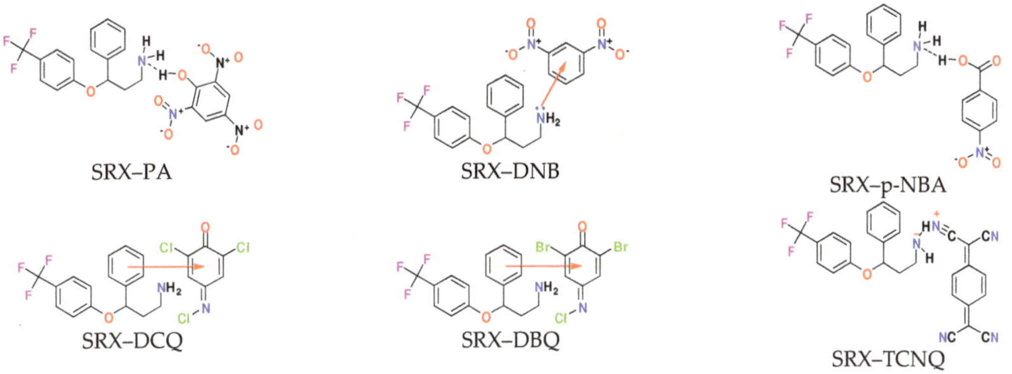

Figure 1. Speculated molecular structures of (1:1) charge-transfer complexes [(SRX)(π-acceptor)].

At room temperature, the mixtures were agitated for about an hour in each case. The precipitate was filtered and washed with the smallest amount of dichloromethane possible before being dried under vacuum over anhydrous $CaCl_2$.

2.2. Instruments and Measurements

With safeguards (platinum pans, nitrogen gas flow, and 30 °C min^{-1} heating rate), thermogravimetric analysis (TGA/DTG) was examined using Shimadzu TGA-50H equipment. A Perkin–Elmer Precisely Lambda 25 UV/Vis Spectrometer was used to scan the electronic absorption spectra of the synthesized charge–transfer complexes in the 200–800 nm region. A Bruker 600 MHz spectrometer was used to measure ^{1}H-NMR spectra in DMSO solvent.

2.3. Molecular Docking

The structures of the SRX drug and CT complexes were handled in PDBQT format via OpenBabelIGUI software (version 2.4.1) [26]. Then, the PyRx-Python prescription 0.8 and MMFF94 force field were used to minimize the energy of the structure for 500 steps [27]. The RCSB Protein Data Bank [28] was used to get the 3D crystal structures of the three receptors. The receptors were arranged using the BIOVIA Discovery Studio Visualizer (v19.1.0.18287). Kollman charges were also measured with the help of the AutoDock Tool [29]. The Geistenger method was used to allocate partial charges. The docking calculations were performed with Autodock Vina [30]. The DS (Discovery Studio) Visualizer was used to examine the docked poses that resulted.

2.4. Molecular Dynamics (MD) Simulation

The optimal receptor–ligand complex pose for SRX and [(SRX)(TCNQ)] with a maximum docking score was acquired through the molecular docking investigation. The GROMACS package version (2019.2) was used to accomplish MD simulation analysis via GROMOS96 43a1 force field. The parameter files and topologies were created with the most recent CGenFF through CHARMM-GUI [31,32]. The SPC water models that prolonged 10 Å from the receptor were utilized to explain receptor–ligand structures [33]. To neutralize the systems, 59 Na$^+$ and 64 Cl$^-$ ions (0.15 M salt) were injected to simulate physiological salt concentrations (Figure 2).

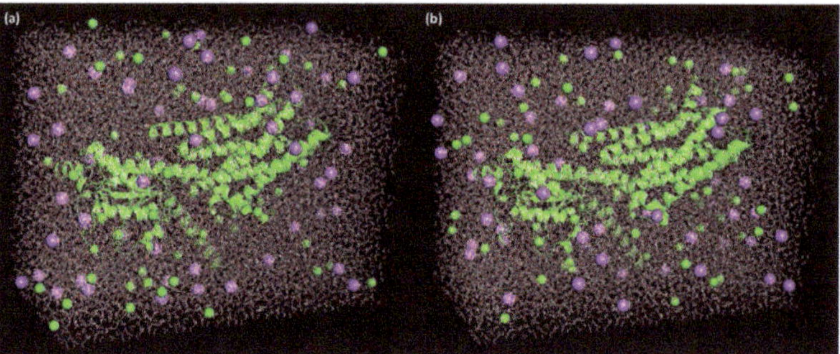

Figure 2. Receptor–ligand complex (**a**) SRXD and (**b**) CTcD in triclinic box solvated with water molecules and neutralized with 59 Na$^+$ and 64 Cl$^-$ ions (0.15 M salt).

Both systems were exposed to periodic boundary conditions at a continuous temperature (300 K) and pressure (1.0 bar) for 100 ns simulation time with a Leap-frog MD integrator [34]. To minimize poor contact inside the system, energy reduction with 5000 steps was performed [35]. The gmx hbond device was used to investigate hydrogen bonding. The gyration radius was measured using gmx gyrate tool, while the solvent-accessible surface area was calculated by gmx sasa. The root mean square deviation (RMSD) of the protein was designed using the gmx rms tools. The GROMACS analytic tools [36] were used to accomplish trajectory analysis. Grace Software was used to compute the plots, while PyMol/VMD was utilized to visualize them [37].

2.5. Computational Structural Analysis

DFT (Density functional theory) computational study was used for structural analysis of CT complexes and optimized geometry with atomic coordinates, strain-free lattice constants, and ground state minimum energy structure are obtained. Gaussian 09RevD.01 program [38] was used for this study. Gradient corrected correlation was applied with Pople's basic set B3LYP/6-311G++ [39]. For visualization of obtained DFT results, ChemCraft 1.5 software [40] was used.

3. Results and Discussion

3.1. Preapprehension

The attachment of the receptor to drugs does not affect the efficiency of its work, in fact, it improves it. However, it should be noted that different drugs have varying efficacy when they are connected with the receptor's site [41–45]. Several reports showed differences in the efficacy of two drugs targeting the same receptor because the activation of the receptor is dependent on the rate of drug interaction with the receptor [43,44].

This drew pharmacologists' attention to the importance of knowing the relationship between drug chemical composition and physiological action. These findings may aid our understanding of the molecular nature of drug–receptor interactions [43,44].

In many cases, the drug's binding to the receptor seems to have low energy, certainly lower than that involved in conventional covalent bonding [45]. Ionic association, particularly hydrogen bonding, and other weaker forces such as charge–transfer forces, or a combination of many of these forces, can produce what is termed "receptor-drug complexing". The capacity of drugs and related compounds to form charge–transfer complexes with well-defined electron acceptors or electron donors, primarily in non-aqueous circumstances, is used as a primary criterion for determining whether charge–transfer forces are manipulated in any way [46–49].

The λ_{max} of UV–Vis spectra of the synthesized charge–transfer complexes were found to be at 340 and 436 nm for (SRX)(PA), 351 nm for (SRX)(DNB), 353 nm for (SRX)(*p*BBA), 528 nm for (SRX)(DCQ), 540 nm for (SRX)(DBQ), and lastly 745 and 833 nm for (SRX)(TCNQ). According to photometric titration measurements, the produced charge–transfer complexes between SRX and corresponding π-acceptors had a 1:1 molar ratio. The dative structure D+–A of charge–transfer complexes in polar solvents were shown to be destabilized by the dissociation of charge–transfer complexes into D+ and A [50–53].

In pharmacokinetics, examining the physical and chemical properties of pharmacological substances in solution, as well as their mechanism of action, is critical. Spectroscopic and thermodynamic approaches are used to assess the binding strength of pharmaceutical compounds to other substances in living systems [41]. In biological and bioelectrochemical energy transfer processes, electron acceptor complexes (EDA) are a common occurrence [42]. The development of highly colored charge–transfer complexes is often related to molecular interactions between electron donors and acceptors, which absorb light in the visible area [48].

Electron acceptor complexes with ionic bands are the most prevalent. Ionic interactions and structural recognition are two crucial mechanisms in biological systems. For example, drug action, enzyme activation, and ion transport across lipophilic membranes are all intricate [45]. Ionic interactions are the fundamental outputs of selectivity, rate control, and reversibility in many biological systems [46].

The most commonly used procedures for assessing various drugs and sophisticated charge transfer investigations include UV direct spectrophotometry [47], colorimetry [48], and HPLC [49]. EDA compounds, as previously reported, have good nonlinear optical properties and electrical conductivity [54].

The six charge–transfer complexes were expected to have particle sizes of 50 nm for (SRX)(PA), 25 nm for (SRX)(DNB), 5 nm for (SRX)(*p*NBA), 10 nm for (SRX)(DCQ), 20 nm for (SRX)(DBQ), and 5 nm for (SRX)(DBQ) (TCNQ). These findings were based on TEM scans, which showed that the particles of the manufactured charge–transfer were nanoscale in size.

The simultaneous thermal stability on the TG/DTG curves of all charge–transfer complexes at a heating rate of 10 °C/min in a static nitrogen atmosphere are shown in Figure 3. The overall mass loss from the TGA curves was 78.17% for SRX–PA, 58.38% for SRX–DNB, 50.45% for SRX-p-NBA, 69.40% for SRX–DCQ, 77.58% for SRX–DBQ, and 75.69% for the SRX–TCNQ complexes. The complexes had mass losses of one to three maxima peaks. The thermal analysis of the curves of the [(SRX)(π-acceptor)] CT complexes clearly shows that the maximum DTG peaks are located at 415, 230, 357, 383, 343, and 370 °C, respectively.

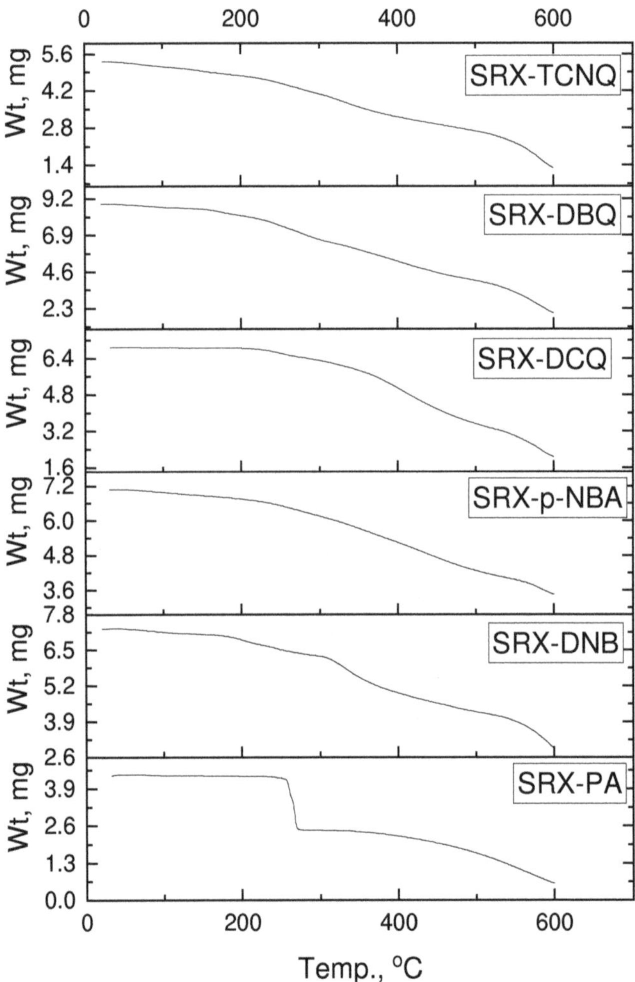

Figure 3. TGA curves of (1:1) charge-transfer complexes [(SRX)(π-acceptor)].

The Coats-Readfern and Horowitez-Metzegar methods [55,56] were used to collect the kinetic thermodynamic data of the maximal DTG peak decomposition steps of all charge–transfer complexes. The kinetic parameters, E, A, ΔS, ΔH, ΔG, and r were calculated, and the data are listed in Table 1 and displayed in Figure 4.

Table 1. Kinetic thermodynamic parameters for the six charge–transfer complexes based on Coats–Redfern (CR) and Horowitz–Metzger (HM) methods.

Complex	Method	Parameter					r
		E (kJol^{-1})	A (s^{-1})	ΔS (J mol^{-1} K^{-1})	ΔH (kJ mol^{-1})	ΔG (kJ mol^{-1})	
(SRX)(PA)	CR	11.5×10^4	4.00×10^8	-8.52×10^1	1.12×10^5	1.54×10^5	0.9990
	HM	11.2×10^4	5.60×10^9	-6.32×10^1	1.12×10^5	1.50×10^5	0.9989
(SRX)(DNB)	CR	7.80×10^4	1.50×10^5	-1.55×10^2	7.25×10^4	1.47×10^5	0.9980
	HM	8.65×10^4	1.34×10^5	-1.30×10^2	8.12×10^4	1.44×10^5	0.9989
(SRX)(*p*NBA)	CR	6.38×10^4	1.32×10^4	-1.72×10^2	5.90×10^4	1.51×10^5	0.9995
	HM	7.23×10^4	1.22×10^4	-1.56×10^2	6.71×10^4	1.54×10^5	0.9985
(SRX)(DCQ)	CR	4.80×10^4	1.25×10^5	-1.45×10^2	4.43×10^4	9.40×10^4	0.9943
	HM	5.22×10^4	1.85×10^6	-1.32×10^2	4.68×10^4	9.22×10^4	0.9987
(SRX)(DBQ)	CR	5.77×10^4	5.12×10^3	-1.85×10^2	5.22×10^4	1.45×10^5	0.9890
	HM	6.35×10^4	2.75×10^4	-1.72×10^2	5.90×10^4	1.40×10^5	0.9994
(SRX)(TCNQ)	CR	11.1×10^4	6.22×10^8	-8.14×10^1	9.72×10^4	1.33×10^5	0.9984
	HM	11.8×10^4	5.50×10^9	-6.35×10^1	1.12×10^5	1.42×10^5	0.9996

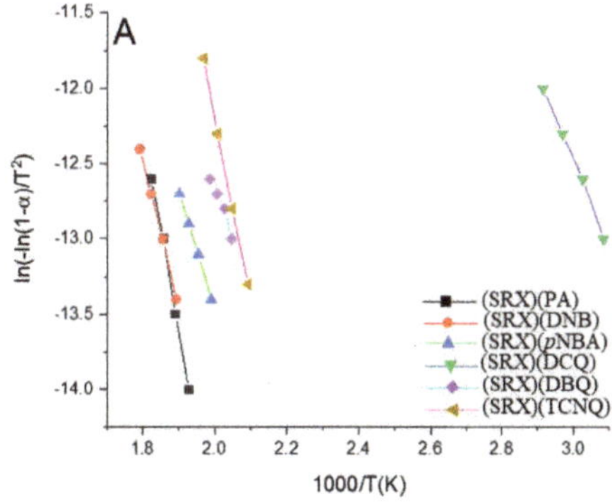

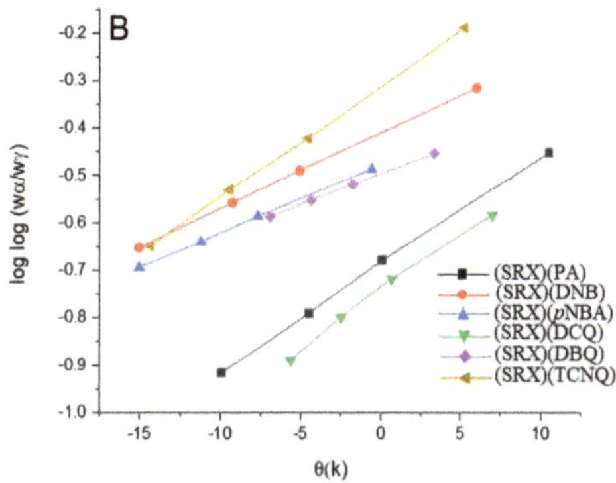

Figure 4. Kinetic curves of (1:1) charge-transfer complexes [(SRX)(π-acceptor)] using (**A**) Coats-Readfern and (**B**) Horowitz-Metzegar methods.

The activation energies of the [(SRX)(π–acceptor)] CT complexes in the case of the maximum DTG peak decomposition step were as follows:

(SRX)(TCNQ) > (SRX)(PA) > (SRX)(DNB) > (SRX)(pNBA) > (SRX)(DBQ) > (SRX)(DCQ).

Among the six π–acceptors, it was found that the SRX–TCNQ and SRX–PA complexes had greater activation energies than the other charge–transfer complexes. This is owing to the presence of cyano and nitro groups in the TCNQ and PA acceptors [57].

3.2. UV–Vis Spectra and Photometric Titration

The UV-Vis spectra of the six charge–transfer complexes in methanol solvent were investigated in the 200–900 nm range (Figure 5) [4]. These charge–transfer complexes are formed by combining 1.00 mL of 0.5 mM from the SRX drug donor with different volumes of the six π-electron acceptors to reach a final concentration of 0.5 mM. With methanol as the solvent, each charge–transfer system had a total volume of 5 mL. Absorption bands for [(SRX)(PA), [(SRX)(DNB)], [(SRX)(p-NBA)], [(SRX)(DCQ)], [(SRX)(DBQ)], and [(SRX)(TCNQ)] donor–acceptor interaction systems appeared at λmax of 436 nm, 351 nm, 353 nm, 528 nm, 540 nm, and 745 nm, respectively. At 25 °C, photometric titrations were performed with the SRX medication as an electron donor and the six π–electron acceptors. The molar ratio of the produced charge–transfer complexes between SRX and the corresponding π–electron was 1:1. The photometric titration curves for the maximal charge–transfer absorption bands (λmax) are shown in Figure 6 [4].

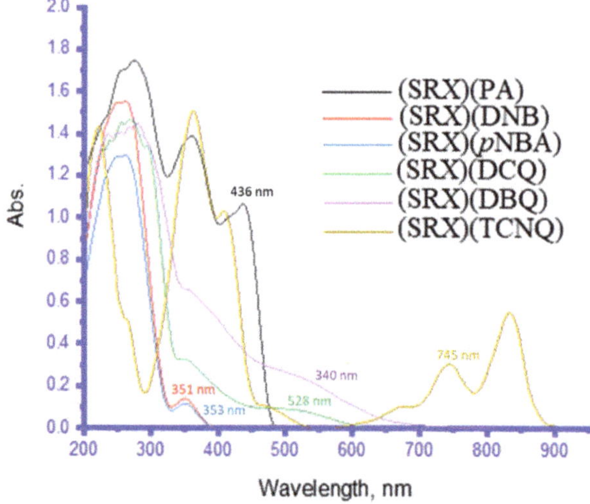

Figure 5. UV–Vis spectra curves of the SRX with the six π–acceptors complex [4].

The photometric titration findings were obtained by graphing the absorbance (Y-axis) against the ratio of indicated acceptors (X-axis) using established procedures [4].

The molar ratio of the produced charge–transfer complexes between SRX medication and identified–acceptors is 1:1 (Figure 6).

3.3. ^{1}H-NMR Spectra

The ^{1}H-NMR spectra of all six π-acceptors complexes are investigated (Figure 7); while the 1H-NMR spectra of SRX only were cited previously [58]. The NH$_2$ protons of the SRX amino group are downfield displaced by 6.87–6.98 ppm as a result of the involvement of one pair of electrons on the amino group towards the six electron π-acceptors. The peaks of other aromatic and methylene protons are similarly pushed downfield to higher ppm values, indicating the formation of six charge–transfer complexes (Supplementary Material Figure S1).

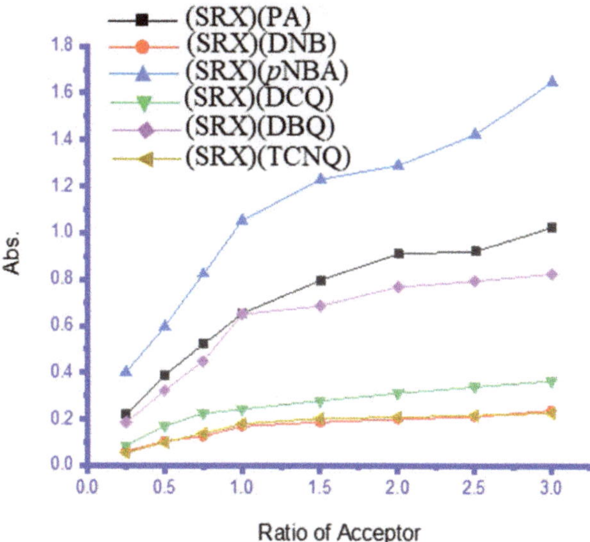

Figure 6. Photometric titration curves of the SRX with the six π–acceptors complex [4].

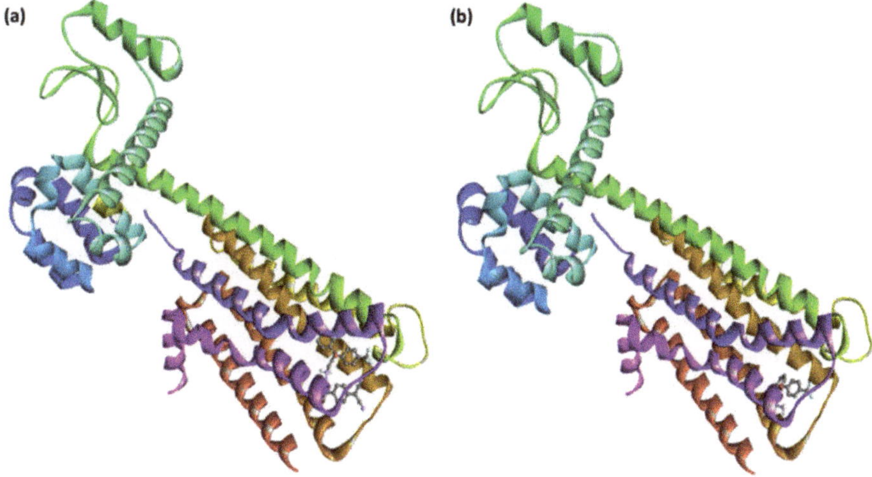

Figure 7. Best docking pose showing a helical model of dopamine docked with (**a**) [(SRX)(TCNQ)] and (**b**) [SRX].

3.4. Molecular Docking Studies

To find the optimal docking pose, the six CT complexes were docked against three protein receptors: serotonin, dopamine, and TrkB kinase. For comparison, the SRX drug (donor moiety) was employed as a control. The potential binding energy of CT complexes was higher than that of SRX in all receptors, according to the molecular docking of these six complexes (Table 2).

Table 2. The docking score of six synthesized CT complexes docked with three receptors [serotonin (PDB ID: 6BQH), dopamine (PDB ID: 6CM4), and TrkB kinase (PDB ID: 4ASZ)].

Receptor	Binding Free Energy (kcal/mol)		
	6BQH	6CM4	4ASZ
SRX-PA	−7.8	−9.2	−8.4
SRX-DNB	−6.8	−8.3	−6.5
SRX-*p*NBA	−8.7	−7.8	−7.0
SRX-DCQ	−7.5	−9.5	−7.4
SRX-DBQ	−7.9	−8.1	−7.5
SRX-TCNQ	−9.4	−9.9	−8.2
SRX	−7.4	−7.3	−6.0

Of the six CT complexes studied, [(SRX)(TCNQ)] exhibited the highest docking energy values. [(SRX)(TCNQ)] had predicted binding energies of −9.3, −9.9, and −8.2 kcal/mol with serotonin, dopamine, and TrkB kinase receptors, respectively. The binding energy of [(SRX)(TCNQ)]-dopamine (CTcD) is higher than that of serotonin and the TrkB kinase receptors, indicating a stronger link. The optimal docking pose of (CTcD) is shown in Figure 5, and the docking data are listed in Table 3.

Table 3. The interactions of SRX-TCNQ and SRX with dopamine (6CM4).

Receptor	Binding Free Energy (kcal/mol)	Interactions	
		H-Bond	Others
SRX-TCNQ	−9.9	Tyr416 and Trp413	Leu94, Trp100 (π-Alkyl); Phe189 (π-Sigma); Asp114 (π-Anion); Ile184 (Halogen-Fluorine)
SRX	−7.3	Ser409 and Thr412	Tpr100, Val91 (π-Alkyl); Tyr416 (π-Sigma)

The [(SRX)(TCNQ)]-dopamine (CTcD) shows that the amino acid residues, including Tyr416 and Trp413, formed hydrogen bond interactions (Figure 8a). There are other interactions between Leu94, Trp100 (π-Alkyl), Phe189 (π-Sigma), Asp114 (π-Anion), and Ile184 (halogen-fluorine) [59]. The theoretical binding energies of SRX with the serotonin, dopamine and TrkB kinase receptors were −7.3, −7.4, and −6.0 kcal/mol, respectively, after molecular docking. The [SRX]-dopamine (SRXD) receptor had a stronger connection than the serotonin and TrkB kinase receptors due to its greater binding energy value.

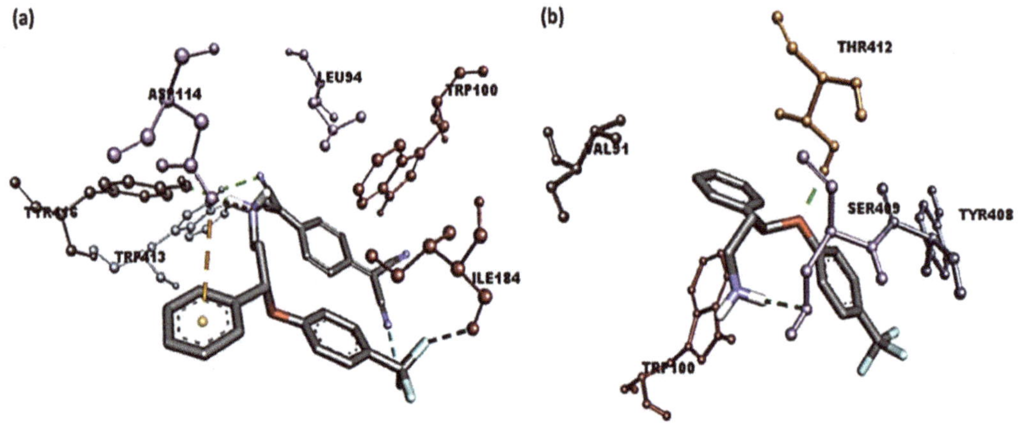

Figure 8. Three-dimesnional representation of interactions for dopamine docked with (**a**) [(SRX)(TCNQ)] and (**b**) [SRX].

The interaction between SRX and dopamine is illustrated in Figure 8b. The amino acid residues, including Ser409 and Thr412, formed hydrogen bond connections between SRX and dopamine. There were also interactions between Tpr100, Val91 (π-alkyl), and Tyr416 (π-sigma). These data indicate that the [(SRX)(TCNQ)] complex binds to the three protein receptors more efficiently than the reactant donor (SRX) alone and that the CTcD has the highest binding energy value. TNCQ is a powerful electron acceptor that forms charge transferring chains due to the existence of its four cyano groups and π-conjugation bonds. This facilities the increase in interactions (such as H-bond, π-Alkyl, π-Sigma, π-Anion, along with SRX) with receptors.

Given the growing evidence that DA transmission assists antidepressant therapeutic goals [60], this augmentation of transmission could have clinical implications. This is because the majority of modern antidepressants do not boost dopamine neurotransmission [60]. One reason for DA's significance is that it regulates motivation, concentration, and pleasure [60]. Figure 9 shows two-dimensional depictions of ligand–receptor interactions. Figure 10 and Figure S2 show the hydrophobic, ionizability, aromatic, and hydrogen bond surfaces at the interaction location of [(SRX)(TCNQ)] and dopamine, respectively.

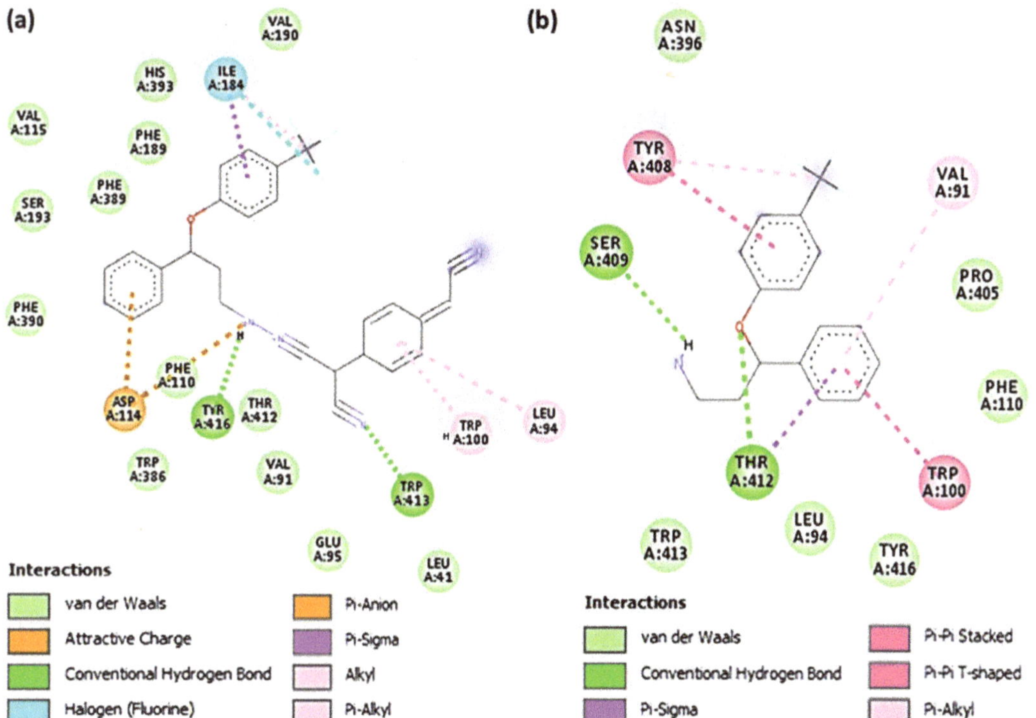

Figure 9. Two-dimensional representation of interactions of dopamine docked with (**a**) CT complex and (**b**) SRX.

3.5. Molecular Dynamics Simulation

For the 100 ns simulation run, the best-docking position for SRXD and CTcD with the highest docking score was used. The RMSD of molecular dynamics data was calculated to investigate structural stability. After 45 ns and 60 ns, respectively, SRXD and CTcD established constant conformation with an appropriate RMSD value of 2.85 and 3.56, respectively (Figure 11).

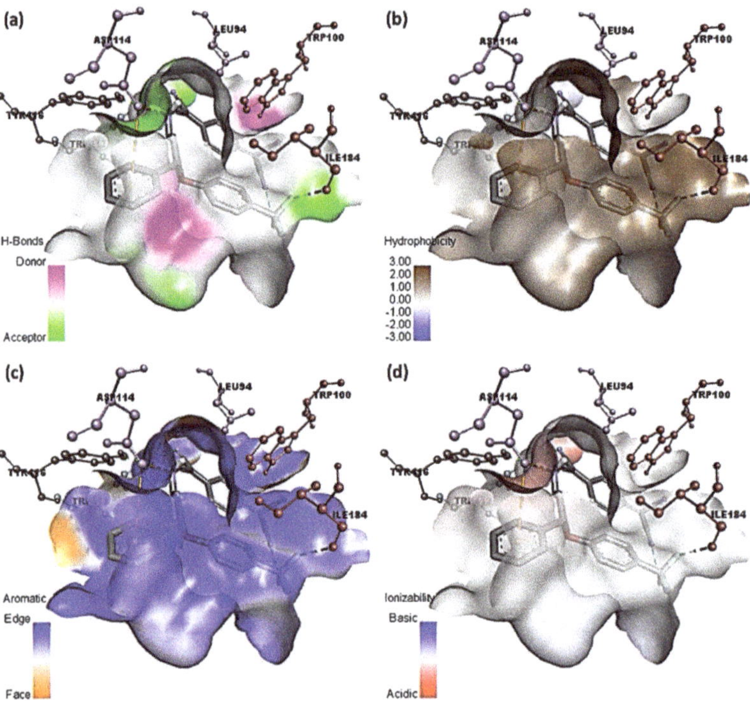

Figure 10. Representation of (**a**) hydrogen binding surface, (**b**) hydrophobic surface, (**c**) aromatic surface, and (**d**) ionizability surface; between dopamine and CT complex.

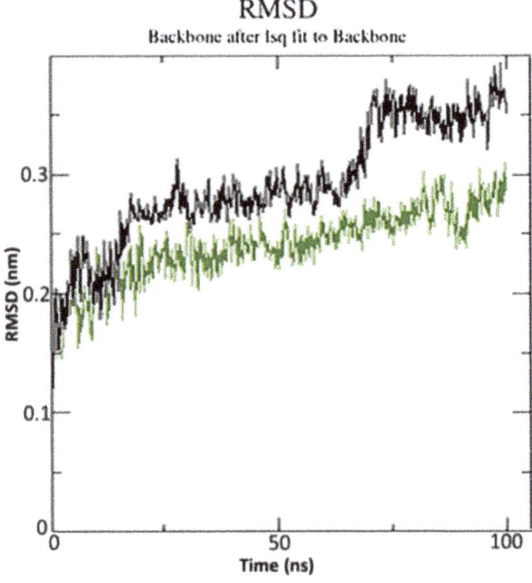

Figure 11. The root mean square deviation (RMSD) of solvated receptor backbone and ligand complex during 100 ns MD simulation [SRXD complex (black) and CTcD complex (green)].

As indicated previously, <3.0 Å is the most acceptable RMSD value range, which indicates better system stability [61]. This finding shows that the CTcD develops a more stable combination. The findings revealed that ligand-receptor interactions bring protein chains closer and reduce the gap between them, as shown in Figure 12 [62].

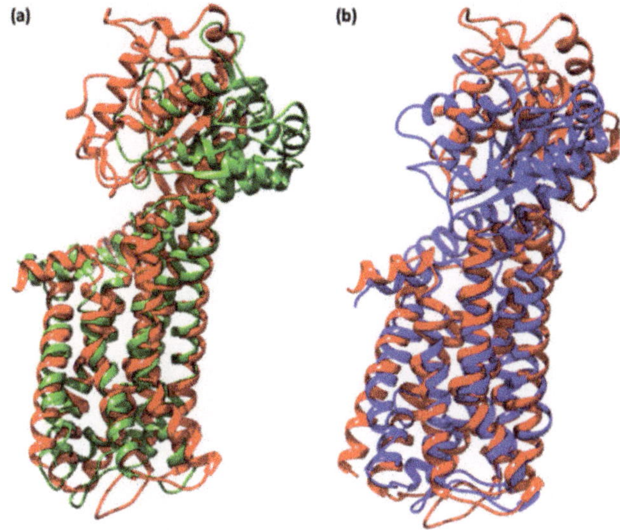

Figure 12. Superimposed structure after simulation of unbound dopamine receptor (red) and (**a**) CTcD (green), and (**b**) SRXD (blue).

The average distance and standard deviation for all amino acid pairs between two conformations were calculated using RR distance maps [63]. In Figure 13, the patterns of spatial interactions are depicted using the RR distance maps [64].

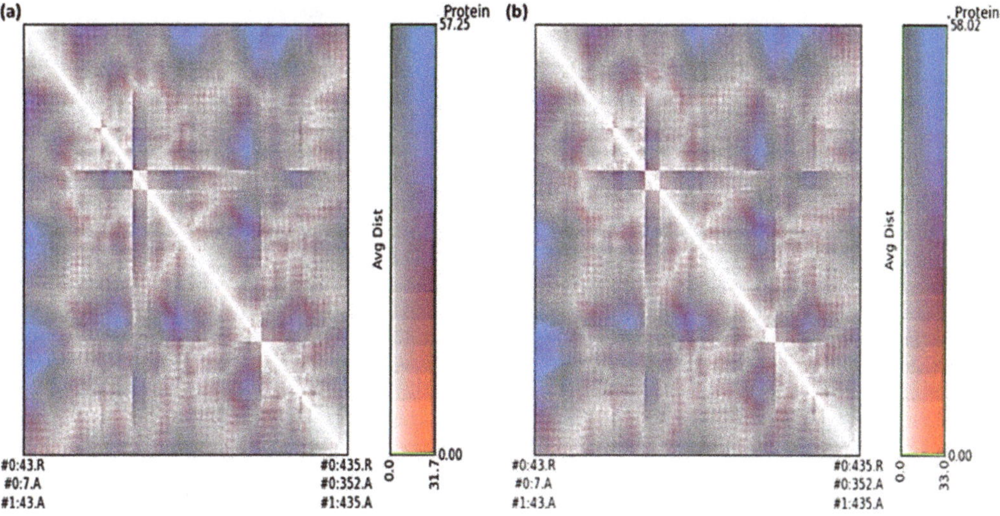

Figure 13. RR distance map between unbound dopamine receptor and after simulation for SRXD (**a**), and CtcD (**b**).

On the map, the white oblique represents the zero distance between two amino acid residues, whereas the red and blue elements depict residue pairs with the biggest distance deviations between the two forms. The average radius of gyration (Rg) value of 28.75 and 28.52 Å was observed for SRXD and CTcD, respectively. Along the simulation time, Rg decreased, indicating that the structures became more compact (Figure 14).

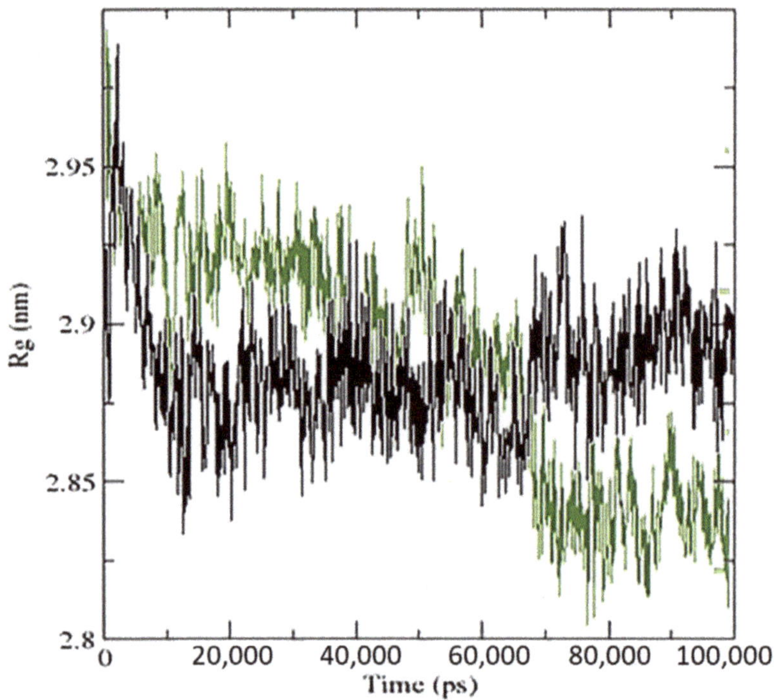

Figure 14. The radius of gyration (Rg) for SRXD complex (black) and CTcD complex (green) during 100 ns simulation time.

The number of hydrogen bond interactions between ligand and receptor combinations (SRXD and CTcD) were displayed against time using a grid search on a $15 \times 20 \times 27$ grid with a rcut = 0.35 value (Figure 15).

The hydrogen bonds between SRX and dopamine were at 33 and 1356 atoms, respectively. While they were between 56 and 5109 atoms for the CT complex and dopamine. However, there were 709 donors for both (SRXD and CTcD), 1356 acceptors for SRXD, and 1426 acceptors for CTcD. For SRXD and CTcD, the average number of hydrogen bonds per time was found to be 0.065 and 0.144 out of 480,702 possible.

Overall, these findings suggest that the receptor–protein interaction increased the number of hydrogen bonds by a significant amount in CTcD. As the ligand attached to the receptor, the values of the solvent-accessible surface area (SASA) changed (Figure 16). When the receptor interacts with the ligand, the SASA is lowered, indicating a change in protein structure and a smaller pocket size with increased hydrophobicity.

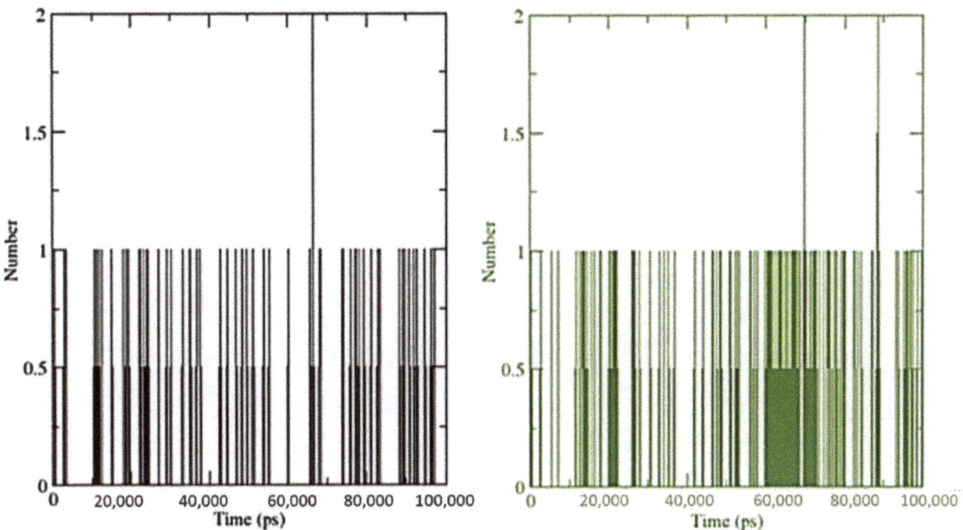

Figure 15. Number of average hydrogen bonding interactions between (**Left**) SRXD complex and (**Right**) CTcD complex during 100 ns simulation time.

Solvent Accessible Surface

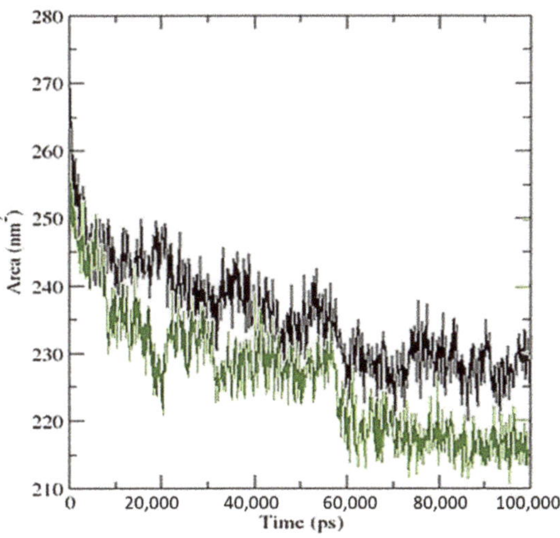

Figure 16. Solvent accessible surface area analysis for the SRXD complex (black) and the CTcD complex (green) during 100 ns simulation time.

3.6. Theoretical Structural Analysis

Density functional theory (DFT) using B-3LYP/6-311G++ (basis set) level of theory and optimized geometry of the CT complexes- [(SRX)(PA}], [(SRX)(DNB), [(SRX)(p-NBA)], [(SRX)(DCQ)], [(SRX)(DBQ)], and [(SRX)(TCNQ)] with atomic coordinates, strain-free lattice constants and ground state minimum energy structure are obtained. The optimized structures of all the CT complexes with the Mulliken numbering scheme are shown in Figure 17. The minimum SCF energy of obtained for [(SRX)(PA}], [(SRX)(DNB), [(SRX)(p-

NBA)], [(SRX)(DCQ)], [(SRX)(DBQ)], and [(SRX)(TCNQ)] is −1958.944644 to, −1689.608194, −1673.728419, −2788.736562, −7011.542112, and −1726.964350 a.u in 87, 90, 38, 176, 34, and 91 steps, respectively (Figure 18). Based on the optimized structure, some molecular parameters (SCF minimum energies, dipole moments, and Electronic spatial extent) were calculated in the gas phase (Table 4). The HOMO–LUMO gap (ΔE) for [(SRX)(PA}], [(SRX)(DNB), [(SRX)(p-NBA)], [(SRX)(DCQ)], [(SRX)(DBQ)], and [(SRX)(TCNQ)] was calculated as 2.78, 3.44, 3.31, 2.29, 2.43, and 1.89 eV, respectively. The overall order of the chemical reactivity of the CT complexes on the bases of ΔE is as follows- [(SRX)(TCNQ)] > [(SRX)(DCQ)] > [(SRX)(DBQ)] > [(SRX)(PA}] > [(SRX)(p-NBA)] > [(SRX)(DNB)].

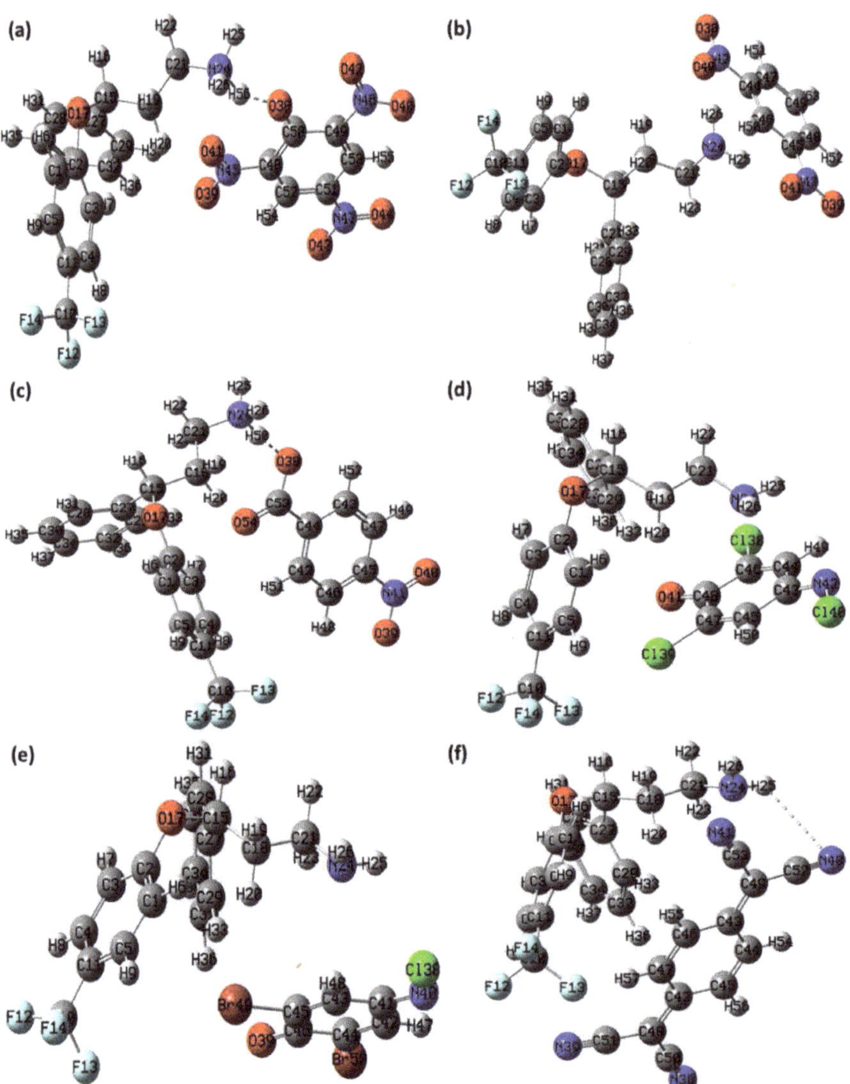

Figure 17. Optimized structure of (**a**) [(SRX)(PA}], (**b**) [(SRX)(DNB), (**c**) [(SRX)(p-NBA)], (**d**) [(SRX)(DCQ)], (**e**) [(SRX)(DBQ)], and (**f**) [(SRX)(TCNQ)] with Mulliken atom numbering scheme.

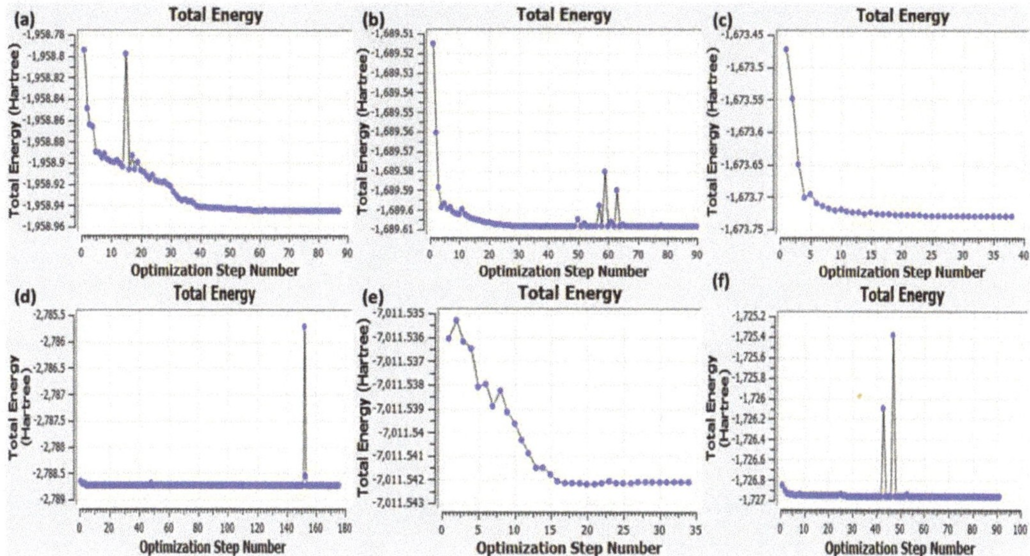

Figure 18. Optimization step graph for (**a**) [(SRX)(PA}], (**b**) [(SRX)(DNB), (**c**) [(SRX)(p-NBA)], (**d**) [(SRX)(DCQ)], (**e**) [(SRX)(DBQ)], and (**f**) [(SRX)(TCNQ)].

Table 4. Theoretical molecular parameters of the CT complexes obtained through DFT.

CT Complex	Minimum SCF Energy (a.u.)	Dipole Moment (Debye)	Electronic Spatial Extent (a.u.)	ΔE (eV)
[(SRX)(PA}]	−1958.944644	10.500053	33,762.8991	2.7845
[(SRX)(DNB)]	−1689.608194	9.644797	20,168.3034	3.4449
[(SRX)(p-NBA)]	−1673.728419	11.524028	26,521.9908	3.3189
[(SRX)(DCQ)]	−2788.736562	5.693616	19,344.1851	2.3924
[(SRX)(DBQ)]	−7011.542112	5.965700	18,542.9710	2.4310
[(SRX)(TCNQ)]	−1726.964350	5.607618	35,156.4199	1.8942

4. Conclusions

The charge transfer complexes between the seproxetine as a donor and picric acid, dinitrobenzene, p-nitrobenzoic acid, 2,6-dichloroquinone-4-chloroimide, 2,6-dibromoquinone-4-chloroimide, and 7,7′,8,8′-tetracyanoquinodi methane as π-electron acceptors were characterized and studied for interaction with three receptors (serotonin, dopamine, and TrkB kinase receptor). The spectrophotometric analysis confirmed that the charge–transfer interactions between the electrons of the donor and acceptors were 1:1 (SRX: π–acceptor). Molecular docking revealed that the CT complex [(SRX)(TCNQ)] interacted with all three receptors more efficiently than the reactant donor (SRX); among all, [(SRX)(TCNQ)]-dopamine (CTcD) had the highest binding energy value. Using AutoDock Vina, the molecular dynamics simulation of the 100 ns run revealed that both the SRX-dopamine and CTcD complexes had a stable conformation; however, the CTcD complex was more stable. DFT calculations provided the optimized geometries of the CT complexes. In the context of mounting evidence for the role of DA transmission, such transmission enhancement might be of potential research and clinical benefit.

Supplementary Materials: The following supporting information can be downloaded at: https://www.mdpi.com/article/10.3390/molecules27103290/s1, Figure S1: ^{1}H-NMR spectrum of all six π–acceptors complexes; Figure S2: Representation of (a) hydrogen binding surface, (b) hydrophobic surface, (c) aromatic surface, and (d) ionizability surface; between dopamine and SRX.

Author Contributions: Conceptualization and visualization, A.A.A., H.H., A.S.A. (Ahmed S. Alzahrani), H.A.A. and R.I.F.; data curation, methodology, S.S., M.S.R. and A.G.; software, S.S. and A.S.A.; validation, W.F.A., M.A. and H.A.; formal analysis, M.A., W.F.A., H.A., S.S., M.S.R. and A.G.; investigation, B.M.R. and A.S.A. (Abdulhakeem S. Alamri); writing—original draft preparation, M.S.R., S.S., W.F.A. and A.G.; writing—review and editing, W.F.A., H.A. and A.G.; project administration, B.M.R.; funding acquisition, W.F.A. All authors have read and agreed to the published version of the manuscript.

Funding: The authors extend their appreciation to the Deputyship for Research and Innovation, Ministry of Education in Saudi Arabia for funding this work through project number 1-441-121.

Data Availability Statement: All data supporting the reported results are available in the manuscript.

Acknowledgments: The authors extend their appreciation to the Deputyship for Research and Innovation, Ministry of Education in Saudi Arabia for funding this work through project number 1-441-121. The authors are also grateful to Christian M. Nefzgar, Institute for Molecular Bioscience, The University of Queensland, Brisbane, QLD, Australia, for his technical support.

Conflicts of Interest: The authors declare no conflict of interest.

Sample Availability: Samples of the compounds are available from the corresponding author.

References

1. Friedrich, M.J. Depression Is the Leading Cause of Disability Around the World. *JAMA* **2017**, *317*, 1517. [CrossRef] [PubMed]
2. Reddy, M.S. Depression: The disorder and the burden. *Indian J. Psychol. Med.* **2010**, *32*, 1–2. [CrossRef] [PubMed]
3. Yohn, C.N.; Gergues, M.M.; Samuels, B.A. The role of 5-HT receptors in depression. *Mol. Brain* **2017**, *10*, 28. [CrossRef] [PubMed]
4. Al-Humaidi, J.Y.; Refat, M.S. Solution, and solid investigations on the charge–transfer complexation between seproxetine as a selective serotonin reuptake inhibitor drug with six kinds of π–electron acceptors. *J. Mol. Liq.* **2021**, *332*, 115831. [CrossRef]
5. Hyttel, J. Pharmacological characterization of selective serotonin reuptake inhibitors (SSRIs). *Int. Clin. Psychopharmacol.* **1994**, *9* (Suppl. S1), 19–26. [CrossRef]
6. Van Oekelen, D.; Luyten, W.H.; Leysen, J.E. 5-HT2A and 5-HT2C receptors and their atypical regulation properties. *Life Sci.* **2003**, *72*, 2429–2449. [CrossRef]
7. Horton, J.R.; Liu, X.; Wu, L.; Zhang, K.; Shanks, J.; Zhang, X.; Rai, G.; Mott, B.T.; Jansen, D.J.; Kales, S.C.; et al. Insights into the Action of Inhibitor Enantiomers against Histone Lysine Demethylase 5A. *J. Med. Chem.* **2018**, *61*, 3193–3208. [CrossRef]
8. Rajamani, S.; Eckhardt, L.L.; Valdivia, C.R.; Klemens, C.A.; Gillman, B.M.; Anderson, C.L.; Holzem, K.M.; Delisle, B.P.; Anson, B.D.; Makielski, J.C.; et al. Drug-induced long QT syndrome: HERG K+ channel block and disruption of protein trafficking by fluoxetine and norfluoxetine. *Br. J. Pharmacol.* **2006**, *149*, 481–489. [CrossRef]
9. Adam, A.A.; Hegab, M.S.; Refat, M.S.; Eldaroti, H.H. Proton-transfer and charge-transfer interactions between the antibiotic trimethoprim and several σ- and π-acceptors: A spectroscopic study. *J. Mol. Struct.* **2021**, *1231*, 129687. [CrossRef]
10. Adam, A.A.; Saad, H.A.; Alsuhaibabi, A.M.; Refat, M.S.; Hegab, M.S. Charge-transfer chemistry of azithromycin, the antibiotic used worldwide to treat the coronavirus disease (COVID-19). Part I: Complexation with iodine in different solvents. *J. Mol. Struct.* **2021**, *325*, 115187. [CrossRef]
11. Al-Humaidi, J.Y.; El-Sayed, M.Y.; Refat, M.S.; Altalhi, T.A.; Eldaroti, H.H. Spectrophotometric studies on the charge transfer interactions between thiazolidine as a donor and three π-acceptors: P-chloranil (CHL), DDQ and TCNQ. *J. Mol. Struct.* **2021**, *333*, 115928. [CrossRef]
12. Khan, I.M.; Islam, M.; Shakya, S.; Alam, K.; Alam, N.; Shahid, M. Synthesis, characterization, antimicrobial and DNA binding properties of an organic charge transfer complex obtained from pyrazole and chloranilic acid. *Bioorg. Chem.* **2020**, *99*, 103779. [CrossRef] [PubMed]
13. Khan, I.M.; Shakya, S.; Akhtar, R.; Alam, K.; Islam, M.; Alam, N. Exploring interaction dynamics of designed organic cocrystal charge transfer complex of 2-hydroxypyridine and oxalic acid with human serum albumin: Single crystal, spectrophotometric, theoretical and antimicrobial studies. *Bioorg. Chem.* **2020**, *100*, 103872. [CrossRef] [PubMed]
14. Karmakar, A.; Bandyopadhyay, P.; Banerjee, S.; Mandal, N.C.; Singh, B. Synthesis, spectroscopic, theoretical and antimicrobial studies on molecular charge-transfer complex of 4-(2-thiazolylazo) resorcinol (TAR) with 3, 5-dinitrosalicylic acid, picric acid, and chloranilic acid. *J. Mol. Liq.* **2020**, *299*, 112217. [CrossRef]
15. Kosower, E.M. The Solvent Sensitivity of the Charge-Transfer Band of Tropylium Iodide. *J. Org. Chem.* **1964**, *29*, 956. [CrossRef]

16. Refat, M.S.; Ibrahim, O.B.; Saad, H.A.; Adam, A.M.A. Usefulness of charge–transfer complexation for the assessment of sympathomimetic drugs: Spectroscopic properties of drug ephedrine hydrochloride complexed with some π-acceptors. *J. Mol. Struct.* **2014**, *1064*, 58–69. [CrossRef]

17. McCormick, D.B.; Li, H.-C.; Mackenzie, R.E. Spectral evidence for the interaction of riboflavin with aromatic hydrocarbons. *Spectrochim. Acta* **1967**, *23*, 2353–2358. [CrossRef]

18. Czent-Gyorgyi, A.; Isenberg, I. On the electron-donating properties of indoles. *Proc. Natl. Acad. Sci. USA* **1960**, *46*, 1334. [CrossRef]

19. Isenberg, I.; Czent-Gyorgyi, A. On charge transfer complexes between substances of biochemical interest. *Proc. Natl. Acad. Sci. USA* **1959**, *45*, 1229. [CrossRef]

20. Fleischman, D.E.; Tollin, G. Molecular complexes of flavins and phenols I. Absorption spectra and properties in solution. *Biochim. Biophys. Acta* **1965**, *94*, 248–257. [CrossRef]

21. Massy, V.; Palmer, G. Charge transfer complexes of lipoyl dehydrogenase and free flavins. *J. Biol. Chem.* **1962**, *237*, 2374. [CrossRef]

22. Cilento, G.; Sanioto, D.L. Electron Transfer from Polycyclic Aromatic Hydrocarbons to Menadione. *Ber. Bunesnges Physik. Chem.* **1963**, *67*, 426. [CrossRef]

23. Cilento, G.; Tedeschi, P.J. Pyridine Coenzymes: IV. Charge Transfer Interaction with the Indole Nucleus. *Biol. Chem.* **1961**, *236*, 907–910. [CrossRef]

24. Wilcox, P.E.; Cohen, E.; Wen Tans, J. Amino acid composition of α-chymotrypsinogen, including estimation of asparagine and glutamine. *Biol. Chem.* **1957**, *228*, 999–1019. [CrossRef]

25. O'Boyle, N.M.; Banck, M.; James, C.A.; Morley, C.; Vandermeersch, T.; Hutchison, G.R. Open Babel: An open chemical toolbox. *J. Cheminf.* **2011**, *3*, 33. [CrossRef]

26. Dallakyan, S. *PyRx-Python Prescription v. 0.8*; The Scripps Research Institute: La Jolla, CA, USA, 2008.

27. Chu, C.-H.; Li, K.-M.; Lin, S.-W.; Chang, M.D.-T.; Jiang, T.-Y.; Sun, Y.-J. Crystal structures of starch binding domain from Rhizopus oryzae glucoamylase in complex with isomaltooligosaccharide: Insights into polysaccharide binding mechanism of CBM21 family. *Proteins Struct. Funct. Bioinform.* **2014**, *82*, 1079–1085. [CrossRef]

28. Morris, G.M.; Goodsell, D.S.; Halliday, R.S.; Huey, R.; Hart, W.E.; Belew, R.K.; Olson, A.J. Automated docking using a Lamarckian genetic algorithm and an empirical binding free energy function. *J. Comput. Chem.* **1998**, *19*, 1639–1662. [CrossRef]

29. Trott, O.; Olson, A.J. AutoDock Vina: Improving the speed and accuracy of docking with a new scoring function, efficient optimization, and multithreading. *J. Comput. Chem.* **2010**, *31*, 455–461. [CrossRef]

30. Vanommeslaeghe, K.; Hatcher, E.; Acharya, C.; Kundu, S.; Zhong, S.; Shim, J.; Darian, E.; Guvench, O.; Lopes, P.; Vorobyov, I.; et al. CHARMM general force field: A force field for drug-like molecules compatible with the CHARMM all-atom additive biological force fields. *J. Comput. Chem.* **2010**, *31*, 671–690. [CrossRef]

31. Yu, W.; He, X.; Vanommeslaeghe, K.; MacKerell, A.D., Jr. Extension of the CHARMM General Force Field to sulfonyl containing compounds and its utility in biomolecular simulations. *J. Comput. Chem.* **2012**, *33*, 2451–2468. [CrossRef]

32. Jorgensen, W.L.; Chandrasekhar, J.; Madura, J.D.; Impey, R.W.; Klein, M.L. Comparison of Simple Potential Functions for Simulating Liquid Water. *J. Chem. Phys.* **1983**, *79*, 926–935. [CrossRef]

33. Allen, M.P.; Tildesley, D.J. *Computer Simulations of Liquids*; Clarendon Press: Oxford, UK, 1987.

34. Essmann, U.; Perera, L.; Berkowitz, M.L.; Darden, T.; Lee, H.; Pedersen, L.G. A Smooth Particle Mesh Ewald Method. *J. Chem. Phys.* **1995**, *103*, 8577–8593. [CrossRef]

35. Steinbach, P.J.; Brooks, B.R. New Spherical-Cutoff Methods for Long-Range Forces in Macromolecular Simulation. *J. Comput. Chem.* **1994**, *15*, 667–683. [CrossRef]

36. Humphrey, W.; Dalke, A.; Schulten, K. VMD: Visual molecular dynamics. *J. Mol. Graph.* **1996**, *14*, 28–33. [CrossRef]

37. DeLano, W.L. *PyMOL*; DeLano Scientific: San Carlos, CA, USA, 2002.

38. Frisch, M.J.; Trucks, G.W.; Schlegel, H.B.; Scuseria, G.E.; Robb, M.A. *Gaussian 09, Revision E.01*; Gaussian, Inc.: Wallingford, CT, USA, 2009.

39. Becke, A.D. Density-functional thermochemistry. III. The role of exact exchange. *J. Chem. Phys.* **1993**, *98*, 5648.

40. Zhurko, G.A.; Zhurko, D.A. *Chemcraft—Graphical Program for Visualization of Quantum Chemistry Computations*; Academic Version 1.5; Chemcraft: Ivanovo, Russia, 2004.

41. Ilangovan, R.; Subha, V.; Ravindran, R.E.; Kirubanandan, S.; Renganathan, S. Chapter 2—Nanomaterials: Synthesis, physicochemical characterization, and biopharmaceutical applications. In *Nanoscale Processing*; Elsevier: Amsterdam, The Netherlands, 2021; pp. 33–70.

42. El-Mossalamy, E.H.; Batouti, M.E.; Fetouh, H.A. The role of natural biological macromolecules: Deoxyribonucleic and ribonucleic acids in the formulation of new stable charge transfer complexes of thiophene Schiff bases for various life applications. *Int. J. Biol. Macromol.* **2021**, *193*, 1572–1586. [CrossRef]

43. Abdallah, A.M.; Frag, E.Y.; Tamam, R.H.; Mohamed, G.G. Gliclazide charge transfer complexes with some benzoquinone acceptors: Synthesis, structural characterization, thermal analyses, DFT studies, evaluation of anticancer activity and utility for determination of gliclazide in pure and dosage forms. *J. Mol. Struct.* **2021**, *1234*, 130153. [CrossRef]

44. Niranjani, S.; Nirmala, C.B.; Rajkumar, P.; Serdaroğlu, G.; Jayaprakash, N.; Venkatachalam, K. Synthesis, characterization, biological and DFT studies of charge-transfer complexes of antihyperlipidemic drug atorvastatin calcium with Iodine, Chloranil, and DDQ. *J. Mol. Liq.* **2022**, *346*, 117862. [CrossRef]

45. Yu, Y.; Liang, G. Interaction mechanism of phenolic acids and zein: A spectrofluorometric and molecular dynamics investigation. *J. Mol. Liq.* **2022**, *348*, 118032. [CrossRef]
46. Samir, A.; Salem, H.; Abdelkawy, M. Optimization of two charge transfer reactions for colorimetric determination of two beta 2 agonist drugs, salmeterol xinafoate and salbutamol, in pharmaceutical and biological samples. *Spectrochim. Acta Mol. Biomol. Spectrosc.* **2022**, *269*, 120747. [CrossRef]
47. Al Rabiah, H.; Yousef, T.A.; Al-Gamal, A.; Homoda, A.M.; Mostafa, G.A. Tamoxifen charge transfer complexes with 2, 3-dichloro-5, 6-dicyano-1, 4-benzoquinone and 7, 7, 8, 8-tetracyanoquinodimethan: Synthesis, spectroscopic characterization and theoretical study. *Bioorg. Chem.* **2022**, *120*, 105603.
48. Khalil, T.E.; Elbadawy, H.A.; Attia, A.A.; El-Sayed, D.S. Synthesis, spectroscopic, and computational studies on molecular charge-transfer complex of 2-((2-hydroxybenzylidene) amino)-2-(hydroxymethyl) propane-1, 3-diol with chloranilic acid: Potential antiviral activity simulation of CT-complex against SARS-CoV-2. *J. Mol. Struct.* **2022**, *1251*, 132010. [CrossRef] [PubMed]
49. Niranjani, S.; Venkatachalam, K. Synthesis, spectroscopic, thermal, structural investigations and biological activity studies of charge-transfer complexes of atorvastatin calcium with dihydroxy-p-benzoquinone, quinalizarin and picric acid. *J. Mol. Struct.* **2020**, *1219*, 128564. [CrossRef]
50. Adam, A.A.; Refat, M.S.; Altalhi, T.A.; Aldawsari, F.S.; Al-Hazmi, G.H. Liquid- and solid-state study of charge-transfer (CT) interaction between drug triamterene as a donor and tetracyanoethylene (TCNE) as an acceptor. *J. Mol. Liq.* **2021**, *336*, 116261. [CrossRef]
51. Adam, A.A.; Refat, M.S. Analysis of charge-transfer complexes caused by the interaction of the antihypertensive drug valsartan with several acceptors in CH_2Cl_2 and $CHCl_3$ solvents and correlations between their spectroscopic parameters. *J. Mol. Liq.* **2022**, *348*, 118466. [CrossRef]
52. Usmana, R.; Khan, A.; Tang, H.; Ma, D.; Alsuhaibani, A.M.; Refat, M.S.; Adnan; Ara, N.; Fan, H.-J.S. Charge Transfer and Hydrogen Bonding Motifs in Organic Cocrystals Derived from Aromatic Diamines and TCNB. *J. Mol. Struct.* **2022**, *1254*, 132360. [CrossRef]
53. Durgadevi, R.; Suvitha, A.; Arumanayagam, T. Growth, optical, electrical properties and DFT studies on piperidinium 4-nitrophenolate NLO single crystal in acetone. *J. Cryst. Growth* **2022**, *582*, 126512. [CrossRef]
54. Coats, A.W.; Redfern, J.P. Kinetic Parameters from Thermogravimetric Data. *Nat. Lett.* **1964**, *201*, 68–69. [CrossRef]
55. Horowitz, H.H.; Metzger, G.A. A New Analysis of Thermogravimetric Traces. *Anal. Chem.* **1963**, *35*, 1464–1468. [CrossRef]
56. Akram, M.; Lal, H.; Shakya, S.; Kabir-ud-Din. Multispectroscopic and Computational Analysis Insight into the Interaction of Cationic Diester-Bonded Gemini Surfactants with Serine Protease α-Chymotrypsin. *ACS Omega* **2020**, *5*, 3624–3637. [CrossRef]
57. Tseng, T.C.; Urban, C.; Wang, Y.; Otero, R.; Tait, S.L.; Alcamí, M.; Ecija, D.; Trelka, M.; Gallego, J.M.; Lin, N.; et al. Charge-transfer-induced structural rearrangements at both sides of organic/metal interfaces. *Nat. Chem.* **2010**, *2*, 374–379. [CrossRef] [PubMed]
58. Garrido, E.M.; Garrido, J.; Calheiros, R.; Marques, M.P.; Borges, F. Fluoxetine and norfluoxetine revisited: New insights into the electrochemical and spectroscopic properties. *J. Phys. Chem. A* **2009**, *113*, 9934–9944. [CrossRef] [PubMed]
59. Khan, I.M.; Islam, M.; Shakya, S.; Alam, N.; Imtiaz, S.; Islam, M.R. Synthesis, spectroscopic characterization, antimicrobial activity, molecular docking and DFT studies of proton transfer (H-bonded) complex of 8-aminoquinoline (donor) with chloranilic acid (acceptor). *J. Biomol. Struct. Dyn.* **2021**, 1–15. [CrossRef]
60. Dunlop, B.W.; Nemeroff, C.B. The role of dopamine in the pathophysiology of depression. *Arch. Gen. Psychiatry* **2007**, *64*, 327–337. [CrossRef] [PubMed]
61. Wu, Q.; Peng, Z.; Anishchenko, I.; Cong, Q.; Baker, D.; Yang, J. Protein contact prediction using metagenome sequence data and residual neural networks. *Bioinformatics* **2020**, *36*, 41–48. [CrossRef]
62. Chen, J.E.; Huang, C.C.; Ferrin, T.E. RRDistMaps: A UCSF Chimera tool for viewing and comparing protein distance maps. *Bioinformatics* **2015**, *31*, 1484–1486. [CrossRef] [PubMed]
63. Marks, D.S.; Colwell, L.J.; Sheridan, R.; Hopf, T.A.; Pagnani, A.; Zecchina, R.; Sander, C. Protein 3D structure computed from evolutionary sequence variation. *PLoS ONE* **2011**, *6*, e28766. [CrossRef]
64. Kavitha, R.; Nirmala, S.; Nithyabalaji, R.; Sribalan, R. Biological evaluation, molecular docking and DFT studies of charge transfer complexes of quinaldic acid with heterocyclic carboxylic acid. *J. Mol. Struct.* **2020**, *1204*, 127508. [CrossRef]

Article

A Comprehensive Investigation of Interactions between Antipsychotic Drug Quetiapine and Human Serum Albumin Using Multi-Spectroscopic, Biochemical, and Molecular Modeling Approaches

Seema Zargar [1], Tanveer A. Wani [2,*], Nawaf A. Alsaif [2] and Arwa Ishaq A. Khayyat [1]

1   Department of Biochemistry, College of Science, King Saud University, P.O. Box 22452, Riyadh 11451, Saudi Arabia; szargar@ksu.edu.sa (S.Z.); aalkhyyat@ksu.edu.sa (A.I.A.K.)
2   Department of Pharmaceutical Chemistry, College of Pharmacy, King Saud University, P.O. Box 2457, Riyadh 11451, Saudi Arabia; nalsaif@ksu.edu.sa
*   Correspondence: twani@ksu.edu.sa

Abstract: Quetiapine (QTP) is a short-acting atypical antipsychotic drug that treats schizophrenia or manic episodes of bipolar disorder. Human serum albumin (HSA) is an essential transport protein that transports hormones and various other ligands to their intended site of action. The interactions of QTP with HSA and their binding mechanism in the HSA-QTP system was studied using spectroscopic and molecular docking techniques. The UV-Vis absorption study shows hyperchromicity in the spectra of HSA on the addition of QTP, suggesting the complex formation and interactions between QTP and HSA. The results of intrinsic fluorescence indicate that QTP quenched the fluorescence of HSA and confirmed the complex formation between HSA and QTP, and this quenching mechanism was a static one. Thermodynamic analysis of the HSA-QTP system confirms the involvement of hydrophobic forces, and this complex formation is spontaneous. The competitive displacement and molecular docking experiments demonstrated that QTP is preferentially bound to HSA subdomain IB. Furthermore, the CD experiment results showed conformational changes in the HSA-QTP system. Besides this, the addition of QTP does not affect the esterase-like activity of HSA. This study will help further understand the credible mechanism of transport and delivery of QTP via HSA and design new QTP-based derivatives with greater efficacy.

Keywords: quetiapine; human serum albumin; hydrophobic interaction; thermodynamic parameters

Citation: Zargar, S.; Wani, T.A.; Alsaif, N.A.; Khayyat, A.I.A. A Comprehensive Investigation of Interactions between Antipsychotic Drug Quetiapine and Human Serum Albumin Using Multi-Spectroscopic, Biochemical, and Molecular Modeling Approaches. *Molecules* **2022**, *27*, 2589. https://doi.org/10.3390/molecules27082589

Academic Editor: Rudy J. Richardson

Received: 14 March 2022
Accepted: 13 April 2022
Published: 18 April 2022

Publisher's Note: MDPI stays neutral with regard to jurisdictional claims in published maps and institutional affiliations.

1. Introduction

In recent years, psychoactive drug usage has increased worldwide due to the increasing incidence of related psychiatric disorders [1]. However, the most commonly prescribed psychoactive drugs, such as antidepressants, antipsychotics, and mood stabilizers, cause unwanted side effects (excessive systemic drug exposure) and toxicity to human systems [2,3].

Quetiapine (QTP Figure 1A) is a second generation (short-acting atypical) antipsychotic drug of dibenzothiazepine (class), which is used to treat schizophrenia, acute bipolar disorder, and major depression in adolescents and adults [4–7]. The exact mechanism of action of QTP is poorly understood. However, QTP is an antagonist of various neurotransmitter receptors in the brain, such as dopamine D_1 and D_2, adrenergic alpha receptors $alpha_1$ and $alpha_2$, histamine H_1, and serotonin 5-HT_{1A} and 5-HT_2, respectively [8,9]. Specifically, the antipsychotic and antidepressant effects of QTP are believed to be due to the interactions of the above-mentioned neurotransmitter receptors dopamine (D_1 and D_2), adrenergic alpha receptors (α_1 and α_2), histamine (H_1), and serotonin (5-HT_{1A} and 5-HT_2) [9].

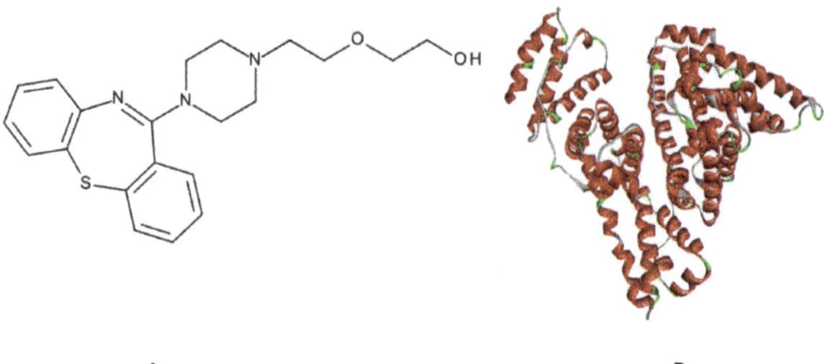

A   B

Figure 1. (**A**) chemical structure of QTP; (**B**) molecular structure of HSA.

Recent studies explore the insight of binding affinity and mechanism of plasma proteins and drug interactions [10–15]. Recently, nanotechnology has helped explore the interaction mechanisms [16,17]. However, the interaction between the drug proteins (plasma) and their mechanism is vital because they directly affect therapeutic drugs' pharmacodynamic and pharmacokinetic properties in the human system [10]. Moreover, the drug proteins (plasma) interactions help to decipher the therapeutic efficacy, distribution, and bioavailability of therapeutic drugs and assist in enhancing solubility in plasma protein, reducing toxicity, and protecting against oxidation [18–20].

Human serum albumin (HSA) is a principal plasma protein with critical physiological functions and facilitates the transportation of many molecules and metabolites (Figure 1B) [21]. It is a monomeric chain globular plasma protein (585 amino acids residues), and its 3D structure consists of three homologous domains (I-III-A and B subdomains). The essential binding regions for drugs in the HSA are Sudlow's site I (subdomains IIA) and Sudlow's site II (subdomains IIIA) [22–25]. However, there is also Site III (subdomain IB), which is also believed to play an essential role in binding various drugs [26]. Therefore, HSA has multiple binding sites and can bind several different drugs, thus making it a fundamental functional drug carrier [27]. Furthermore, the binding of therapeutic drugs within HSA is commonly reversible via weak interactions such as hydrogen bonding, hydrophobic forces, ionic interactions, and van der Waal's interactions [28].

To the best of our knowledge, the interaction binding mechanism of QTP and HSA has still not been investigated. Here, multi-spectroscopic techniques and biochemical and molecular docking approaches were applied to scrutinize the binding properties of QTP with HSA under physiological conditions. However, we considered the possibility of complexation between QTP-HSA, which would explore the pharmacodynamics and pharmacokinetics of QTP. The QTP-HSA interactions reported here would explain the binding mechanism at the molecular level and facilitate efforts to modify new therapeutic drugs that optimize their distribution within the human body.

2. Results and Discussion

2.1. UV-Vis Absorption Spectroscopy

UV-Vis spectral analyses are carried out to observe the structural and conformational changes in the protein molecule induced by the binding ligands and thus to obtain information about their interaction mechanism. [29]. The UV-Vis absorption spectra of the HSA and HSA-QTP complex are shown in Figure 2. It is apparent from the spectra that HSA exhibits an absorption peak at 280 nm coming from the π-π^* transition of the aromatic amino acids (tryptophan (W), tyrosine (Y), phenylalanine) [30]. An increase in QTP concentration was accompanied by a slight shift in the absorption wavelength. This blue shift indicates that QTP binding is associated with changes in the local environment of HSA. In addition, there is an increase in UV-absorption intensities of HSA at around 280 nm at increasing concentrations of QTP, and this hyperchromicity suggests the HSA-QTP system formation.

Hyperchromicity at around 280 nm in HSA after QTP addition also confirms that the aromatic amino acid (W and Y) microenvironment changes due to the HSA-QTP complex formation [31,32].

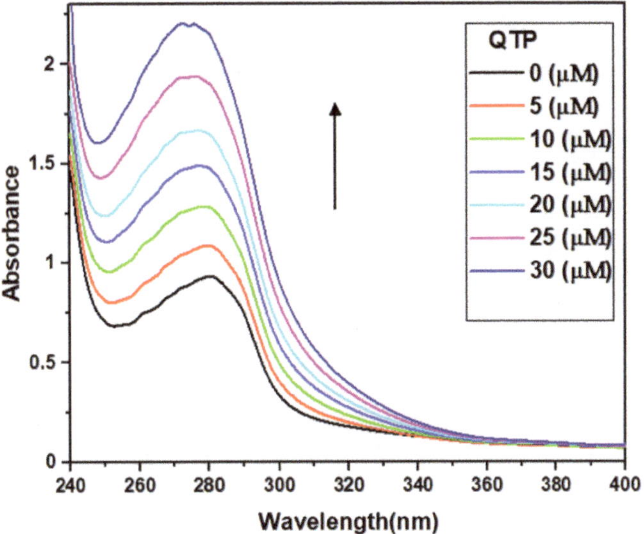

Figure 2. UV absorption spectra of HSA (5 µM) in the absence and presence of increasing concentrations of QTP (5–30 µM) in the wavelength range 240–410 nm.

2.2. Fluorescence Emission Spectroscopy of the HSA-QTP Complex

Fluorescence emission spectroscopy is a multipurpose biophysical technique used to study the binding mechanism of protein-ligand interactions and to evaluate the binding parameters [10,12,26]. The fluorescence emission spectra of HSA alone and the HSA-QTP complex are given in Figure 3A. It is apparent from Figure 3A that HSA exhibits a strong emission peak at 340 nm upon excitation at 295 nm due to W-214 residue. Further, the addition of different concentrations of QTP (0–35 µM) leads to the quenching of HSA fluorescence intensity without changing the peak shape. This fluorescence quenching suggests the formation of the HSA-QTP system and suggests a possible microenvironmental alteration in HSA upon treatment with QTP [33,34].

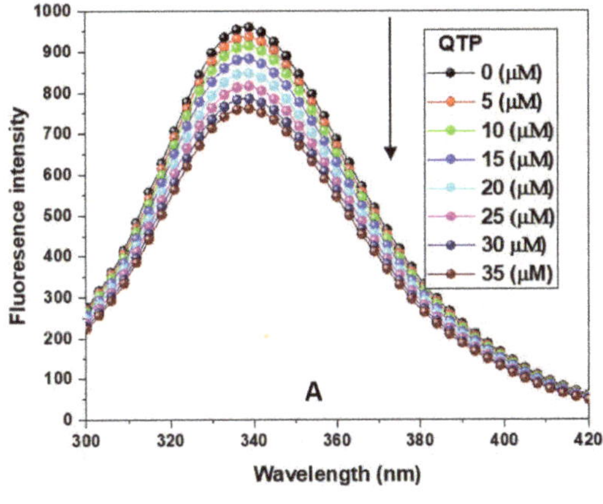

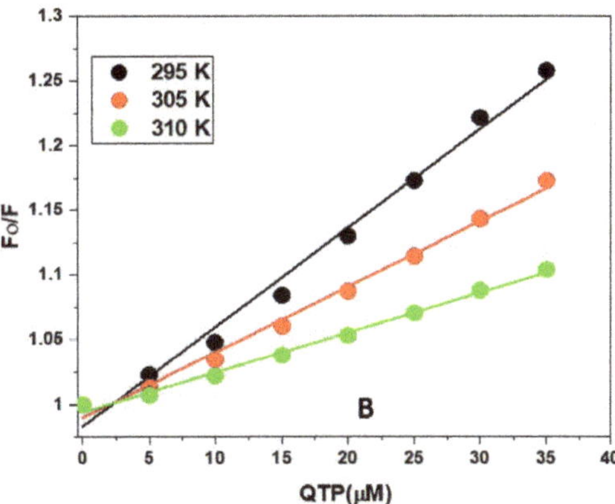

Figure 3. *Cont.*

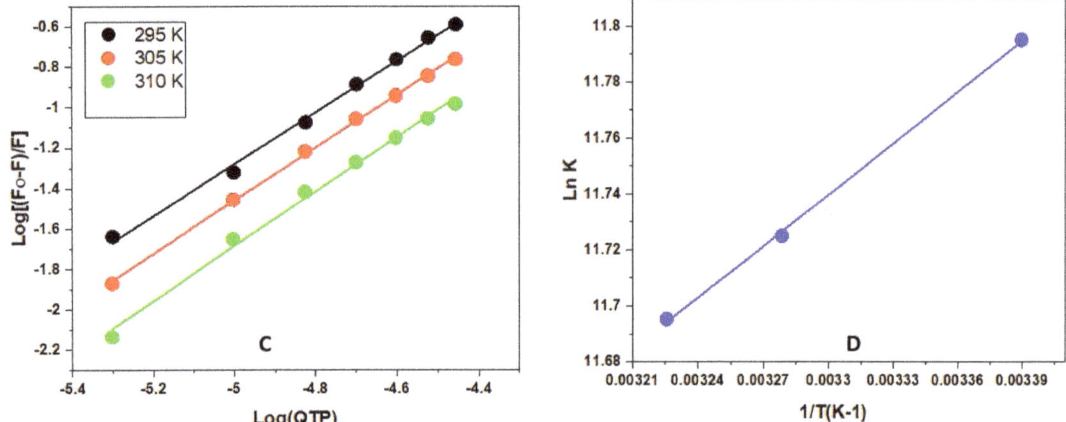

Figure 3. (**A**) Steady-state fluorescence emission spectra of HSA were recorded in the absence and presence of increasing concentrations of QTP. The intrinsic fluorescence of the HSA was measured at 295 K in the wavelength range of 300–420 nm after exciting at 295 nm. The black arrow represents fluorescence quenching of HSA on titration with QTP (**B**) Stern-Volmer plot for QTP-HSA interaction (295, 300, 310 K). (**C**) Double log plot for the QTP-HSA interaction at different temperatures (295, 300, 310 K). (**D**) van 't Hoff plot (lnK vs. 1/T) for the binding of QTP to HSA. The concentration of HSA was 5 µM and was titrated with QTP (0–35 µM) in all the experiments (**A–D**).

2.2.1. Fluorescence Quenching Mechanism (FQM) of the Interactions of the HSA-QTP System

According to the literature, the protein's fluorescence quenching mechanism (FQM) consists mainly of dynamic quenching and static quenching. In the case of dynamic quenching, the interaction of the fluorophore with the quencher is indirect. In contrast, in the case of static quenching, a ground state complex formation exists between the fluorophore and quencher [30]. Therefore, the FQM can be sorted out based on their temperature dependence. Furthermore, in the case of static quenching, K_{sv} values are inversely proportional to temperature, whereas in dynamic K_{sv}, the values are directly proportional to temperature. Therefore, the FQM of the HSA-QTP system was evaluated by recording the fluorescence spectra of HSA-QTP at different temperatures (295, 300, and 305 K), and the fluorescence quenching data of the HSA-QTP system was analyzed using the Stern-Volmer equation [30]:

$$\frac{F_0}{F} = 1 + K_{sv}[Q] \tag{1}$$

where F_0 and F represent the steady-state fluorescence of HSA and the HSA-QTP complex, respectively. [Q] represents the quencher concentration (QTP), and K_{sv} represents the Stern-Volmer constant. The K_{sv} plot for the HSA-QTP system obtained at various temperatures (295, 300, and 305 K) is given in Figure 3B. It is found that the K_{sv} values for the HSA-QTP system decreased with a temperature rise, confirming the static quenching mechanism for the HSA-QTP system (Table 1). In addition, the fluorescence mechanism (quenching) was also analyzed according to the bimolecular rate constant values using the equation:

$$k_q = K_{sv}/\tau_0 \tag{2}$$

where k_q is the bimolecular rate constant, and τ_0 is the average lifetime of the protein in the absence of the quencher and is valued at 10^{-8} for biopolymers [35]. The calculated bimolecular quenching rate constant value for the HSA-QTP system is presented in Table 1. The k_q values were found to be higher than the value of the scattering collision constant (2×10^{10} M^{-1} s^{-1}), which again suggests the involvement of a static quenching mechanism between the HSA-QTP system [36].

Table 1. The values of the Stern-Volmer constant and quenching rate constant for the QTP-HSA system.

pH	Temp (K)	K_{sv} ($\times 10^4$ M^{-1})	K_q ($\times 10^{12}$ M^{-1} s^{-1})	R^2
	295	0.7	0.7	0.987
7.4	305	0.5	0.5	0.992
	310	0.3	0.3	0.993

2.2.2. Evaluation of the Binding Constants and the Number of Binding Sites in the HSA-QTP System

Intrinsic fluorescence data at different temperatures (295, 300, and 305 K) were used to determine the binding constant (K_b) and binding stoichiometry (n) of the HSA-QTP system by using the following equation [10,30]:

$$\log \frac{(F_0 - F)}{F} = \log K_b + n \log[Q] \tag{3}$$

where F_0 and F represent fluorescence intensities of HSA with or without the quencher (QTP), respectively. K_b and n represent the binding constant and binding stoichiometry in the HSA-QTP system. The double log plot of log $[(F_0 - F)/F]$ vs. log $[Q]$ (Figure 3C) was used for the determination of the binding constant and binding stoichiometry. The values of K_b and n were calculated from the intercept and slope of the plot, as shown in Figure 3C. As per Figure 3C, K_b and n at different temperatures for the HSA-QTP system are presented in Table 2. A decrease in the binding constant was observed at higher temperatures for the HSA-QTP system. Further, the binding constants were ~10^4, suggesting a moderate binding between HSA and QTP.

Table 2. The binding constant values and the number of binding sites for the interaction of QTP with HSA.

pH	Temp (K)	K_b ($\times 10^4$ M^{-1})	N	R^2
	295	1.326	1.28	0.996
7.4	305	1.236	1.31	0.994
	310	1.200	1.35	0.994

2.2.3. Determination of the Binding Forces between HSA and QTP-Thermodynamic Analysis

The primary binding intermolecular forces that are involved in the drug-protein interactions were estimated via thermodynamic parameters. The protein-drug interactions are held together by hydrophobic interactions, hydrogen bonds, electrostatic forces, and van der Waal interactions. Moreover, the sign and magnitude of the enthalpy (ΔH^0) and entropy (ΔS^0) change to determine the nature of binding forces in the drug-protein complex. For the hydrophobic interactions, the sign and magnitude must have a positive value for ΔH^0 and ΔS^0. At the same time, in the case of van der Waals forces and hydrogen bonding, it must be negative for ΔH^0 and ΔS^0 [36,37]. Additionally, for the electrostatic interaction, ΔH^0 should be negative and ΔS^0 positive. The free energy (ΔG^0) change of the HSA-QTP system can be determined by using the van 't Hoff equation and the thermodynamic equation given below:

$$\ln K_b = -\frac{\Delta H^0}{RT} + \frac{\Delta S^0}{R} \tag{4}$$

$$\Delta G^0 = \Delta H^0 - T\Delta S^0 \tag{5}$$

where R represents the gas constant (8.314 J mol^{-1} K^{-1}), T is the temperature in kelvins, and K_b represents the binding constant at the studied different temperatures. ΔH^0 and ΔS^0 are obtained from the slope and intercept of the plot between lnK and 1/T (Figure 3D). The results of ΔG^0, ΔH^0, and ΔS^0 obtained from HSA-QTP interactions are summarized in Table 3. The positive values of ΔH^0 and ΔS^0 for the HSA-QTP system suggest that hydrophobic interactions played a significant role in the binding process of QTP to HSA. Thus, the formation of the HSA-QTP complex was exothermic and spontaneous [38].

Table 3. Various thermodynamic parameters for QTP-HSA complex formation at various temperatures.

Temp (K)	ΔH^0 (KJ mol^{-1})	ΔS^0 (JK^{-1} mol^{-1})	$T\Delta S^0$ (KJ mol^{-1})	ΔG^0 (KJ mol^{-1})
295			23.89	−18.8
305	5.087	81	24.705	−19.61
310			25.11	−20.11

2.2.4. Synchronous Fluorescence Spectroscopy (SFS) Experiment

The synchronous fluorescence spectrometry helps to provide information about the local environment of proteins around W and Y residues upon interaction with ligands [30,39]. In this experiment, the fluorescence difference between excitation and emission wavelengths reflects the nature of the spectra. A difference of wavelength ($\Delta\lambda$) of 15 nm is characteristic for (Y), and 60 nm is typical of (W) residues. Therefore, any shift in the maximum emission wavelength reflects the local environment changes around aromatic amino acid residue (Y and W) [40]. The SFS emission spectra of the HSA-QTP complex are given in Figure 4A,B. It was clear from Figure 4 that the HSA fluorescence intensity of both (W and Y) regularly decreases with the addition of QTP. Further, no shift in the emission wavelength was observed for either of the spectra at $\Delta\lambda$ = 15 nm or 60 nm. The HSA-QTP interaction did not lead to any microenvironmental change in the protein molecule upon interaction.

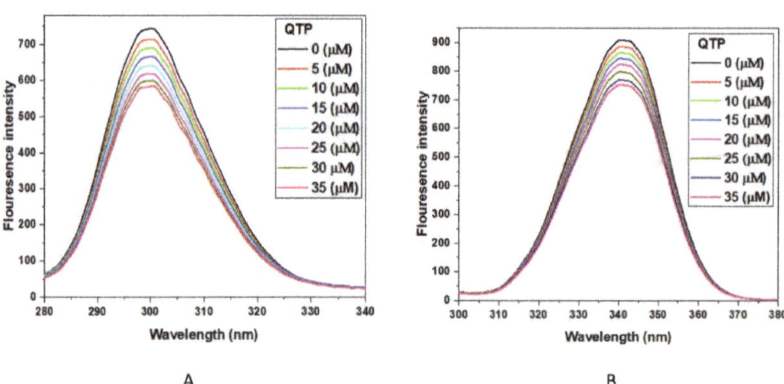

Figure 4. Synchronous fluorescence spectra at $\Delta\lambda$ = 15 nm (**A**) and $\Delta\lambda$ = 60 nm (**B**) of HSA (5 μM) in the absence and presence of increasing concentrations of QTP (0-35 μM). At $\Delta\lambda$ = 15 nm (for Y), the excitation wavelength of HSA was fixed at 240 nm, and the emission range was 255–400 nm, whereas at $\Delta\lambda$ = 60 nm (for W), the excitation wavelength was taken at 240 nm and the emission range was 300–400 nm.

2.2.5. Binding and Prediction of Site Markers in the HSA-QTP System

A site marker displacement experiment was investigated to identify QTP binding site on HSA. In this experiment, here warfarin (WAR) for Sudlow's site I (subdomain IIA), ibuprofen (IBU) for Sudlow's Site II (subdomain IIIA), and hemin (HEM) for binding site III (subdomain IB) were used as HSA site marker probes; [10,26]. As a result, fluorescence spectra were recorded HSA-QTP system in the presence of site marker probes (0–30 μM). Moreover, the displacement percentage (I%) of QTP with the site markers is estimated by the following methods [40,41]:

$$I(\%) = \frac{F_2}{F_1} \times 100 \tag{6}$$

F_1 and F_2 represent the fluorescence emission intensities of the HSA-QTP system in the absence and presence of different site markers, respectively. However, the percentage of displacement values of the HSA-QTP complex against the different concentrations of site markers is shown in Figure 5. It is apparent from Figure 5 that the displacement percentage of QTP from HSA by hemin is appreciably higher than WAR and IBU. Thus, the binding site of QTP is predicted to be in site III (subdomain IB) of HSA.

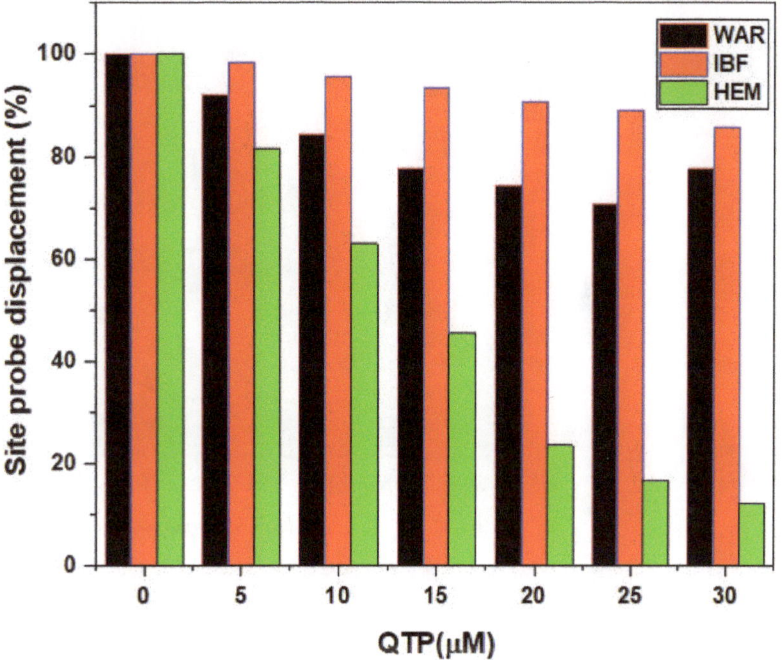

Figure 5. Effect of site probes on the fluorescence emission intensities of the HSA-QTP system. The experiments were carried out using three site probes (warfarin, ibuprofen, and hemin). (HSA = 5 μM, QTP = 10 μM, C = 0-30 μM), λex = 295 nm, T = 295 K.

2.2.6. Circular Dichroism Spectra Changes in HSA upon QTP Binding

Circular dichroism (CD) spectroscopy is a versatile technique mainly used to detect structural and conformational changes in protein structure. The CD spectra of HSA have two negative peaks in the UV region, which reflect α-helix at around 208 and 222 nm of the protein [42]. Figure 6 represents the CD spectra of HSA alone and the HSA-QTP system at different molar ratios of 1:0–1:2. The addition of QTP leads to a decrease in the ellipticity of HSA, suggesting the loss of α-helical content. The CD results showed that the α-helix content of the HSA and QTP-HSA system was 55.92% and 48.88%, respectively. Therefore, these results suggest that the addition of QTP leads to secondary structure change of HSA α-helix content.

2.2.7. QTP-Induced Thermal Stabilization of HSA

The binding of drugs to plasma proteins can increase the protein's thermal stability [43]. Various studies have shown that drugs induced thermal stabilization to HSA [44,45]. Therefore, the thermal stability measurements of HSA were carried out at different temperatures in the absence and presence of QTP binding. The temperature-dependent titrations measurements were performed on HSA (5 μM) without or with QTP (50 μM) in different temperature range, 25–80 °C (5 °C intervals). Figure 7 shows the influence of temperature on the fluorescence intensity of the HSA and HSA-QTP system at 343 nm. In the presence of QTP at 45 °C, the decrease in FI of the HSA-QTP system was lesser than HSA alone. However, our thermal stability results demonstrated QTP-induced stability to HSA via QTP-HSA system formation (coupling of binding and unfolding equilibrium) [46].

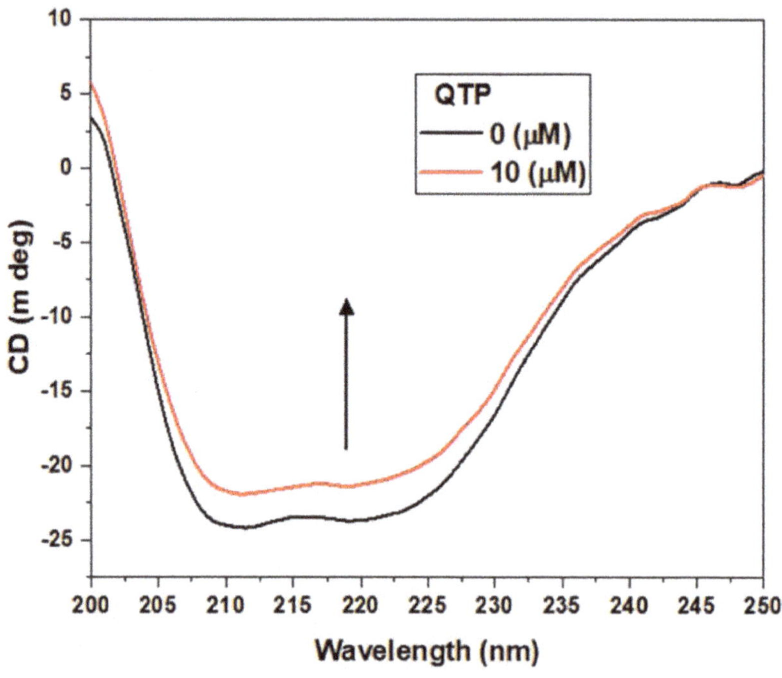

Figure 6. Circular dichroism spectra of HSA (5 µM) in the absence and presence of QTP (10 µM).

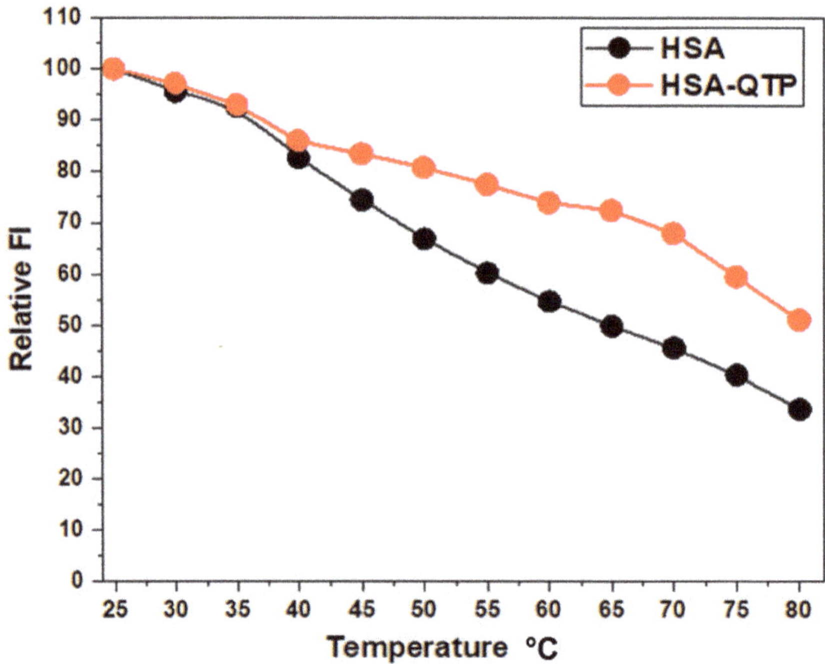

Figure 7. Thermal stability profiles of HSA and the QTP-HSA (1:10) system in the temperature range, 25–80 °C, as monitored by fluorescence intensity measurements at 343 nm (FI 343 nm) using a protein concentration of 5 µM in 60 mM sodium phosphate buffer, pH 7.4.

2.2.8. Effect of QTP Binding on the Esterase-Like Activity of HSA

HSA is the most abundant protein in the blood plasma and possesses catalytic functions such as esterase-like activity [47]. Amino acid residues such as Arg-410 Tyr-41 (Sudlow's site II (subdomain IIIA)) of HSA play a predominant function in esterase activity (Watanabe et al., 2000) [48]. However, the effect of QTP binding on the esterase-like activity of HSA is shown in Figure 8. It was observed that upon the addition of QTP (0–75 μM), there is no inhibiting effect on the esterase-like activity of HSA.

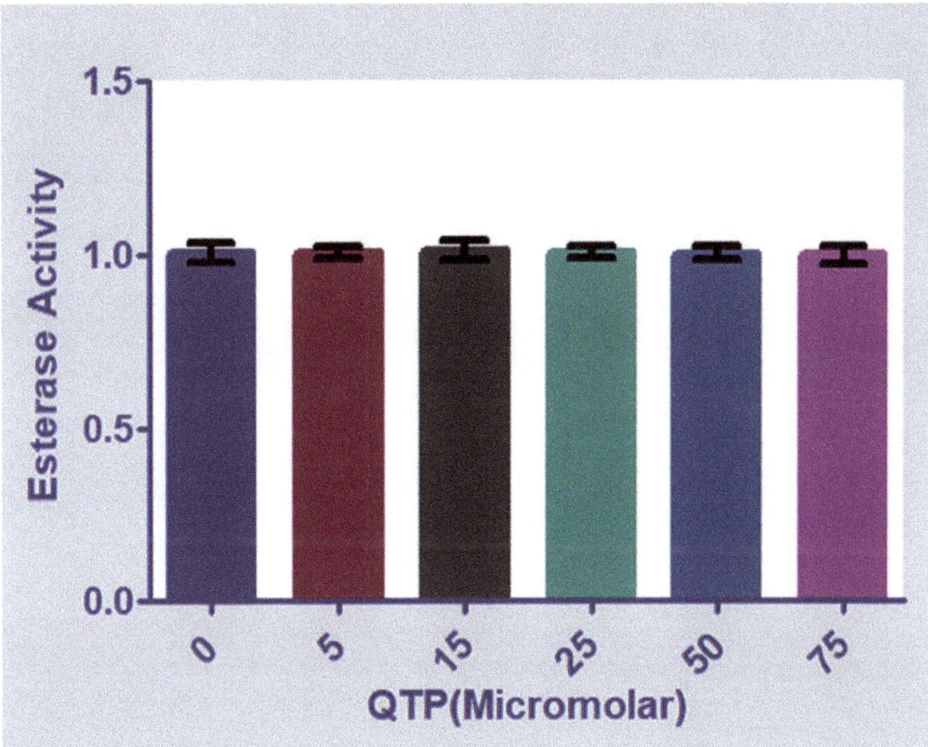

Figure 8. Estimated esterase activity in HSA (5 μM) in the absence and presence of increasing concentrations of QTP (0–75 μM).

2.2.9. Computational Modeling of the HSA-QTP Complex

The binding region and amino acid residues involved in the interaction of QTP with HSA were evaluated by molecular docking analysis [27,45,47]. The most suitable confirmation of the HSA-QTP system is given in Figure 9A,B. The molecular docking results suggested the QTP binding region at subdomain IB (Site III) of HSA (Figure 9A). Further, QTP binds to HSA and forms two hydrogen bonds with VAL120 and ASP173 amino acid residues of HSA (Figure 10A). In addition to the two hydrogen bonds, the QTP molecule is surrounded by LEU-179, ARG-117, ALA-176, ASP-121, LEU-179, PRO-118, ALA-172, VAL-120, ASP-173, GLU-119, LYS-174, and ALA-175 through different interactions (Figure 10). The autodock results also showed that the binding affinity of QTP to HSA was −8.2 kcal mol^{-1}. Thus, we can conclude that the molecular docking results agree with the site displacement markers experiments (Figure 5).

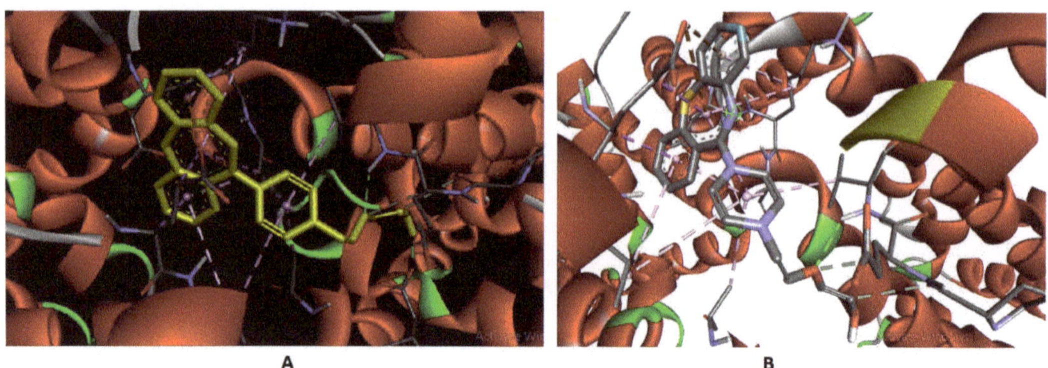

Figure 9. (**A**) Molecular models of HSA complex with QTP. (**B**) Detailed view of the docking poses of the HSA-QTP complex. Selected protein side-chains are shown as ribbons.

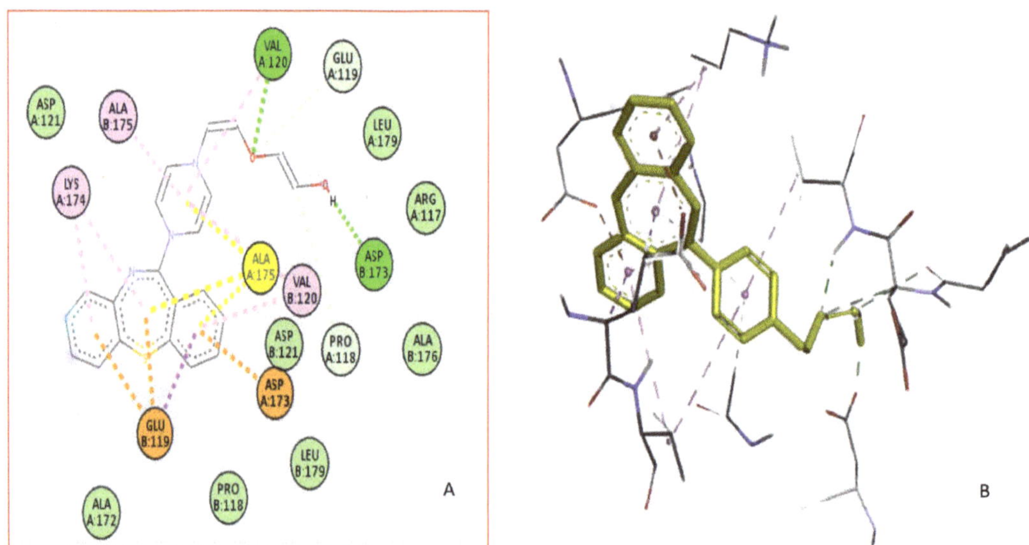

Figure 10. (**A**) The 2D binding site was magnified to show the surrounding amino acid residue of HSA interacting with QTP. (**B**) Three-dimensional structure of interactions of HSA with QTP.

3. Materials and Methods

3.1. Chemical Reagents

HSA (A1887, fatty acid and globulin free) and QTP (purity, 90%) were obtained from Sigma Chemical Co. (St. Louis, Mo, USA) and GLR. Scientific. Co. (Delhi, India), warfarin, ibuprofen through the National Scientific company (Riyadh, KSA) and hemin were obtained from SRL Pvt. Ltd. (Mumbai, India). All other chemicals and reagents for this study were of high analytical grade.

3.2. Sample Preparation

HSA stock solution (200 µM) was prepared in Tris-HCI buffers (0.2 M) pH 7.4. In addition, the stock of QTP (10 mM) was prepared in methanol and then diluted with Tris-HCI buffers (0.2 M), pH 7.4, to prepare the working standard samples of QTP. Finally, the buffer was prepared using Type I Millipore water (Burlington, MA, USA).

3.3. Instrumentations

The UV-Vis absorption spectra were recorded on a UV-1800 spectrophotometer (Shimadzu, Kyoto, Japan) using a 1.0 × 1.0 cm cell. The fluorescence experiments were recorded on an RF-5301PC spectrofluorometer (Shimadzu, Kyoto, Japan) fitted with a xenon-flash lamp with quartz-cuvettes. The circular dichroism experiments were recorded on a JASCO J-1500-CD spectrophotometer (Mary's Court, Easton, MD, USA) equipped with a Peltier temperature controller with a quartz cuvette.

3.4. Methods

3.4.1. UV-Visible Absorption Spectroscopy

UV-Visible absorbance spectra of HSA (5 μM) in the absence and presence of QTP (0–30 μM) at 298 k were recorded at wavelengths from 240 to 410 nm, and baseline correction was performed using an appropriate buffer.

3.4.2. Steady-State Fluorescence Measurements

The intrinsic fluorescence spectra of HSA were recorded at an emission wavelength (300–420 nm) upon excitement at 295 nm. The HSA (5 μM) samples were titrated with QTP (0–35 μM) at three different temperatures (295, 300, 305 K) to estimate thermodynamic parameters. The obtained fluorescence data were corrected for inner filter effects.

3.4.3. Synchronous Fluorescence Spectroscopy (SFS) Experiments

For this experiment, SFS measurements of HSA (5 μM) titrated with different concentrations of QTP (0–35 μM) were performed in different experiments by setting wavelength intervals ($\Delta\lambda$) at 15 nm for tyrosine residue (Y) and 60 nm for tryptophan residue (W) in the same experimental conditions as the fluorescence measurements.

3.4.4. Competitive Site Probe Displacement (CSPD) Experiments

Briefly, in these experiments, CSPD experiments were carried out to locate the binding site of QTP on the HSA. Warfarin (WAR) (Sudlow's site I), ibuprofen (IBU) (Sudlow's site II), and hemin (HEM) (site III) site markers were used to locate the binding region of QTP in HSA. Initially, fluorescence spectra were performed by titrating a solution of 5 μM HSA and QTP 10 μM with increasing site marker (0–30 μM) concentrations in separate experiments. All other parameters (excitation and emission wavelength) were uniform for the fluorescence measurements.

3.4.5. Circular Dichroism (CD) Spectroscopy Measurements

The far CD spectra of HSA and the HSA-QTP complex were recorded at a wavelength between 200 and 260 nm on the spectropolarimeter with a scan speed of 100 nm min^{-1}. HSA (5 μM) was titrated with 10 μM QTP. The measured ellipticity values were expressed as the mean residue ellipticity (MRE) in deg cm^2 dmol^{-1}, defined by equation [48]:

$$MRE = \frac{\text{Observed CD }(\theta obs)}{c \times n \times l \times 10} \tag{7}$$

where θobs is the measured ellipticity in millidegree, "n" is the number of amino acids residues, "l" is the path length of the cuvette (cm), and "c" is the molar concentration of protein. The α-helical content of HSA was determined by equation [48]:

$$\alpha\text{-helical content}(\%) = \frac{[MRE_{208} - 4000]}{[33,000 - 4000]} \times 100 \tag{8}$$

where MRE_{208} is the mean residue elasticity (MRE) at 208 nm.

3.4.6. Thermal Stability Studies of HSA and the HSA-QTP System

The thermal stability of HSA without and with QTP was investigated using fluorescence measurements. The fluorescence spectra of the HSA (5 μM) and HSA-QTP (50 μM) complex were recorded (300–400 nm upon excitation at 295 nm) in the temperature range 25–80 °C (with 5 °C intervals). The solution mixture (HSA-QTP) was incubated for 1 h at 25 °C before fluorescence measurements.

3.4.7. HSA Esterase Activity (E.A.) Assay

The influence of QTP on the esterase activity of HSA was investigated by estimating the formation of p-nitrophenol [40]. The E.A. analysis is based on the fact that 4-nitrophenyl acetate (P-NPA) interacts with HSA and generates 4-nitrophenol (maximum absorption at 400 nm) [49,50]. For this experiment, the concentration of P-NPA (5 μM) and HSA (5 μM) was fixed, and the concentration of QTP increased (0–75 μM).

3.4.8. Molecular Docking between HSA and QTP

The mechanism of QTP binding with HSA has been predicted by molecular docking using AutoDock Vina [51]. The molecular structure of HSA (PDB ID: 1AO6) and QTP (Chem-Spider ID 4827) was obtained from Protein Data Bank (PDB) and Chem-spider, respectively. In the docking protocol, a grid box size of $60 \times 60 \times 60$ with coordinates set to $x = 45$, $y = 12$, and $z = 18$ was built to cover the entire protein. All other parameters were maintained to the default setting. The docked structure of the HSA-QTP system was analyzed with Discovery studio.

4. Conclusions

In the present study, the antipsychotic drug QTP was characterized for its binding interaction to HSA using spectroscopic and biochemical methods and computational approaches. The results obtained from the QTP-HSA binding interactions showed moderate binding affinity of QTP toward HSA. In addition, the involvement of hydrogen bonding and hydrophobic interactions was observed. The spectroscopic studies suggest a complex formation between QTP and HSA, and the system follows a static quenching mechanism. Conversely, the thermodynamic parameters of the HSA-QTP system calculated via fluorescence spectroscopy at different temperatures indicate a spontaneous and exothermic process and indicate the predominant forces to be hydrophobic interactions.

Further, the site-displacement assay and molecular docking results confirm the QTP binding region at subdomain IB of HSA. The CD spectra and UV-Vis spectroscopy identified changes in the secondary structure of HSA upon its interaction with QTP. In addition, QTP did not inhibit the esterase-like activity of HSA. This study is essential and is expected to help understand the drug's mechanisms and pharmacokinetics for further clinical research and novel drug delivery systems.

Author Contributions: Conceptualization: T.A.W., S.Z.; Methodology: T.A.W., S.Z.; Software: T.A.W., N.A.A.; Formal analysis: A.I.A.K., S.Z.; Investigation: N.A.A., T.A.W., S.Z, A.I.A.K.; Resources: T.A.W.; Writing: S.Z., T.A.W.; Review & Editing N.A.A., T.A.W. Project administration: T.A.W. All authors have read and agreed to the published version of the manuscript.

Funding: Researchers Supporting Project (RSP-2021/357), King Saud University, Riyadh, Saudi Arabia.

Institutional Review Board Statement: Not Applicable.

Informed Consent Statement: Not Applicable.

Data Availability Statement: Data will be available on request to corresponding author.

Acknowledgments: The authors extend their appreciation to the Researchers Supporting Project number (RSP-2021/357), King Saud University, Riyadh, Saudi Arabia, for funding this work.

Conflicts of Interest: The authors declare no conflict of interest.

References

1. Rinaldi, R.; Bersani, G.; Marinelli, E.; Zaami, S. The rise of new psychoactive substances and psychiatric implications: A wide-ranging, multifaceted challenge that needs far-reaching common legislative strategies. *Hum. Psychopharmacol. Clin. Exp.* **2020**, *35*, e2727. [CrossRef] [PubMed]
2. Harbell, M.W.; Dumitrascu, C.; Bettini, L.; Yu, S.; Thiele, C.M.; Koyyalamudi, V. Anesthetic Considerations for Patients on Psychotropic Drug Therapies. *Neurol. Int.* **2021**, *13*, 640–658. [CrossRef] [PubMed]
3. Pringsheim, T.; Barnes, T.R. Antipsychotic drug-induced movement disorders: A forgotten problem? *Can. J. Psychiatry* **2018**, *63*, 717–718. [CrossRef] [PubMed]
4. Gupta, A.; Kumar, Y.; Zaheer, M.R.; Roohi; Iqbal, S.; Iqbal, J. Electron Transfer-Mediated Photodegradation of Phototoxic Antipsychotic Drug Quetiapine. *ACS Omega* **2021**, *6*, 30834–30840. [CrossRef] [PubMed]
5. Tran, J.; Gervase, M.A.; Evans, J.; Deville, R.; Dong, X. The stability of quetiapine oral suspension compounded from commercially available tablets. *PLoS ONE* **2021**, *16*, e0255963. [CrossRef]
6. Evoy, K.E.; Teng, C.; Encarnacion, V.G.; Frescas, B.; Hakim, J.; Saklad, S.; Frei, C.R. Comparison of quetiapine abuse and misuse reports to the FDA Adverse Event Reporting System with other second-generation antipsychotics. *Subst. Abus. Res. Treat.* **2019**, *13*. [CrossRef]
7. Akamine, Y.; Yasui-Furukori, N.; Uno, T. Drug-drug interactions of P-gp substrates unrelated to CYP metabolism. *Curr. Drug Metab.* **2019**, *20*, 124–129. [CrossRef]
8. Ignacio, Z.M.; Calixto, A.V.; da Silva, R.H.; Quevedo, J.; Reus, G.Z. The use of quetiapine in the treatment of major depressive disorder: Evidence from clinical and experimental studies. *Neurosci. Biobehav. Rev.* **2018**, *86*, 36–50. [CrossRef]
9. de Miranda, A.S.; Ferreira, R.N.; Teixeira, A.L.; de Miranda, A.S. Mood Stabilizers: Quetiapine. *NeuroPsychopharmacotherapy* **2020**, 1–23.
10. Wani, T.A.; Alsaif, N.; Alanazi, M.M.; Bakheit, A.H.; Zargar, S.; Bhat, M.A. A potential anticancer dihydropyrimidine derivative and its protein binding mechanism by multispectroscopic, molecular docking and molecular dynamic simulation along with its in-silico toxicity and metabolic profile. *Eur. J. Pharm. Sci.* **2021**, *158*, 105686. [CrossRef]
11. Wani, T.A.; Alsaif, N.A.; Alanazi, M.M.; Bakheit, A.H.; Khan, A.A.; Zargar, S. Binding of colchicine and ascorbic acid (vitamin C) to bovine serum albumin: An in-vitro interaction study using multispectroscopic, molecular docking and molecular dynamics simulation study. *J. Mol. Liq.* **2021**, *342*, 117542. [CrossRef]
12. Wani, T.A.; Bakheit, A.H.; Al-Majed, A.A.; Altwaijry, N.; Baquaysh, A.; Aljuraisy, A.; Zargar, S. Binding and drug displacement study of colchicine and bovine serum albumin in presence of azithromycin using multispectroscopic techniques and molecular dynamic simulation. *J. Mol. Liq.* **2021**, *333*, 115934. [CrossRef] [PubMed]
13. Wani, T.A.; Bakheit, A.H.; Zargar, S.; Alamery, S. Mechanistic competitive binding interaction study between olmutinib and colchicine with model transport protein using spectroscopic and computer simulation approaches. *J. Photochem. Photobiol. A Chem.* **2022**, *426*, 113794. [CrossRef]
14. Wani, T.A.; Alanazi, M.M.; Alsaif, N.A.; Bakheit, A.H.; Zargar, S.; Alsalami, O.M.; Khan, A.A. Interaction Characterization of a Tyrosine Kinase Inhibitor Erlotinib with a Model Transport Protein in the Presence of Quercetin: A Drug–Protein and Drug–Drug Interaction Investigation Using Multi-Spectroscopic and Computational Approaches. *Molecules* **2022**, *27*, 1265. [CrossRef]
15. Zargar, S.; Wani, T.A. Protective Role of Quercetin in Carbon Tetrachloride Induced Toxicity in Rat Brain: Biochemical, Spectrophotometric Assays and Computational Approach. *Molecules* **2021**, *26*, 7526. [CrossRef]
16. Tebeta, R.; Ahmed, N.; Fattahi, A. Experimental study on the effect of compression load on the elastic properties of HDPE/SWCNTs nanocomposites. *Microsyst. Technol.* **2021**, *27*, 3513–3522. [CrossRef]
17. Alemi Parvin, S.; Ahmed, N.; Fattahi, A. Numerical prediction of elastic properties for carbon nanotubes reinforced composites using a multi-scale method. *Eng. Comput.* **2021**, *37*, 1961–1972. [CrossRef]
18. Chamani, J.; Heshmati, M. Mechanism for stabilization of the molten globule state of papain by sodium n-alkyl sulfates: Spectroscopic and calorimetric approaches. *J. Colloid Interface Sci.* **2008**, *322*, 119–127. [CrossRef]
19. Marouzi, S.; Rad, A.S.; Beigoli, S.; Baghaee, P.T.; Darban, R.A.; Chamani, J. Study on effect of lomefloxacin on human holo-transferrin in the presence of essential and nonessential amino acids: Spectroscopic and molecular modeling approaches. *Int. J. Biol. Macromol.* **2017**, *97*, 688–699. [CrossRef]
20. Wani, T.A.; Bakheit, A.H.; Zargar, S.; Bhat, M.A.; Al-Majed, A.A. Molecular docking and experimental investigation of new indole derivative cyclooxygenase inhibitor to probe its binding mechanism with bovine serum albumin. *Bioorg. Chem.* **2019**, *89*, 103010. [CrossRef]
21. He, X.M.; Carter, D.C. Atomic structure and chemistry of human serum albumin. *Nature* **1992**, *358*, 209–215. [CrossRef] [PubMed]
22. Carballal, S.; Radi, R.; Kirk, M.C.; Barnes, S.; Freeman, B.A.; Alvarez, B. Sulfenic acid formation in human serum albumin by hydrogen peroxide and peroxynitrite. *Biochemistry* **2003**, *42*, 9906–9914. [CrossRef] [PubMed]
23. Qi, J.; Zhang, Y.; Gou, Y.; Zhang, Z.; Zhou, Z.; Wu, X.; Yang, F.; Liang, H. Developing an anticancer copper (II) pro-drug based on the His242 residue of the human serum albumin carrier IIA subdomain. *Mol. Pharm.* **2016**, *13*, 1501–1507. [CrossRef] [PubMed]
24. Alsaif, N.A.; Al-Mehizia, A.A.; Bakheit, A.H.; Zargar, S.; Wani, T.A. A spectroscopic, thermodynamic and molecular docking study of the binding mechanism of dapoxetine with calf thymus DNA. *S. Afr. J. Chem.* **2020**, *73*, 44–50. [CrossRef]

25. Wani, T.A.; Bakheit, A.H.; Zargar, S.; Alanazi, Z.S.; Al-Majed, A.A. Influence of antioxidant flavonoids quercetin and rutin on the in-vitro binding of neratinib to human serum albumin. *Spectrochim. Acta Part A Mol. Biomol. Spectrosc.* **2021**, *246*, 118977. [CrossRef] [PubMed]
26. Kou, S.-B.; Lin, Z.-Y.; Wang, B.-L.; Shi, J.-H.; Liu, Y.-X. Evaluation of the binding behavior of olmutinib (HM61713) with model transport protein: Insights from spectroscopic and molecular docking studies. *J. Mol. Struct.* **2021**, *1224*, 129024. [CrossRef]
27. Rabbani, G.; Lee, E.J.; Ahmad, K.; Baig, M.H.; Choi, I. Binding of tolperisone hydrochloride with human serum albumin: Effects on the conformation, thermodynamics, and activity of HSA. *Mol. Pharm.* **2018**, *15*, 1445–1456. [CrossRef]
28. Alam, M.M.; Abul Qais, F.; Ahmad, I.; Alam, P.; Hasan Khan, R.; Naseem, I. Multi-spectroscopic and molecular modelling approach to investigate the interaction of riboflavin with human serum albumin. *J. Biomol. Struct. Dyn.* **2018**, *36*, 795–809. [CrossRef]
29. Zhao, X.; Liu, R.; Chi, Z.; Teng, Y.; Qin, P. New insights into the behavior of bovine serum albumin adsorbed onto carbon nanotubes: Comprehensive spectroscopic studies. *J. Phys. Chem. B* **2010**, *114*, 5625–5631. [CrossRef]
30. Lakowicz, J.R. *Principles of Fluorescence Spectroscopy*; Springer Science & Business Media: Berlin, Germany, 2013.
31. Kandagal, P.; Ashoka, S.; Seetharamappa, J.; Shaikh, S.; Jadegoud, Y.; Ijare, O. Study of the interaction of an anticancer drug with human and bovine serum albumin: Spectroscopic approach. *J. Pharm. Biomed. Anal.* **2006**, *41*, 393–399. [CrossRef]
32. Peng, X.; Wang, X.; Qi, W.; Huang, R.; Su, R.; He, Z. Deciphering the binding patterns and conformation changes upon the bovine serum albumin-rosmarinic acid complex. *Food Funct.* **2015**, *6*, 2712–2726. [CrossRef] [PubMed]
33. Kameníková, M.; Furtmüller, P.G.; Klacsová, M.; Lopez-Guzman, A.; Toca-Herrera, J.L.; Vitkovská, A.; Devínsky, F.; Mučaji, P.; Nagy, M. Influence of quercetin on the interaction of gliclazide with human serum albumin-spectroscopic and docking approaches. *Luminescence* **2017**, *32*, 1203–1211. [CrossRef] [PubMed]
34. Tayyab, S.; Min, L.H.; Kabir, M.Z.; Kandandapani, S.; Ridzwan, N.F.W.; Mohamad, S.B. Exploring the interaction mechanism of a dicarboxamide fungicide, iprodione with bovine serum albumin. *Chem. Pap.* **2020**, *74*, 1633–1646. [CrossRef]
35. Lakowicz, J.R.; Weber, G. Quenching of protein fluorescence by oxygen. Detection of structural fluctuations in proteins on the nanosecond time scale. *Biochemistry* **1973**, *12*, 4171–4179. [CrossRef]
36. Shahabadi, N.; Hadidi, S.; Feizi, F. Study on the interaction of antiviral drug 'Tenofovir'with human serum albumin by spectral and molecular modeling methods. *Spectrochim. Acta Part A Mol. Biomol. Spectrosc.* **2015**, *138*, 169–175. [CrossRef]
37. Jafari, F.; Samadi, S.; Nowroozi, A.; Sadrjavadi, K.; Moradi, S.; Ashrafi-Kooshk, M.R.; Shahlaei, M. Experimental and computational studies on the binding of diazinon to human serum albumin. *J. Biomol. Struct. Dyn.* **2018**, *36*, 1490–1510. [CrossRef]
38. Ross, P.D.; Subramanian, S. Thermodynamics of protein association reactions: Forces contributing to stability. *Biochemistry* **1981**, *20*, 3096–3102. [CrossRef]
39. Lloyd, J. The nature and evidential value of the luminescence of automobile engine oils and related materials: Part I. Synchronous excitation of fluorescence emission. *J. Forensic Sci. Soc.* **1971**, *11*, 83–94. [CrossRef]
40. Ahmad, A.; Ahmad, M. Understanding the fate of human serum albumin upon interaction with edifenphos: Biophysical and biochemical approaches. *Pestic. Biochem. Physiol.* **2018**, *145*, 46–55. [CrossRef]
41. Wani, T.A.; Bakheit, A.H.; Zargar, S.; Khayyat, A.I.A.; Al-Majed, A.A. Influence of Rutin, Sinapic Acid, and Naringenin on Binding of Tyrosine Kinase Inhibitor Erlotinib to Bovine Serum Albumin Using Analytical Techniques Along with Computational Approach. *Appl. Sci.* **2022**, *12*, 3575. [CrossRef]
42. Greenfield, N.J. Using circular dichroism spectra to estimate protein secondary structure. *Nat. Protoc.* **2006**, *1*, 2876–2890. [CrossRef] [PubMed]
43. Sancataldo, G.; Vetri, V.; Fodera, V.; Di Cara, G.; Militello, V.; Leone, M. Oxidation enhances human serum albumin thermal stability and changes the routes of amyloid fibril formation. *PLoS ONE* **2014**, *9*, e84552. [CrossRef] [PubMed]
44. Farruggia, B.; Nerli, B.; Di Nuci, H.; Rigatusso, R.; Picó, G. Thermal features of the bovine serum albumin unfolding by polyethylene glycols. *Int. J. Biol. Macromol.* **1999**, *26*, 23–33. [CrossRef]
45. Rahman, S.; Rehman, M.T.; Rabbani, G.; Khan, P.; AlAjmi, M.F.; Hassan, M.; Muteeb, G.; Kim, J. Insight of the interaction between 2, 4-thiazolidinedione and human serum albumin: A spectroscopic, thermodynamic and molecular docking study. *Int. J. Mol. Sci.* **2019**, *20*, 2727. [CrossRef]
46. Celej, M.S.; Montich, G.G.; Fidelio, G.D. Protein stability induced by ligand binding correlates with changes in protein flexibility. *Protein Sci.* **2003**, *12*, 1496–1506. [CrossRef]
47. Moradi, N.; Ashrafi-Kooshk, M.R.; Ghobadi, S.; Shahlaei, M.; Khodarahmi, R. Spectroscopic study of drug-binding characteristics of unmodified and pNPA-based acetylated human serum albumin: Does esterase activity affect microenvironment of drug binding sites on the protein? *J. Lumin.* **2015**, *160*, 351–361. [CrossRef]
48. Zargar, S.; Wani, T.A. Exploring the binding mechanism and adverse toxic effects of persistent organic pollutant (dicofol) to human serum albumin: A biophysical, biochemical and computational approach. *Chem.-Biol. Interact.* **2021**, *350*, 109707. [CrossRef]
49. Ahmad, B.; Khan, M.K.A.; Haq, S.K.; Khan, R.H. Intermediate formation at lower urea concentration in 'B'isomer of human serum albumin: A case study using domain specific ligands. *Biochem. Biophys. Res. Commun.* **2004**, *314*, 166–173. [CrossRef]
50. Zhao, L.; Song, W.; Wang, J.; Yan, Y.; Chen, J.; Liu, R. Mechanism of dimercaptosuccinic acid coated superparamagnetic iron oxide nanoparticles with human serum albumin. *J. Biochem. Mol. Toxicol.* **2015**, *29*, 579–586. [CrossRef]
51. Trott, O.; Olson, A.J. AutoDock Vina: Improving the speed and accuracy of docking with a new scoring function, efficient optimization, and multithreading. *J. Comput. Chem.* **2010**, *31*, 455–461. [CrossRef]

2022, 27, 3295

molecules

MDPI

Enhancement of Haloperidol Binding Affinity to Dopamine Receptor via Forming a Charge-Transfer Complex with Picric Acid and 7,7,8,8-Tetracyanoquinodimethane for Improvement of the Antipsychotic Efficacy

Abdulhakeem S. Alamri [1,2], Majid Alhomrani [1,2], Walaa F. Alsanie [1,2], Hussain Alyami [3], Sonam Shakya [4], Hamza Habeeballah [5], Abdulwahab Alamri [6], Omar Alzahrani [7], Ahmed S. Alzahrani [2], Heba A. Alkhatabi [8,9,10], Raed I. Felimban [8,11], Abdulhameed Abdullah Alhabeeb [12], Bassem M. Raafat [13], Moamen S. Refat [14,*] and Ahmed Gaber [2,15,*]

[1] Department of Clinical Laboratories Sciences, The Faculty of Applied Medical Sciences, Taif University, P.O. Box 11099, Taif 21944, Saudi Arabia; a.alamri@tu.edu.sa (A.S.A.); m.alhomrani@tu.edu.sa (M.A.); w.alsanie@tu.edu.sa (W.F.A.)
[2] Centre of Biomedical Sciences Research (CBSR), Deanship of Scientific Research, Taif University, P.O. Box 11099, Taif 21944, Saudi Arabia; a.s.zahrani@tu.edu.sa
[3] College of Medicine, Taif University, P.O. Box 11099, Taif 21944, Saudi Arabia; hmyami@tu.edu.sa
[4] Department of Chemistry, Faculty of Science, Aligarh Muslim University, Aligarh 202002, India; sonamshakya08@gmail.com
[5] Department of Medical Laboratory Technology, Faculty of Applied Medical Sciences in Rabigh, King Abdulaziz University, Jeddah 21589, Saudi Arabia; hhabeeballah@kau.edu.sa
[6] Department of Pharmacology and Toxicology, College of Pharmacy, University of Hail, Hail 55211, Saudi Arabia; a.alamry@uoh.edu.sa
[7] School of Health and Biomedical Sciences, RMIT University, Melbourne 3001, Australia; o_s_z@hotmail.com
[8] Department of Medical Laboratory Sciences, Faculty of Applied Medical Sciences, King Abdulaziz University, Jeddah 21589, Saudi Arabia; halkhattabi@kau.edu.sa (H.A.A.); faraed@kau.edu.sa (R.I.F.)
[9] Center of Excellence in Genomic Medicine Research (CEGMR), King Abdulaziz University, Jeddah 21589, Saudi Arabia
[10] Hematology Research Unit, King Fahd Medical Research Centre, King Abdulaziz University, Jeddah 21589, Saudi Arabia
[11] 3D Bioprinting Unit, Center of Innovation in Personalized Medicine (CIPM), King Abdulaziz University, Jeddah 21589, Saudi Arabia
[12] National Centre for Mental Health Promotion, P.O. Box 95459, Riyadh 11525, Saudi Arabia; aalhabeeb@ncmh.org.sa
[13] Department of Radiological Sciences, College of Applied Medical Sciences, Taif University, P.O. Box 11099, Taif 21944, Saudi Arabia; bassemraafat@tu.edu.sa
[14] Department of Chemistry, College of Science, Taif University, P.O. Box 11099, Taif 21944, Saudi Arabia
[15] Department of Biology, College of Science, Taif University, P.O. Box 11099, Taif 21944, Saudi Arabia
* Correspondence: moamen@tu.edu.sa (M.S.R.); a.gaber@tu.edu.sa (A.G.)

Citation: Alamri, A.S.; Alhomrani, M.; Alsanie, W.F.; Alyami, H.; Shakya, S.; Habeeballah, H.; Alamri, A.; Alzahrani, O.; Alzahrani, A.S.; Alkhatabi, H.A.; et al. Enhancement of Haloperidol Binding Affinity to Dopamine Receptor via Forming a Charge-Transfer Complex with Picric Acid and 7,7,8,8-Tetracyanoquinodimethane for Improvement of the Antipsychotic Efficacy. *Molecules* **2022**, *27*, 3295. https://doi.org/10.3390/molecules27103295

Academic Editors: Tanveer A. Wani, Seema Zargar and Afzal Hussain

Received: 3 April 2022
Accepted: 18 May 2022
Published: 20 May 2022

Publisher's Note: MDPI stays neutral with regard to jurisdictional claims in published maps and institutional affiliations.

Abstract: Haloperidol (HPL) is a typical antipsychotic drug used to treat acute psychotic conditions, delirium, and schizophrenia. Solid charge transfer (CT) products of HPL with 7,7,8,8-tetracyanoquinodimethane (TCNQ) and picric acid (PA) have not been reported till date. Therefore, we conducted this study to investigate the donor–acceptor CT interactions between HPL (donor) and TCNQ and PA (π-acceptors) in liquid and solid states. The complete spectroscopic and analytical analyses deduced that the stoichiometry of these synthesized complexes was 1:1 molar ratio. Molecular docking calculations were performed for HPL as a donor and the resulting CT complexes with TCNQ and PA as acceptors with two protein receptors, serotonin and dopamine, to study the comparative interactions among them, as they are important neurotransmitters that play a large role in mental health. A molecular dynamics simulation was ran for 100 ns with the output from AutoDock Vina to refine docking results and better examine the molecular processes of receptor–ligand interactions. When compared to the reactant donor, the CT complex [(HPL)(TCNQ)] interacted with serotonin and dopamine more efficiently than HPL only. CT complex [(HPL)(TCNQ)] with dopamine (CTtD) showed the greatest binding energy value among all. Additionally, CTtD complex established more a stable interaction with dopamine than HPL–dopamine.

Keywords: charge transfer; haloperidol; π-acceptors; antipsychotics

---

1. Introduction

Atypical antipsychotics or serotonin–dopaminergic antagonists, the fourth class of antipsychotic drugs, can improve the so-called positive symptoms of schizophrenia, such as hallucinations, delusions, and agitation, as well as negative symptoms, such as catatonia and flattening of the ability to feel emotion. Each agent in this group has a unique profile of receptor interactions. Almost all antipsychotics block dopamine receptors and reduce dopamine transmission in the forebrain. Moreover, atypical antipsychotics have an affinity for serotonin receptors. Atypical antipsychotics are related to chlorpromazine and haloperidol (HPL), and HPL is used to treat acute psychotic conditions, delirium, and schizophrenia [1].

The formation of highly colored charge-transfer (CT) complexes that absorb light in the visible region is often related with molecular interactions between electron donors and acceptors [2,3]. CT complexes have become more important in the fields of drug receptor binding; DNA binding; and antibacterial, antifungal, and anticancer applications [4,5]. A weak interaction between donors and acceptors causes the reaction [6].

Donor–acceptor complexation plays an important role especially in the field of biochemical energy transfer process [7]. The formation of brilliantly colored CT complexes that absorb visible light is frequently linked to charge transfer interactions between electron acceptors and donors [2]. In biological systems, mechanisms requiring molecular complexation and structural recognition include drug design, enzyme catalysis, and ion exchanges via lipophilic membranes [3,4]. Mulliken postulated that an electron transfer from a Lewis base's π-molecular orbital to a Lewis acid's vacant λ-molecular orbital causes the development of molecular complexes from two aromatic molecules, with the resonance between this dative structure and the no-band structure maintaining the complex.

Solid CT products of HPL with 7,7,8,8-tetracyanoquinodimethane (TCNQ) and picric acid (PA) have not been reported till date. Therefore, we conducted this study to investigate such reactions. The molecular docking software AutoDock Vina was used to investigate the interactions between ligands (HPL and synthesized CT complexes) and receptors (serotonin and dopamine). Hydrophobic, ionizability, aromatic, and hydrogen bond surfaces were studied as well as binding energy. The best molecular docking data were submitted to molecular dynamic simulation at 300 K for 100 ns to give a more effective mechanism for illustrating receptor–ligand interactions. In terms of residue flexibility, structural stability, solvent accessible surface area, structure compactness, and hydrogen bond interactions, the dynamic properties of the complexes were compared.

2. Results and Discussion

2.1. Preface

The micro analytical technique confirmed that the molar ratio between HPL donor and PA and TCNQ (π–acceptors) was 1:1. The conductivities of HPL-PA and HPL-TCNQ CT complexes were 45 and 53 Ω^{-1} cm^{-1} mol^{-1}, respectively. The low conductance values of the synthesized CT complexes deduced the formation of D$^+$ and A$^-$ datives anions based on the association of donor–acceptor chelation. The electronic spectra of synthesized CT complexes of HPL-PA and HPL-TCNQ refer to the association of new electronic absorption bands (447 nm, 738, and 837 nm), which did not exist in the spectra of free reactants. The infrared spectrum of HPL-PA complex was assigned upon intermolecular hydrogen bonding between the –OH group of the PA acceptor and basic oxygen atom center of HPL donor (Figure 1). In the case of the infrared spectrum of HPL-TCNQ solid CT complex

(Figure 2), the –OH stretching band of HPL shifted to higher frequencies. This was assigned to the increase in polarity status, $-^-O\text{-}C\equiv NH^+$, during the complexation process.

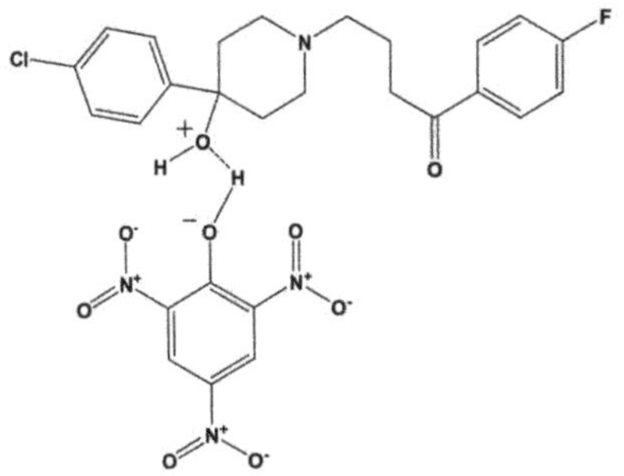

Figure 1. Charge-transfer (CT) complex of [(HPL)(PA)].

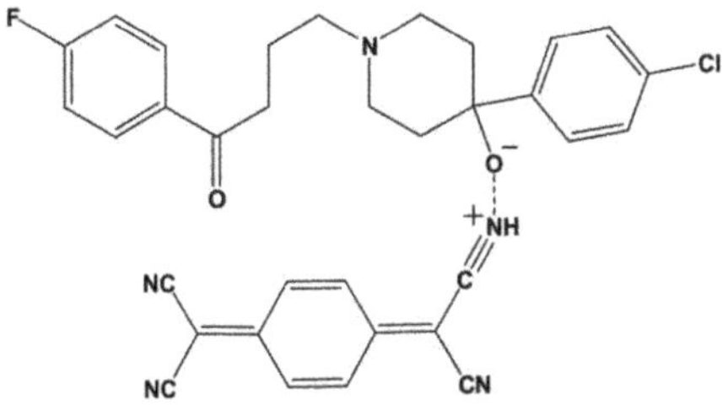

Figure 2. CT complex of [(HPL)(TCNQ)].

The bonding of the –OH group of HPL and the –OH group of PA to the –CN of TCNQ acceptors via intermolecular hydrogen bonding was confirmed by proton NMR spectra of the free HPL donor and its HPL-PA and HPL-TCNQ complexes. The activation energy (E) was used to calculate the thermal stability of both HPL-PA and HPL-TCNQ complexes using Coats–Redfern and Horowitz–Metzger techniques [8–10].

The average activation energies for the [(HPL)(PA)] complex and the [(HPL)(TCNQ)] complex were 132 kJ mol^{-1} and 98 kJ mol^{-1}, respectively, and the variant data might be influenced by the acceptor type. The activated complexes had a more ordered structure than the reactants, and the activation of entropy (ΔS *) had negative values, indicating that the reaction rates were slower than normal.

The optical band gap (E_g), which refers to the minimum transition energy, was determined based on the electronic absorption spectra. Optical absorption near the edge of the absorption band can be used to estimate E_g and confirm the formation of CT complexes. The absorption coefficient (α) can be estimated from the transmittance (T) of the complex according to the following equation:

$$\alpha = 1/d \, \text{lin} \, (1/T) \tag{1}$$

where d is the sample thickness. The bandgap of CT complexes can be calculated from the relationship between α and E_g based on the following equation [11]:

$$\alpha h\nu = A(h\nu - E_g)^m \tag{2}$$

where (m) equals to $1/2$ and 2 for direct and indirect transitions, respectively, whereas (A) is an energy-independent constant. The values of $(\alpha h\nu)2$ were plotted against hν. The direct optical bandgap Eg was determined from the linear relationship of the plots at the absorption edge where $(\alpha h\nu)2 = 0$ [12]. Eg values for HPL-PA and HPL-TCNQ CT complexes were 2.483 and 2.895, respectively (Figure 3), and the values were dependent on the nature of the acceptor. These data indicate the conducting behavior of HPL-PA and HPL-TCNQ complexes [13,14].

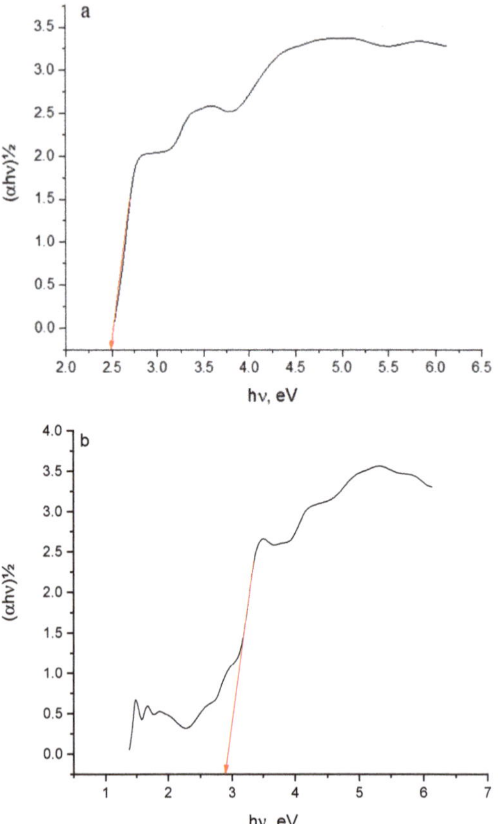

Figure 3. Plots of optical energy (hv) against $(\alpha h\nu)^{\frac{1}{2}}$ for (**a**) haloperidol–picric acid (HPL-PA) and (**b**) HPL-7,7,8,8-tetracyanoquinodimethane CT complexes.

2.2. Molecular Docking Studies

The docking positions of the synthesized CT complexes [(HPL)(PA)] and [(HPL)(TCNQ)] against serotonin (PDB ID: 6A94) and dopamine (PDB ID: 6CM4) were determined. For comparison, HPL was employed as the control. CT complexes have a larger potential binding energy than HPL in both receptors (Tables 1 and 2). Among them, [(HPL)(TCNQ)] had the greatest docking energy. The theoretical binding energies of [(HPL)(TCNQ)] with serotonin and dopamine were −10.2, and −11.8 kcal/mol, respectively. Additionally, the higher binding energy value of [(HPL)(TCNQ)]–dopamine (CTtD) signifies a stronger interaction with dopamine compared to that with serotonin. The best docking position of CTtD is shown in Figure 4, and the docking data are given in Table 2.

Table 1. Docking score of the ligands and their interactions with serotonin (6A94).

Receptor	Binding Free Energy (kcal/Mol)	Interactions	
		H-Bond	Others
HPL-PA	−9.4	Thr160	Leu229, Phe243, Val366 (π-Alkyl); Trp336, Phe340 (π-π T-Shape)
HPL-TCNQ	−10.2	Asn363 and Tyr139	Ala321, Val324 (π-Alkyl); Ala360 (π-Alkyl); Val366 (π-Sigma)
HPL	−10.0	Leu229	Phe332, Phe243, Val366 (π-Alkyl); Phe340, Trp336, Ser159 (π-π T-Shape)

Table 1. Target: PDB: 6A94

Table 2. Docking score of the ligands and their interactions with dopamine (6CM4).

Receptor	Binding Free Energy (kcal/Mol)	Interactions	
		H-Bond	Others
HPL-PA	−9.6	Trp100 and Tro419	Cys118, Val115, Leu94 (π-Alkyl); Phe390, Trp389, Tyr480 (π-π T-Shape)
HPL-TCNQ	−11.8	His393, Ser193, and Tyr416	Val91 and Trp413 (π-Alkyl); Tyr408, (π-π T-Shaped); Cys118 (π-Alkyl); Thr412 and Leu94 (π-Sigma); Asp114 (Attractive charge)
HPL	−10.9	Asp114	Phe198, Phe382, Cys118, Val91 (π-Alkyl); Trp100, Trp386, Phe390 (π-π T-shaped); Lue94, Thr412 (π-Sigma)

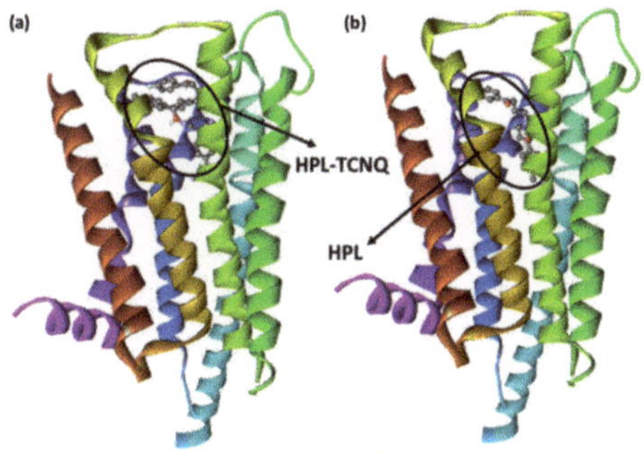

Figure 4. The best-docked position showing a helical model of dopamine docked with (**a**) CT complex and (**b**) HPL drug only.

The illustration of molecular docking for ligand–receptor interactions depicted in Figure 5a,b. As shown in Figure 5a, CT complex [(HPL)(TCNQ)] with dopamine (CTtD) revealed that the amino acid residues, including His393, Ser193, and Tyr416, formed hydrogen bond interactions. Additionally, Val91 and Trp413 (π-Alkyl); Tyr408 (π-π T-Shaped); Cys118 (π-Alkyl); Thr412 and Leu94 (π-Sigma); and Asp114 (Attractive charge) interactions were present [15,16].

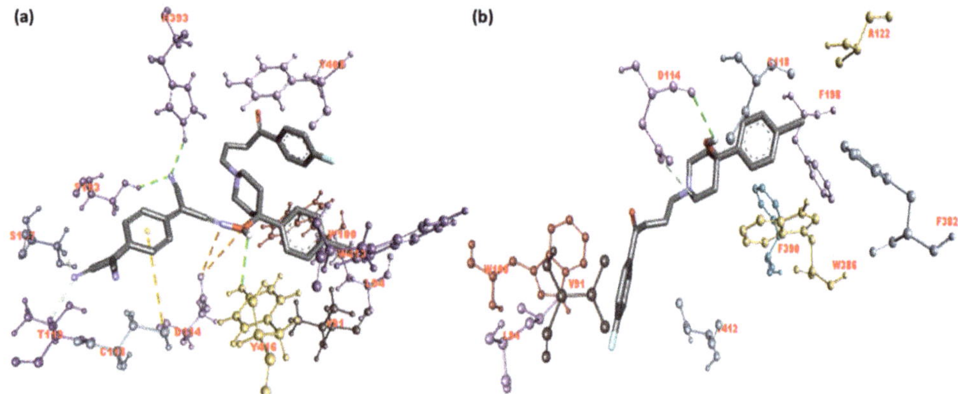

Figure 5. 3D representation of the interactions for dopamine docked with (**a**) CT complex and (**b**) HPL drug only.

Molecular docking of HPL drug with serotonin and dopamine revealed the potential binding energies as −10.0 and −10.9 kcal/mol, respectively. The higher binding energy value of HPL–dopamine (HPLD) signifies stronger interaction with dopamine compared to that with serotonin. The best docking position with dopamine (HPLD) is shown in Figure 4, and the docking data are shown in Table 2. Figure 5b shows the interaction between HPL and dopamine, which reveals that the amino acid residue Asp114 formed hydrogen bond interactions. Additionally, Phe198, Phe382, Cys118, and Val91 (π-Alkyl); Trp100, Trp386, and Phe390 (π-π T-shaped); and Lue94 and Thr412 (π-Sigma) interactions were present. This shows that the CT complex [(HPL)(TCNQ)] binds to both receptors more efficiently as compared toHPL alone, and among them, CTtD had the highest binding energy value. 2D representations of ligand–receptor interactions are shown in Figure 6. Hydrophobic, ionizability, aromatic, and hydrogen bond surfaces at the interaction site of [(HPL)(TCNQ)] and dopamine are represented in Figure 7, and those for HPL and dopamine are shown in Figure 8.

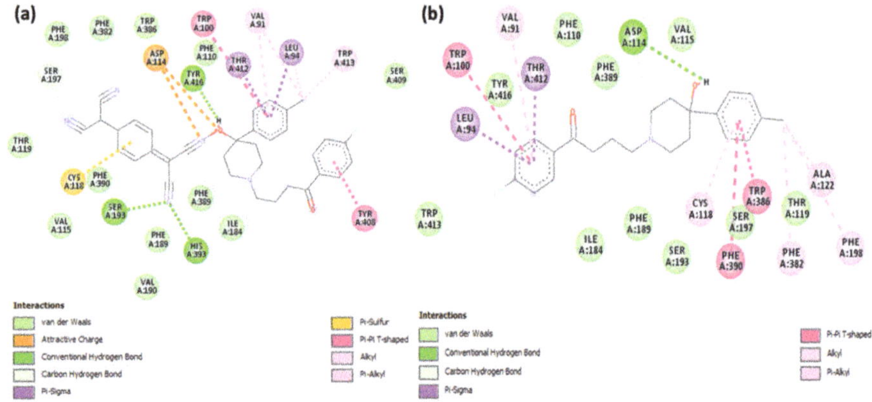

Figure 6. 2D representation of the interactions for dopamine docked with (**a**) CT complex and (**b**) HPL drug only.

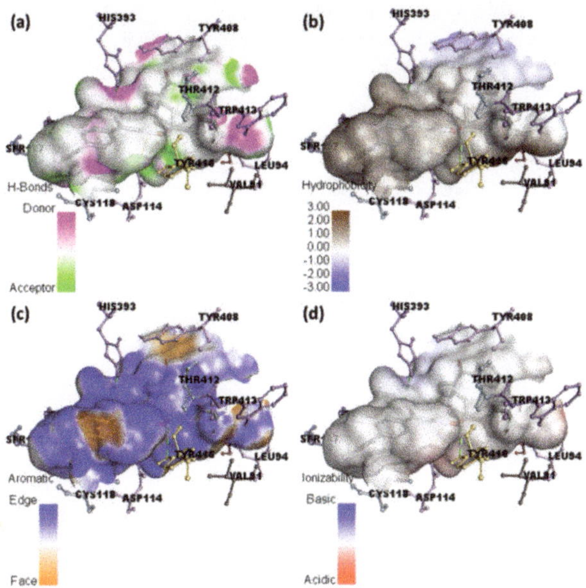

Figure 7. Representation of (**a**) hydrogen binding, (**b**) hydrophobic, (**c**) aromatic, and (**d**) ionizability surfaces between dopamine and CT complex.

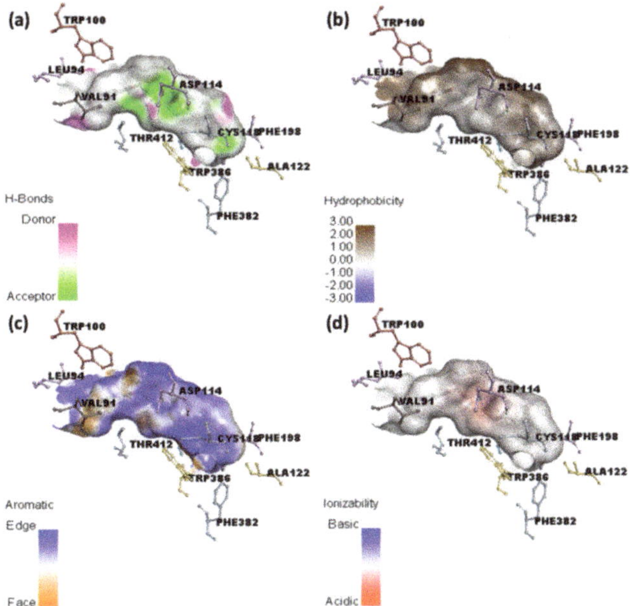

Figure 8. Representation of (**a**) hydrogen binding, (**b**) hydrophobic, (**c**) aromatic, and (**d**) ionizability surfaces between dopamine and HPL drug only.

2.3. Molecular Dynamics Simulation

The best docked location of HPLD and CTtD with the highest docking score at 100 ns molecular dynamics (MD) generated by AutoDock Vina was used. Only the best docking output was used to build up this method in a high-throughput manner for analyzing the binding mechanism of the ligand at the active site of protein under clearly defined aqueous conditions. The root mean square deviation was computed to determine structural stability from MD data (RMSD). HPLD and CTtD formed stable conformation after ~80 ns and ~65 ns, respectively, with an appropriate RMSD value of 3.04 and 2.41 Å, respectively, as seen in the RMSD plot (Figure 9).

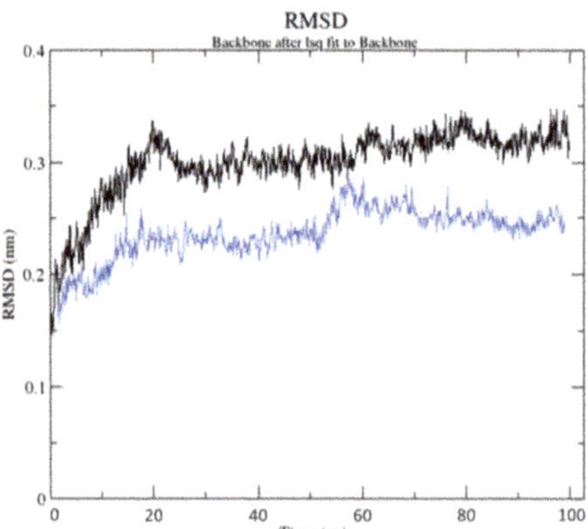

Figure 9. Root mean square deviation of solvated receptor backbone and ligand complex during the 100 ns MD simulation [HPLD complex (black) and CTtD complex (blue)].

The RMSD value range of <3.0 Å is the most acceptable [17]. CTtD produces a more stable combination as a result of this discovery. The MD findings of ligand–receptor interaction, as shown in Figure 10, bring protein chains closer together and close the distance between them [18]. Chimera 1.15 software was used to create the superimposed structures by employing the tool–structure comparison followed by the MatchMaker feature. Pairing uses both sequence and secondary structure to superimpose comparable structures.

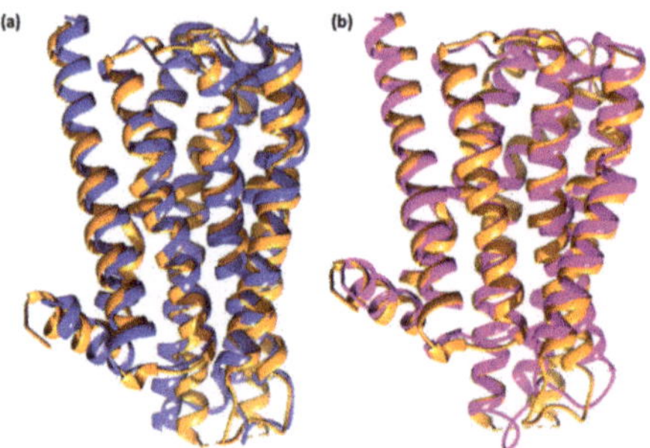

Figure 10. Superimposed structures of (**a**) unbounded dopamine receptor (orange) and dopamine receptor after simulation (blue) for HPLD and (**b**) unbounded dopamine receptor (orange) and dopamine receptor after simulation (pink) for CTtD.

RR Distance Maps creates a distance map through Chimera 1.15 software by a structural comparison tool. The map can display Cα-Cα distances within a single protein chain, as well as averages and standard deviations for many chains. The white diagonal on the map represents no distance between the two residues, but the red and blue on the map reflect residue pairings with the biggest distance differences between the two conformations (Figure 11).

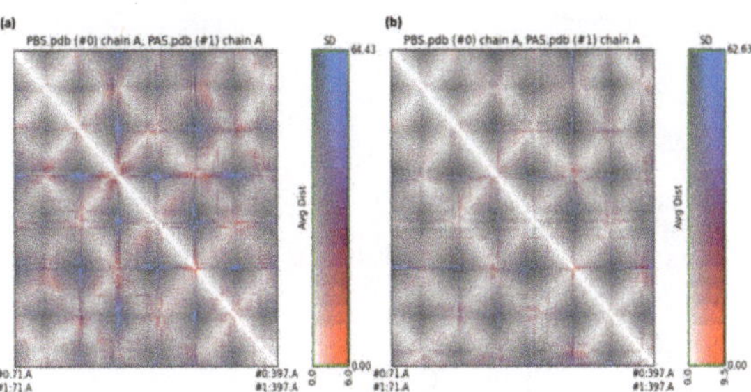

Figure 11. RR distance map: (**a**) unbounded dopamine receptor and dopamine receptor after simulation for HPLD and (**b**) unbounded dopamine receptor and dopamine receptor after simulation for CTtD, [PAS (#1) value = protein after simulation, SD = Standard deviation].

The average radius of gyration (Rg) values for HPLD and CTtD were 22.647 and 22.586 Å, respectively. Along the simulation time, Rg decreased, indicating that the structures became more compact (Figure 12).

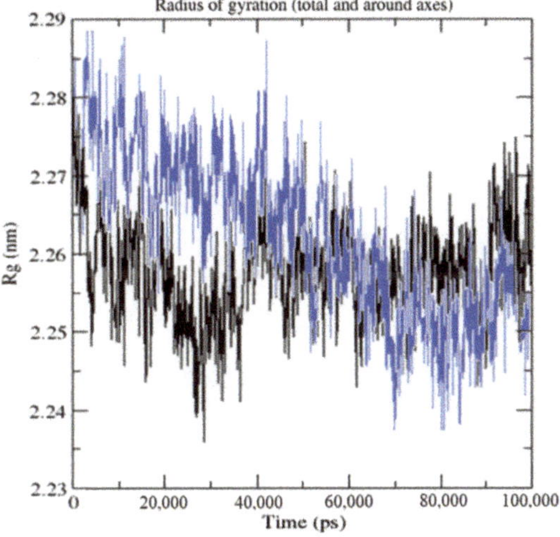

Figure 12. Radius of gyration for HPLD (black) and CTtD complexes (blue) during the 100-ns simulation time.

The number of hydrogen bond interactions between HPLD and CTtD was displayed against time using a grid-search on a $27 \times 18 \times 21$ grid with rcut = 0.35 (Figure 13). There were 396 atoms of donors and 753 atoms of acceptors detected when the hydrogen bonds were between the ligand at 36 and 53 atoms for HPL and CT complex, respectively, and 2941 atoms of the dopamine receptor were calculated. Out of a total of 201,657 potential bonds, the average number of hydrogen bonds per period for HPLD and CTtD was 1.738 and 2.481, respectively.

Overall, the receptor–protein interaction enhanced the number of hydrogen bonds substantially, and it was more in CTtD.

The values of solvent accessible surface area (SASA) altered as the ligand bound to the receptor (Figure 14). When the receptor binds to a ligand, the SASA value drops, indicating a change in conformation in the protein structure and a smaller pocket with more hydrophobicity around it.

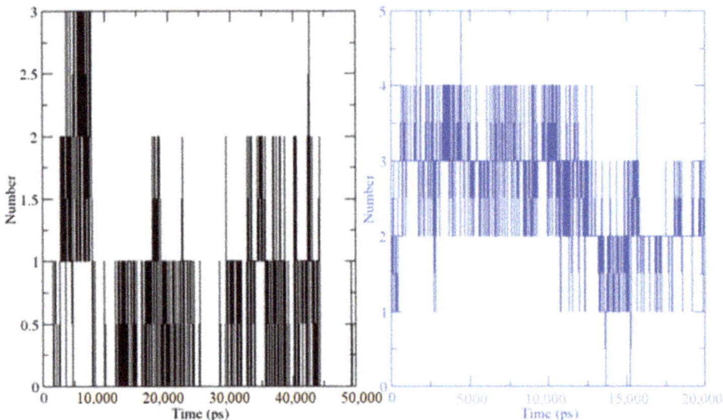

Figure 13. Number of average hydrogen bonding interactions between (left) HPLD complex and (right) **CTtD** complex during the 100 ns simulation time.

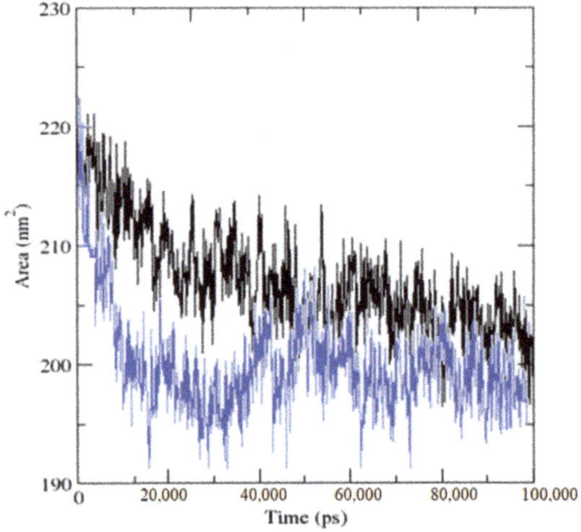

Figure 14. Solvent-accessible surface area analysis for HPLD complex (black) and CTtD complex (blue) during the 100 ns simulation time.

3. Materials and Methods

3.1. Synthesis of [(HPL)(PA)] and [(HPL)(TCN)] CT Complexes

Previously, synthesis and characterizations (elemental analyses, conductivities, electronic absorption spectra, infrared spectra, Raman laser spectra, 1H-NMR, DSC-TG thermograms, scanning electron microscopy, energy dispersive X-ray detection, and X-ray diffraction patterns) of the two solid HPL CT complexes were performed [19]. For preparation, 3 mmol of pure HPL drugin 20 mL CH_3OH was allowed to react with 3 mmol of each acceptor (PA and TCNQ) in 10 mL CHCl3 solvent, and both mixtures were stirred at room temperature for 45 min. Following this, the yellow and green solid complexes were isolated, washed three times with a minimum amount of CHCl3 solvent, and dried under vacuum over anhydrous $CaCl_2$.

3.2. Physical Measurements

A Perkin–Elmer Precisely Lambda 25 UV/Vis spectrometer using a 1 cm quartz cell was used to measure the electronic absorption spectra of charge transfer complexes produced in the presence of methanol at the range of 200–800 nm. The Jenway 4010 conductivity device was also used to measure the molar conductivity in the presence of newly prepared dimethyl sulfoxide solutions.

3.3. Molecular Docking

The structures of [(HPL)(PA)] and [(HPL)(TCNQ)] were obtained in PDBQT format using the OpenBabelIGUI tool (http://openbabel.org/wiki/Main Page; accessed 1 March 2022) [20]. The structure's energy was reduced by 500 steps utilizing the MMFF94 force field and conjugation of gradient optimization procedure using PyRx-Python prescription 0.8 [21]. The RCSB protein data library was used to get the 3D crystal structures of both receptors [22]. Natural bonding and other heterogeneous atoms were removed from both acceptors by BIOVIA Discovery Studio Visualizer. The Kollman charges of the receptor were calculated, and polar hydrogen atoms were placed into the receptor using the AutoDock tool [23]. The Geistenger method was used to assign partial charges. Docking calculations were performed using AutoDock Vina [24]. The resultant docked positions were examined to check the interactions using DS Visualizer.

3.4. MD Simulations

Simulations were conducted using processor Intel(R) Xeon(R) CPU E5-2680 v4 @ 2.40GHz, 64 bit. MD simulation was performed using the optimal receptor–ligand complex position and an evaluation of the conformational space and inhibitory potential. MD simulation analysis with the GROMOS96 43a1 force field was performed using the Gronningen machine for chemical simulations (GROMACS, version 2019.2 package).

Both ligands' parameter files and topologies were created using the latest CGenFF via CHARMM-GUI [25,26]. Online server CHARMM-gui was used to insert the dipalmitoylphosphatidylcholine (DPPC) membrane. Seventy two DPPC molecules were added to upperleaflet and lowerleaflet. SPC water models that extended 10 Å from the receptor were used to examine the receptor–ligand configurations in a rectangular box [27]. 37 K$^+$ and 46 Cl$^-$ ions (0.15 M salt) were administered to neutralize the systems and reproduce physiological salt concentrations (Figure 15).

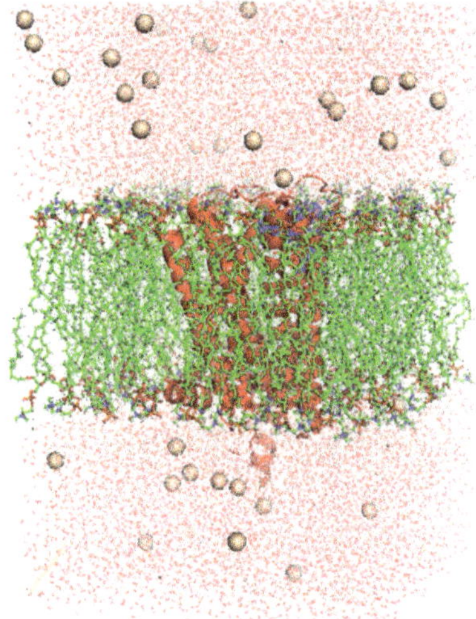

Figure 15. Lateral view of CTtD complex incorporated in dipalmitoylphosphatidylcholine (DPPC) membrane in a rectangular box solvated with water molecules and neutralized with 37 K$^+$ and 46 Cl$^-$ ions (0.15 M salt).

A constant temperature (300 K) and constant pressure (1.0 bar) over 100 ns were used for simulations using the leap-frog MD complement in the NPT/NVT equilibration run [28]. The steepest descent technique with 5000 steps was also used to reduce improper contact within the system [29]. Hydrogen bonds checked with the gmx hbond tool. Rg and SASA were calculated using the gmx gyrate and gmx sasa programs, respectively. The RMSD of protein was calculated using the gmx rms tools. Trajectory examination was accomplished using GROMACS program [30]. The plots were made with Grace Software, and the visualization was done with PyMol/VMD [8,31,32].

4. Conclusions

In the present research, we looked into how haloperidol (HPL) interacts with two key neurotransmitters (serotonin and dopamine) that are vital in mental health. These findings were compared with the synthesized charge transfer complexes of TCNQ and PA with HPL. The [(HPL)(TCNQ)] coupled with serotonin and dopamine more efficiently than HPL alone. Also, [(HPL)(TCNQ)]–dopamine has a higher binding energy value than HPL–dopamine. The molecular dynamic simulation at 100 ns demonstrated that the [(HPL)(TCNQ)]–dopamine complex had a more stable interaction with the dopamine receptor than the HPL–dopamine complex.

Author Contributions: Conceptualization, H.H., A.S.A. (Ahmed S. Alzahrani), H.A.A. and R.I.F.; methodology, S.S., M.S.R. and A.G.; software, S.S. and A.S.A. (Abdulhakeem S. Alamri); validation, M.A., W.F.A., A.A.A. and H.A.; formal analysis, M.A., W.F.A., S.S., M.S.R. and A.G.; investigation, A.A., O.A., A.S.A. (Abdulhakeem S. Alamri), B.M.R. and A.S.A. (Abdulhakeem S. Alamri); resources, M.S.R. and A.G.; data curation, M.A., S.S., M.S.R. and A.G.; writing—original draft preparation, M.S.R., S.S., M.A. and A.G.; writing—review and editing, H.A., A.A.A., M.A. and A.G.; visualization, H.H., H.A.A., A.S.A. (Ahmed S. Alzahrani) and R.I.F.; supervision, A.G.; project administration, B.M.R.; funding acquisition, M.A. and W.F.A. All authors have read and agreed to the published version of the manuscript.

Funding: This research was funded by the Deputyship for Research and Innovation, Ministry of Education in Saudi Arabia, grant number 1-441-120.

Institutional Review Board Statement: Not applicable.

Informed Consent Statement: Not applicable.

Data Availability Statement: All the data supporting reported results are available in the manuscript.

Acknowledgments: The authors extend their appreciation to the Deputyship for Research & Innovation, Ministry of Education in Saudi Arabia for funding this work through project number 1-441-120. Additionally, authors are appreciatively to Christian M. Nefzgar; Institute for Molecular Bioscience, The University of Queensland, Brisbane, QLD, Australia, for his technical assistance.

Conflicts of Interest: The authors declare no conflict of interest.

Sample Availability: Samples of all compounds are available from the corresponding authors.

References

1. Irving, C.B.; Adams, C.E.; Lawrie, S.M. Haloperidol versus placebo for schizophrenia. *Cochrane Database Syst. Rev.* **2004**, *4*, CD003082.
2. Mulliken, R.S. Structures of complexes formed by halogen molecules with aromatic and with oxygenated solvents. *J. Am. Chem. Soc.* **1950**, *72*, 600–608. [CrossRef]
3. Vinay, K.B.; Revanasiddappa, H.D.; Raghu, M.S.; Sameer, A.M.A.; Rajendraprasad, N. Spectrophotometric determination of mycophenolate mofetil as its charge-transfer complexes with two π-acceptors. *J. Anal. Methods Chem.* **2012**, *2012*, 8. [CrossRef] [PubMed]
4. Murugesan, V.; Saravanabhavan, M.; Sekar, M. Synthesis, spectroscopic characterization and structural investigation of a new CT complex of 2, 6-diaminopyridine with 4-nitrophenylacetic acid: Antimicrobial, DNA binding/cleavage and antioxidant studies. *Spectrochim. Acta A* **2015**, *147*, 99–106. [CrossRef]

5. Miyan, L.; Ahmad, A.; Alam, M.F.; Younus, H. Synthesis, single-crystal, DNA interaction, spectrophotometric and spectroscopic characterization of the hydrogen-bonded CT complex of 2-aminopyrimidine with π-acceptor chloranilic acid at different temperature in acetonitrile. *J. Photochem. Photobiol. B* **2017**, *174*, 195–208.
6. Singh, N.; Khan, I.M.; Ahmad, A.; Javed, S. Synthesis and dynamics of a novel proton transfer complex containing 3,5-dimethylpyrazole as a donor and 2,4-dinitro-1-naphthol as an acceptor: Crystallographic, UV-visible spectrophotometric, molecular docking and Hirshfeld surface analyses. *New J. Chem.* **2017**, *41*, 6810–6821. [CrossRef]
7. Bai, H.; Wang, Y.; Cheng, P.; Li, Y.; Zhu, D.; Zhan, X. Acceptor-donoracceptor small molecules based on indacenodithiophene for efficient organic solar cells. *ACS Appl. Mater. Interfaces* **2014**, *6*, 8426–8433. [CrossRef]
8. DeLano, W.L. The PyMOL Molecular Graphics System PyMOL DeLano Scientific. 2002. Available online: http://www.pymol.org (accessed on 1 March 2022).
9. Horowitz, H.H.; Metzger, G. A new analysis of thermogravimetric traces. *Anal. Chem.* **1963**, *35*, 1464–1468. [CrossRef]
10. Coats, A.W.; Redfern, J.P. Kinetic parameters from thermogravimetric data. *Nature* **1964**, *201*, 68–69. [CrossRef]
11. Mott, N.F.; Davis, E.A. *Electronic Process in Non-Crystalline Materials*; Calendron Press: Oxford, UK, 1972.
12. Hoffmann, M.R.; Martin, S.T.; Choi, W.; Bahnemann, D.W. Environmental applications of semiconductor photocatalysis. *Chem. Rev.* **1995**, *95*, 69–96. [CrossRef]
13. Fu, M.L.; Guo, G.C.; Liu, X.; Cai, L.Z.; Huang, J.S. Syntheses, structures, and properties of three selenoarsenates templated by transition metal complexes. *Inorg. Chem. Commun.* **2005**, *8*, 18–21. [CrossRef]
14. Adam, A.M.A.; Refat, M.S.; Hegab, M.S.; Saad, H.A. Spectrophotometric and thermodynamic studies on the 1:1 charge transfer interaction of several clinically important drugs with tetracyanoethylene in solution-state: Part one. *J. Mol. Liq.* **2016**, *224*, 311–321. [CrossRef]
15. Akram, M.; Lal, H.; Shakya, S.; Kabir-ud-Din. Multispectroscopic and computational analysis insight into the interaction of cationic diester-bonded gemini surfactants with serine protease α-chymotrypsin. *ACS Omega* **2020**, *5*, 3624–3637. [CrossRef]
16. Khan, I.M.; Shakya, S.; Islam, M.; Khan, S.; Najnin, H. Synthesis and spectrophotometric studies of CT complex between 1, 2-dimethylimidazole and picric acid in different polar solvents: Exploring antimicrobial activities and molecular (DNA) docking. *Phys. Chem. Liq.* **2021**, *59*, 753–769. [CrossRef]
17. Kufareva, I.; Abagyan, R. Methods of protein structure comparison. *Methods Mol. Biol.* **2012**, *857*, 231–257.
18. Wu, S.; Zhang, Y. A comprehensive assessment of sequence-based and template-based methods for protein contact prediction. *Bioinformatics* **2008**, *24*, 924–931. [CrossRef]
19. El-Habeeb, A.A.; Al-Saif, F.A.; Refat, M.S. CT complex of some nervous and brain drugs—Part 1: Synthesis, spectroscopic, analytical and biological studies on the reaction between haloperidol antipsychotic drugs with π-acceptors. *J. Mol. Struct.* **2013**, *1034*, 1–18. [CrossRef]
20. O'Boyle, N.M.; Banck, M.; James, C.A.; Morley, C.; Vandermeersch, T.; Hutchison, G.R. Open Babel: An open chemical toolbox. *J. Cheminform.* **2011**, *3*, 33. [CrossRef]
21. Dallakyan, S. *PyRx-Python Prescription*, Version 0.8; The Scripps Research Institute: La Jolla, CA, USA, 2008.
22. Chu, C.H.; Li, K.M.; Lin, S.W.; Chang, M.D.T.; Jiang, T.Y.; Sun, Y.J. Crystal structures of starch binding domain from *Rhizopus oryzae* glucoamylase in complex with isomaltooligosaccharide: Insights into polysaccharide binding mechanism of CBM21 family. *Proteins* **2014**, *82*, 1079–1085. [CrossRef]
23. Morris, G.M.; Goodsell, D.S.; Halliday, R.S.; Huey, R.; Hart, W.E.; Belew, R.K.; Olson, A.J. Automated docking using a Lamarckian genetic algorithm and an empirical binding free energy function. *J. Comput. Chem.* **1998**, *19*, 1639–1662. [CrossRef]
24. Trott, O.; Olson, A.J. AutoDock Vina: Improving the speed and accuracy of docking with a new scoring function, efficient optimization, and multithreading. *J. Comput. Chem.* **2010**, *31*, 455–461. [CrossRef]
25. Vanommeslaeghe, K.; Hatcher, E.; Acharya, C.; Kundu, S.; Zhong, S.; Shim, J.; Darian, E.; Guvench, O.; Lopes, P.; Vorobyov, I.; et al. CHARMM general force field: A force field for drug-like molecules compatible with the CHARMM all-atom additivebiological force fields. *J. Comput. Chem.* **2010**, *31*, 671–690. [CrossRef]
26. Yu, W.; He, X.; Vanommeslaeghe, K.; MacKerell, A.D., Jr. Extension of the CHARMM General Force Field to sulfonylcontaining compounds and its utility in biomolecular simulations. *J. Comput. Chem.* **2012**, *33*, 2451–2468. [CrossRef]
27. Jorgensen, W.L.; Chandrasekhar, J.; Madura, J.D.; Impey, R.W.; Klein, M.L. Comparison of simple potential functions for simulating liquid water. *J. Chem. Phys.* **1983**, *79*, 926–935. [CrossRef]
28. Allen, M.P.; Tildesley, D.J. *Computer Simulations of Liquids*; Clarendon Press: Oxford, UK, 1987.
29. Essmann, U.; Perera, L.; Berkowitz, M.L.; Darden, T.; Lee, H.; Pedersen, L.G. A smooth particle mesh Ewald method. *J. Chem. Phys.* **1995**, *103*, 8577–8593. [CrossRef]
30. Steinbach, P.J.; Brooks, B.R. New spherical-cutoff methods for long-range forces in macromolecular simulation. *J. Comput. Chem.* **1994**, *15*, 667–683. [CrossRef]
31. Humphrey, W.; Dalke, A.; Schulten, K. VMD: Visual molecular dynamics. *J. Mol. Graph.* **1996**, *14*, 33–38. [CrossRef]
32. Alhomrani, M.; Alsanie, W.F.; Alamri, A.S.; Alyami, H.; Habeeballah, H.; Alkhatabi, H.A.; Felimban, R.I.; Haynes, J.M.; Shakya, S.; Raafat, B.M.; et al. Enhancing the Antipsychotic Effect of Risperidone by Increasing Its Binding Affinity to Serotonin Receptor via Picric Acid: A Molecular Dynamics Simulation. *Pharmaceuticals* **2022**, *15*, 285. [CrossRef]

Article

Biophysical, Biochemical, and Molecular Docking Investigations of Anti-Glycating, Antioxidant, and Protein Structural Stability Potential of Garlic

Mohd W. A. Khan [1,*], Ahmed A. Otaibi [1], Abdulmohsen K. D. Alsukaibi [1], Eida M. Alshammari [1], Salma A. Al-Zahrani [1], Subuhi Sherwani [2], Wahid A. Khan [3], Ritika Saha [4], Smita R. Verma [4] and Nessar Ahmed [5]

[1] Department of Chemistry, College of Sciences, University of Ha'il, Ha'il 2440, Saudi Arabia; ahmed.alotaibi@uoh.edu.sa (A.A.O.); a.alsukaibi@uoh.edu.sa (A.K.D.A.); eida.alshammari@uoh.edu.sa (E.M.A.); s.alzahrane@uoh.edu.sa (S.A.A.-Z.)
[2] Department of Biology, College of Sciences, University of Ha'il, Ha'il 2440, Saudi Arabia; s.sherwani@uoh.edu.sa
[3] Department of Clinical Biochemistry, College of Medicine, King Khalid University, Abha 61412, Saudi Arabia; wkhan@kku.edu.sa
[4] Department of Biotechnology, Delhi Technological University, Delhi 110042, India; ritikasaha_bt20a16_19@dtu.ac.in (R.S.); smitar@dtu.ac.in (S.R.V.)
[5] Department of Life Sciences, Manchester Metropolitan University, Manchester M1 5GD, UK; n.ahmed@mmu.ac.uk
* Correspondence: mw.khan@uoh.edu.sa or wajidkhan11@gmail.com

Citation: Khan, M.W.A.; Otaibi, A.A.; Alsukaibi, A.K.D.; Alshammari, E.M.; Al-Zahrani, S.A.; Sherwani, S.; Khan, W.A.; Saha, R.; Verma, S.R.; Ahmed, N. Biophysical, Biochemical, and Molecular Docking Investigations of Anti-Glycating, Antioxidant, and Protein Structural Stability Potential of Garlic. *Molecules* **2022**, 27, 1868. https://doi.org/10.3390/molecules27061868

Academic Editors: Tanveer A. Wani, Seema Zargar and Afzal Hussain

Received: 8 February 2022
Accepted: 9 March 2022
Published: 14 March 2022

Publisher's Note: MDPI stays neutral with regard to jurisdictional claims in published maps and institutional affiliations.

Abstract: Garlic has been reported to inhibit protein glycation, a process that underlies several disease processes, including chronic complications of diabetes mellitus. Biophysical, biochemical, and molecular docking investigations were conducted to assess anti-glycating, antioxidant, and protein structural protection activities of garlic. Results from spectral (UV and fluorescence) and circular dichroism (CD) analysis helped ascertain protein conformation and secondary structure protection against glycation to a significant extent. Further, garlic showed heat-induced protein denaturation inhibition activity (52.17%). It also inhibited glycation, advanced glycation end products (AGEs) formation as well as lent human serum albumin (HSA) protein structural stability, as revealed by reduction in browning intensity (65.23%), decrease in protein aggregation index (67.77%), and overall reduction in cross amyloid structure formation (33.26%) compared with positive controls (100%). The significant antioxidant nature of garlic was revealed by FRAP assay (58.23%) and DPPH assay (66.18%). Using molecular docking analysis, some of the important garlic metabolites were investigated for their interactions with the HSA molecule. Molecular docking analysis showed quercetin, a phenolic compound present in garlic, appears to be the most promising inhibitor of glucose interaction with the HSA molecule. Our findings show that garlic can prevent oxidative stress and glycation-induced biomolecular damage and that it can potentially be used in the treatment of several health conditions, including diabetes and other inflammatory diseases.

Keywords: garlic; antioxidant; anti-glycation; glycation; AGEs; HSA

1. Introduction

AGEs are a complex and highly reactive group of heterogeneous compounds produced by glycation of proteins by reducing sugars [1,2]. Protein glycation has been shown to increase the accumulation of AGEs, the production and release of reactive oxygen species (ROS), and the structural and functional alteration of proteins, as well as resulting in tissue damage [3,4]. AGEs interact with particular receptors or bind proteins, activating a number of signaling pathways implicated in diabetic complications such as nephropathy, cataracts, Alzheimer's disease, and atherosclerosis, among others [5,6]. Glycation, AGEs, and oxidative stress have all been linked to a variety of health problems [7]. The states of total metabolic burden, chronic hyperglycemia, oxidative stress, and inflammation have been linked to an excessive buildup of AGEs [8].

In chronic inflammation, such as that observed in rheumatoid arthritis (RA), AGEs are produced and can accumulate in tissues [9]. Besides, AGEs build with age and promote pathologic stiffening of cartilage and extracellular matrix. Pentosidine is an AGE, and it is found in blood, synovial fluid, and articular cartilage of osteoarthritis patients [9]. AGEs are linked with endothelial activation and endothelial dysfunction [10]. As a result, AGEs have been suggested as an early biomarker for cardiovascular disease [11].

Intracellular ROS inhibit glucose consumption and cause oxidative alteration of intracellular proteins. Such exposure of protein can result in fragmentation, aggregation, oxidative phosphorylation, and unwanted interactions with ion channel-coupled receptors [12]. An excessive generation of free radicals/ROS leads to oxidative stress, which is involved in the development of inflammatory illnesses such as diabetes, cancer, cardiovascular disease, Parkinson's disease, Alzheimer's disease, and aging. Chronic inflammation is involved in tumor development. Tissue damage and endothelial dysfunction arise from increased ROS generation at the site of inflammation, leading to inflammatory diseases [13].

Inhibition of glycation has also been shown to be beneficial in the treatment of diabetic complications [14]. Synthetic chemicals are powerful anti-glycating agents, but they can also have serious side effects such as gastrointestinal problems, uncommon vasculitis, anemia, flu-like symptoms, nausea, and diarrhea. Thus, a lot of interest has been focused on finding natural plant phytochemicals that efficiently prevent glycation and produce fewer adverse side effects. Traditional medicine practitioners frequently employ medicinal plants and natural products in their everyday practice to treat a variety of illnesses because they are generally non-toxic, cheap, ingestible, and have fewer adverse effects. [4]. The medicinal properties of garlic (*Allium sativum*) were well documented in Sanskrit literature 5000 years ago, and its use in Chinese medicine was also reported as far back as 3000 years ago. The healing properties of garlic were well known in the ancient world and utilized by the Egyptians, Babylonians, Greeks, and Romans [15]. Louis Pasteur documented the antibacterial activity of garlic in 1858. Due to its healing potential and antiseptic properties, it was used to treat and prevent gangrene during the Second World War [16]. The overall benefits of garlic in maintaining good health and in preventing a range of health issues has shifted the focus of modern-day medicine back to a time-tested natural remedy. The therapeutic roles of garlic are supported by modern-day epidemiologic evidence, with studies indicating the benefits of garlic preparations in having antioxidant and antimicrobial effects and in reducing diabetes, cardiovascular disease, and cancer [17,18]. This study focused on investigating the anti-glycation and antioxidative stress activities of garlic extract, as well as conducting an in-depth study of the secondary structural alterations of HSA proteins and how garlic extract metabolites might inhibit the glycation reactions in addition to inhibiting secondary structure alterations. This study provides a platform for the future studies in this direction, which may potentially elucidate physiological and immunological imbalances.

2. Results

2.1. Biochemical Analysis

Phytochemical Screening

The chemical and biological properties detected in preliminary screening of aqueous solution of garlic extract are shown in Table 1. The presence of flavonoids was determined using an alkaline reagent test. $FeCl_3$ test was used to estimate phenolic content. The total phenolic compounds in garlic extract were determined to be 21.45 ± 0.02 mg gallic acid equivalent/g dry weight of the extract. Phenolic content serves as a means of protection against both infection and oxidative stress. Calculation of total flavonoid content of the extract was performed by $AlCl_3$ test, with quercetin as a reference. Thus, total flavonoid content was found to be 16.58 ± 0.03 mg quercetin equivalents (QE)/g dry weight of the extract.

Table 1. Chemical and biological studies of aqueous garlic extract.

Preliminary Screening	Garlic Extract
Weight of dry powder	50 g
Yield	5.19%
Extract	Aqueous
Flavonoids	+
Polyphenolic compounds	+
Total phenolic compounds	21.45 ± 0.02 mg gallic acid equivalent/g dry weight of extract
Total flavonoid content	16.58 ± 0.03 mg quercetin equivalent/g dry weight of the extract

+ sign indicates presence of flavonoids and polyphenolic compounds.

2.2. Antioxidant and Free Radical Scavenging Activities of Garlic Extract

2.2.1. Assay for Ferric Reducing Antioxidant Power (FRAP)

Using the FRAP test and ascorbic acid as a standard reference, we evaluated the reducing capability of garlic extract. The reduction of Fe^{+3} to Fe^{+2} by the extract is the basis for this test. At 700 nm, the solution of ascorbic acid (0–100 µg/mL) followed Beer's Law with a regression coefficient (R^2) of 0.9973 and a slope (m) of 0.004. For this figure, the intercept was 0.0216. The standard curve's equation is y = 0.004x + 0.0216 (not given). Garlic extract had FRAP values of 32.41 ± 0.86 g ascorbic acid/100 mg dry weight of extract. Garlic extract was found to have improved ferric reducing power in a dosage-dependent manner (Figure 1).

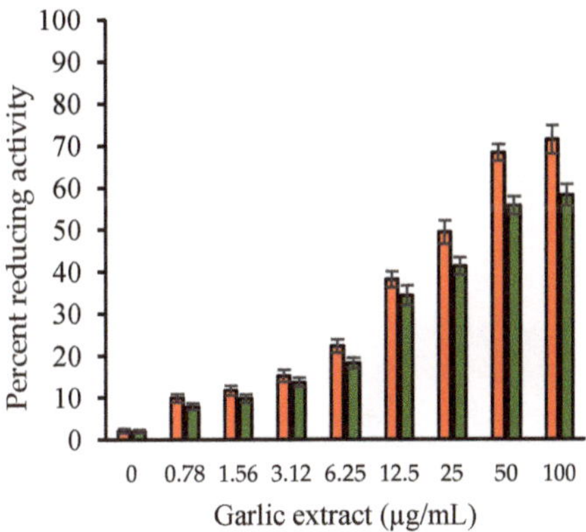

Figure 1. Percentage reducing power of ascorbic acid (red) and aqueous garlic extract (green). Samples in the histogram showed varying concentrations of ascorbic acid and garlic extract (0–100 µg/mL). The *y* axis shows the corresponding percentage reducing power. The results are presented as means ± SEM (*n* = 3). All the results (0.78–100 µg/mL garlic extract) were statistically significant compared with the sample without extract (0 µg/mL). Comparison between two groups was performed based on *t* test, and significance was defined as $p < 0.05$.

2.2.2. 2,2-Diphenyl-1-Picrylhydrazyl (DPPH) Radical Scavenging Assay

The DPPH free radical scavenging method is a widely used method for determining antioxidant capacity of various compounds. Figure 2 depicts the DPPH radical scavenging capabilities of aqueous garlic extract. Garlic extract exhibits a substantial DPPH scavenging activity that was observed to rise as extract concentrations were raised from 0–100 µg/mL. The DPPH scavenging activity of 100 µg/mL extract was 66.18%. However, ascorbic acid showed 39.59% at 100 µg/mL.

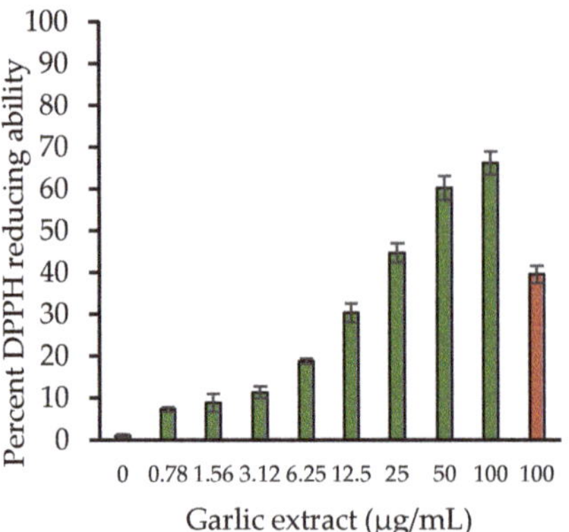

Figure 2. Percentage of free radical reduced vs. aqueous garlic extract concentrations. Various concentrations (0–100 µg/mL) of garlic extracts (green) and 100 µg/mL of ascorbic acid (red). The results are presented as means ± SEM ($n = 3$). All the results (0.78–100 µg/mL garlic extract) were statistically significant compared with the sample without extract (0 µg/mL). Comparison between two groups was performed based on t test, and significance was defined as $p < 0.05$.

2.3. Inhibition of Structural Changes by Garlic Extract

2.3.1. Protein Denaturation Inhibition

Inhibition in in vitro HSA protein denaturation was examined using garlic extracts. Tissue proteins have been reported to be denatured by inflammatory and oxidative reactions. Natural products were used to evaluate their potential to protect proteins from denaturation and would be employed as an anti-inflammatory dietary source. Garlic extract (50 µg/mL) inhibited heat-induced albumin denaturation at a higher percentage of 50.66% (Figure 3).

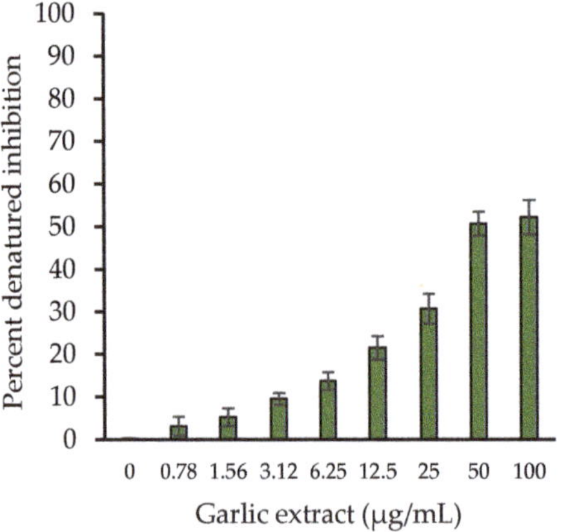

Figure 3. Percentage protection from denaturation induced by heat vs. garlic extract concentration (0–100 µg/mL). The results are presented as means ± SEM ($n = 3$). All the percentage denaturation inhibition results (0.78–100 µg/mL garlic extract) were statistically significant compared with the sample without extract (0 µg/mL). Comparison between two groups was performed based on t test, and significance was defined as $p < 0.05$.

2.3.2. Protein Browning Inhibition

Browning intensity is used to assess a product's capacity to defend against glycation. The effectiveness of extract to protect against glycation was assessed by measuring the percentage intensity of browning of HSA treated with glucose. Protein glycation causes an increase in browning intensity under hyperglycemic conditions. At 420 nm, we examined the degree of browning in samples. HSA incubated with glucose without extract had the maximum browning intensity. For this sample, we used a browning intensity of 100 percent. However, when HSA was incubated with glucose in the presence of extract, there was a noticeable reduction in browning intensity (65.23%) at a concentration of 50 µg/mL (Figure 4). In addition, in the case of HSA kept without glucose and extract, some browning (5%) was seen. Internal structural changes that occur over time might be one reason for this sample. Our hypothesis that extracts prevent the development of glycated products is supported by these findings.

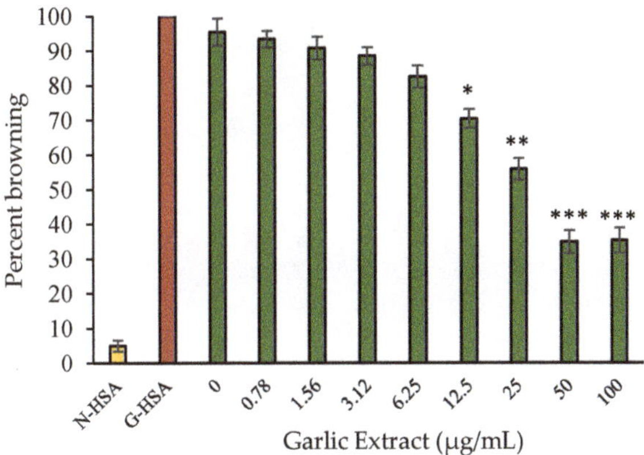

Figure 4. Percentage browning vs. concentration of garlic extract. Native HSA (N-HSA) and glycated HSA (G-HSA) were incubated for 10 weeks under similar conditions and are considered as negative and positive controls, respectively. G-HSA was co-incubated with garlic extract at various concentrations (0–100 µg/mL) of garlic extract (green). The results are presented as means ± SEM (n = 3). All the percentage browning inhibition results (0.78–100 µg/mL garlic extract) were compared with G-HSA sample. Comparison between two groups was performed based on t test, and significance was defined as * $p < 0.05$, ** $p < 0.01$, *** $p < 0.001$.

2.3.3. Inhibition in Protein Aggregate Formation

Glycation causes the binding of carbonyl groups to proteins, resulting in the creation of protein molecules clustering, which is also known as protein aggregation. As a result of glycation, protein aggregates formation occurs. In one of our previous studies, we showed aggregation occurs due to protein glycation [1]. Garlic extract had a beneficial effect on the inhibition of protein aggregation. The addition of extract lowered the aggregation index of G-HSA in a concentration-dependent manner (Figure 5). At a concentration of 100 µg/mL, the extract exhibited the levels of aggregation (67.77%).

2.3.4. Amyloid Structure Inhibition

Congo red (CR) dye is used to determine how much of a protein's secondary structure has been altered. CR contacts hydrophobic clefts located in between beta fibrils and has a unique ability to bind to the protein sheet structure. After binding, the CR dye has a specific absorbance of 530 nm. In Figure 6, the findings of the CR binding experiment showed reduction in cross amyloid structure formation (33.26%) compared with positive controls (G-HSA). By masking the sites of glycation and limiting the surface area accessible by solvent, garlic extract reduced HSA fibrillation and perhaps prevented the transition from α-helix to β-sheet.

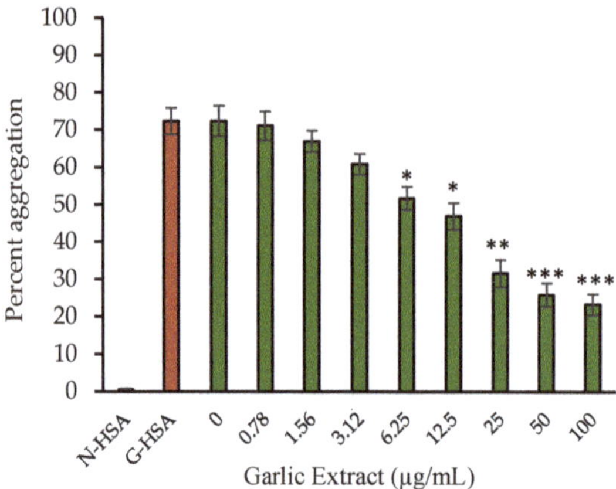

Figure 5. Percentage aggregation vs. concentration of garlic extract. Native HSA and G-HSA were incubated for 10 weeks and served as negative and positive controls, respectively. G-HSA was incubated with various concentrations (0–100 µg/mL) of garlic extract (green). The results are presented as means ± SEM (n = 3). All the percentage inhibition of protein aggregation results (0.78–100 µg/mL garlic extract) were compared with G-HSA sample. Comparison between two groups was performed based on t test, and significance was defined as * $p < 0.05$, ** $p < 0.01$, *** $p < 0.001$.

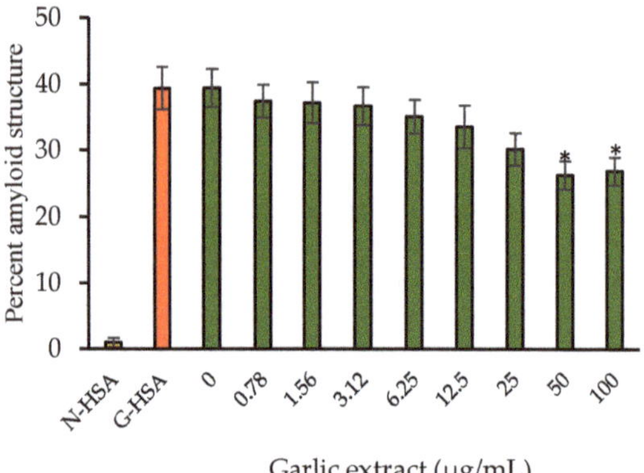

Figure 6. Percentage amyloid structure vs. concentration of garlic extract. Native HSA and G-HSA were incubated for 10 weeks as negative and positive controls, respectively. G-HSA was incubated with various concentrations (0–100 µg/mL) of garlic extract (green). The results are presented as means ± SEM (n = 3). All the percentage amyloid structure inhibition results (0.78–100 µg/mL garlic extract) were compared with G-HSA sample. Comparison between two groups was performed based on t test, and significance was defined as * $p < 0.05$.

2.3.5. Spectral Studies

UV spectral studies were conducted for all the protein samples. HSA incubated with glucose has been reported to have significant decrease in absorption peak at 280 nm as compared with the native HSA, indicating considerable hypochromicity. The observed hypochromicity of glycated sample at 280 nm might be attributable to an alteration in the protein microenvironment and the change in aromatic amino acids. Glycated HSA samples incubated with the extract (50 µg/mL) showed significantly increased UV absorption (Figure 7).

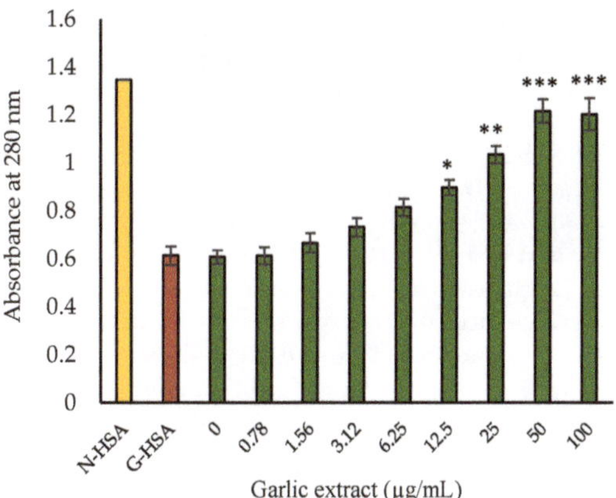

Figure 7. Absorbance of G-HSA vs. concentration of garlic extract. Native HSA and G-HSA were incubated for 10 weeks as negative and positive controls, respectively. G-HSA was incubated with various concentrations (0–100 µg/mL) of garlic extract (green). The results are presented as means ± SEM (*n* = 3). Change in absorbance results (0.78–100 µg/mL garlic extract) were compared with G-HSA sample. Comparison between two groups was performed based on *t* test, and significance was defined as * $p < 0.05$, ** $p < 0.01$, *** $p < 0.001$.

The formation of AGEs in the glycated samples was detected using AGE pentosidine-specific autofluorescence. For each sample, the specific fluorescence of AGEs was observed to be in the range of 400–480 nm. The fluorescence intensity vs. wavelength (400–480 nm) spectra were found to be rather wide (data not shown), which might be attributable to the variety of fluorescent molecules created during the glycation process. Extract containing samples showed a steady reduction in AGE-specific fluorescence at 450 nm as the extract concentration increased in the incubated samples (Figure 8). The fluorescence intensity of HSA treated with glucose was maximum at 450 nm. Our data show that garlic extract exhibited protection against AGEs synthesis by glycation. Protection level increased as the concentration of garlic extract increased.

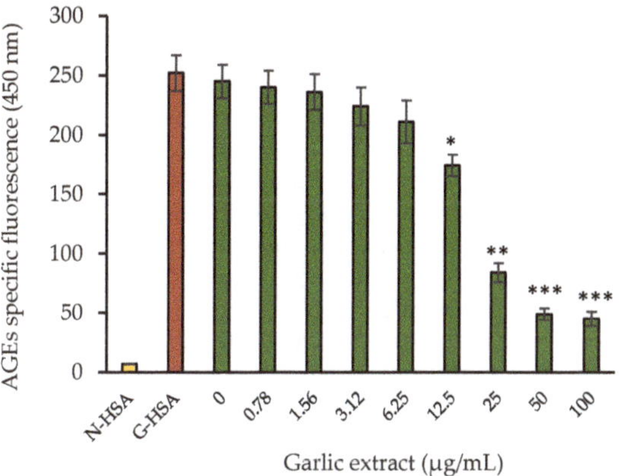

Figure 8. AGE-specific fluorescence intensity at 450 nm vs. concentration of garlic extract. Native HSA and G-HSA were incubated for 10 weeks as negative and positive controls, respectively. G-HSA was incubated with various concentrations (0–100 µg/mL) of garlic extract (green). The results are presented as means ± SEM (*n* = 3). Reduction in AGE fluorescence results (0.78–100 µg/mL garlic extract) were compared with G-HSA sample. Comparison between two groups was performed based on *t* test, and significance was defined as * $p < 0.05$, ** $p < 0.01$, *** $p < 0.001$.

2.3.6. Circular Dichroism

Analysis of secondary structure change with glycating agent 'glucose' alone or together with garlic extract with varying concentrations (0–100 µg/mL) was investigated using a Jasco J 810 spectropolarimeter (Table 2). The presence of secondary structural elements was estimated as relative percentages by using the Chen and Yang equation [19] via a computer data processor. CD spectra were recorded for all the samples at a similar wavelength range (200 to 280 nm). As we have shown previously, there is a significant change in α-helix (-10.5%) and β-sheet ($+14.9\%$) structures upon glycation of HSA [1]. Inhibition in the change of both α-helix (-2.8%) and β-sheet ($+4.6\%$) structures of glycated samples were observed when incubated with garlic extract of 100 (µg/mL). Significant decreases in all the secondary structure (α-helix ($p < 0.05$), β-sheet ($p < 0.01$), β-turns ($p < 0.01$), and random coils ($p < 0.05$)) changes in glycated HSA samples were observed at 25 µg/mL concentration of garlic extract. Furthermore, the highest reduction in all the secondary structures was achieved at 100 µg/mL concentration of extract used (Table 2).

Table 2. Secondary structure composition of N-HSA, G-HSA, and G-HSA incubated with different concentrations of garlic extract (0–100 µg/mL).

Conformation	N-HSA	G-HSA	G-HSA with Garlic Extracts (µg/mL)								
	-		0.78	1.56	3.12	6.25	12.5	25	50	100	AG
α-helix	42.7 ± 0.6	38.2 ± 0.3 (-10.5%)	38.2 ± 0.3 (-10.5%)	38.3 ± 0.3 (-10.3%)	38.4 ± 0.3 (-10.0%)	38.8 ± 0.3 (-9.1%)	39.8 ± 0.3 * (-6.8%)	40.5 ± 0.3 * (-5.2%)	40.9 ± 0.3 ** (-4.2%)	41.5 ± 0.3 *** (-2.8%)	40.8 ± 0.3 ** (-4.4%)
β-sheet	26.2 ± 0.5	30.1 ± 0.2 ($+14.9\%$)	30.1 ± 0.2 ($+14.9\%$)	30.0 ± 0.2 ($+14.5\%$)	29.9 ± 0.2 ($+14.1\%$)	29.6 ± 0.2 ($+13.0\%$)	28.8 ± 0.2 * ($+9.9\%$)	28.1 ± 0.2 ** ($+7.3\%$)	27.9 ± 0.2 ** ($+6.5\%$)	27.4 ± 0.2 *** ($+4.6\%$)	27.8 ± 0.2 ** ($+6.1\%$)
β-turn	18.5 ± 0.2	19.4 ± 0.3 ($+4.9\%$)	19.4 ± 0.3 ($+4.9\%$)	19.4 ± 0.3 ($+4.9\%$)	19.4 ± 0.3 ($+4.9\%$)	19.2 ± 0.3 ($+4.3\%$)	19.1 ± 0.3 * ($+3.2\%$)	19.0 ± 0.3 ** ($+2.7\%$)	18.8 ± 0.3 ** ($+1.6\%$)	18.6 ± 0.3 ** ($+0.5\%$)	19.0 ± 0.3 ** ($+2.7\%$)
Random coil	12.6 ± 0.5	12.3 ± 0.4 (-2.4%)	12.3 ± 0.4 (-2.4%)	12.3 ± 0.4 (-2.4%)	12.3 ± 0.4 (-2.4%)	12.3 ± 0.4 (-2.4%)	12.3 ± 0.4 (-2.4%)	12.4 ± 0.4 * (-1.6%)	12.4 ± 0.4 * (-1.6%)	12.5 ± 0.4 *** (-0.8%)	12.4 ± 0.4 * (-1.6%)

The values are in percentage. Each sample was read in triplicate. Data are mean ± standard deviation. * $p < 0.05$, ** $p < 0.01$, and *** $p < 0.001$ vs. control (N-HSA). Values in parentheses represent the percentage change in the secondary structure from N-HSA. Percentage decrease and increase are denoted by "−" and "+" signs. Different GE concentrations were used in µg/mL. The 5 mm of AG was used as negative control. The t test was adopted for the comparison between the groups.

2.4. Molecular Docking Studies for Potential Natural Product Metabolites as Inhibitors of Glycation

Some of the important phenolic compounds present in garlic extracts were analyzed for their inhibitory role in the glycation reaction. Sudlow Site I is the primary binding site for glucose in pyranose form. The interaction of the glucose ring form and other ligands (considered in this study) with the SAL subsite of the Sudlow Site I are summarized in Table 1. Figures 9 and 10 provide insight into the most stable 2D and 3D docked conformations of the ligands at the SAL subsite obtained via exhaustive molecular docking application. The orientation of the glucopyranose at the SAL subsite was obtained from the crystal of glucose-bound has submitted by Wang et. al. [20] to the protein data bank (PDB ID: 4iw2). The binding efficiency reported in Table 3 is a result of multiple interactions of the ligands with the target active site.

The blue dotted interaction of Arg257 with glucopyranose (Figure 9a) represents an altered confirmation state of glucopyranose. The 2D interaction images in Figures 9 and 10 represent the type and count of interactions specific ligand undergoes. The 3D images of docked ligands in Figures 9 and 10 represent the polarity distribution at the docking site. The pink environment in the close vicinity of the ligands represents hydrogen donors or an electronegative environment; however, the green cloud near the ligands is due to amino acid with a side chain rich in hydrogen acceptors and results in an electropositive environment. The white cloudy areas around the ligands constitute the neutral space due to the presence of hydrophobic amino acids.

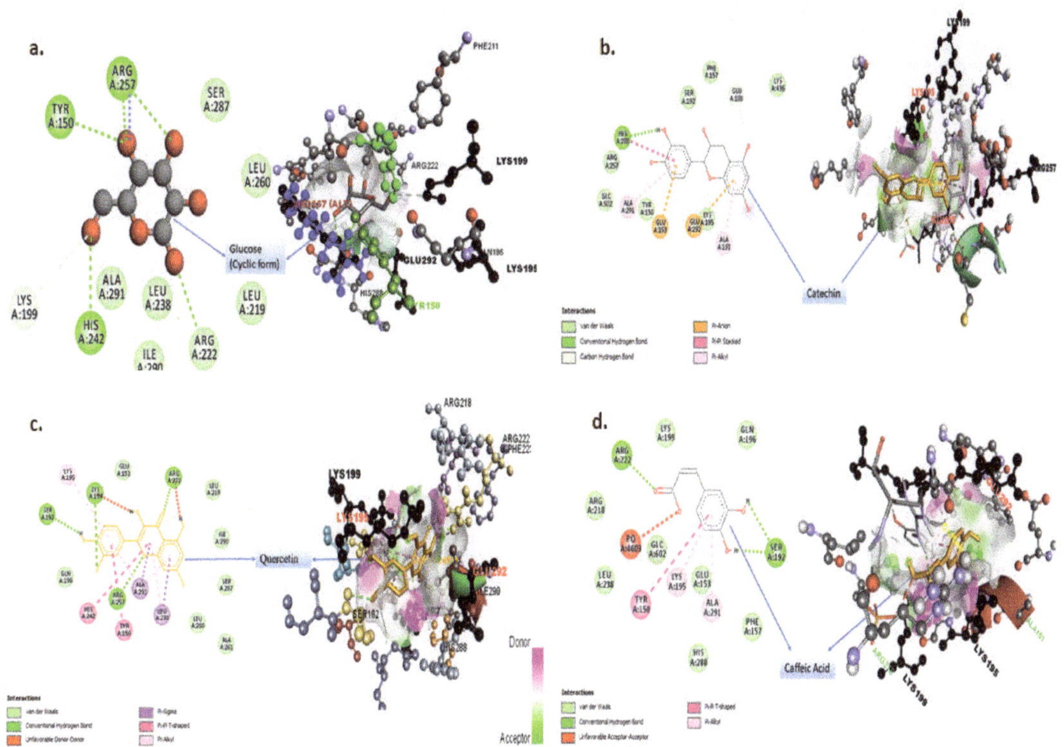

Figure 9. 2D and 3D interaction models of ligands with HSA. Ligands used are (**a**) glucose, (**b**) catechin, (**c**) quercetin, and (**d**) caffeic acid.

Table 3. Ligands. Common name refers to the compound name used in the study.

Ligand/Inhibitor Name (Common Name)	IUPAC Name, Mol Formula, and PubChem ID	Binding Affinity Kcal/Mol	Number of Hydrogen Bonds	Other Interactions *
Glucose (Cyclic form)	(3R,4S,5S,6R)-6-(hydroxymethyl)oxane-2,3,4,5-tetrol Mol formula: $C_6H_{12}O_6$ PubChem ID: 5793	−6.2	6 (3 × Arg257, 1 × Arg221, 1 × Tyr150, 1 × His242)	1 (1 × Lys199) Leu238 and Ala291 (Hydrophobic interactions)
Catechin	(2S,3R)-2-(3,4-dihydroxyphenyl)-3,4-dihydro-2H-chromene-3,5,7-triol Mol formula: $C_{15}H_{14}O_6$ PubChem ID: 73160	−6.7	1 (1 × His 288)	13 (1 × Arg 257, 1 × His 288, 1 × Glc 602, 1 × Ala 291, 1 × Tyr 150, 1 × Glu 153, 1 × Glu 292, 1 × Lys 195, 1 × Ala 191, 1 × Phe 157, 1 × Ser 192, 1 × Lys 436, 1 × Glu 188 Carbon H Bond)

Table 3. *Cont.*

Ligand/Inhibitor Name (Common Name)	IUPAC Name, Mol Formula, and PubChem ID	Binding Affinity Kcal/Mol	Number of Hydrogen Bonds	Other Interactions *
Caffeic Acid	(E)-3-(3,4-dihydroxyphenyl)prop-2-enoic acid Mol formula: $C_9H_8O_4$ PubChem ID:689043	−6.6	3 (1 × Arg 222, 2 × Ser 192)	12 (1 × Arg 218, 1 × Po 4603, 1 × Glc 602, 1 × Leu 238, 1 × Tyr 150, 1 × Lys 195, 1 × Glu 153, 1 × Ala 291, 1 × His 288, 1 × Phe 157,1 × Gln 196, 1 × Lys 199)
Gallic Acid	3,4,5-trihydroxybenzoic acid Mol formula: $C_7H_6O_5$ PubChem ID: 370	−6.4	4 (2 × Lys 199, 1 × Gln 196, 1 × Glu 153)	10 (1 × His 242, 1 × Gln 196, 1 × Ser 192, 1 × Lys 195, 1 × Tyr 150, 1 × Ala 291, 1 × Phe 157, 1 × His 288, 1 × Glu 292, 1 × Arg 257)
m-Coumaric Acid	E)-3-(3-hydroxyphenyl)prop-2-enoic acid Mol formula: $C_9H_8O_3$ PubChem ID: 637541	−6.3	4 (1 × Arg 222, 1 × PO 4603, 1 × Glc 602, 1 × Ser 192)	6 (1 × Arg 222, 1 × PO 4603, 1 × Glc 602, 1 × Ser 192, 1 × Gln 196 Unfavorable donor–donor, 1 × Gln 196 Unfavorable donor–donor)
Quercetin	2-(3,4-dihydroxyphenyl)-3,5,7-trihydroxychromen-4-one Mol formula: $C_{15}H_{10}O_7$ PubChem ID: 5280343	−8.1	4 (1 × Ser 192, 1 × Lys 199, 1 × Arg 257, 1 × Arg 222)	17 (1 × Lys 195, 1 × Lys 199, 1 × Gln 196, 1 × His 242, 1 × Tyr 150, 3 × Ala 291, 2 × Leu 238, 1 × Leu 260, 1 × Ala 261, 1 × Ser 287, 1 × Ile 290, 1 × Leu 219,1 × Arg 222, 1 × Glu 153)
Pyrogallol	benzene-1,2,3-triol Mol formula: $C_6H_6O_3$ PubChem ID: 1057	−5.4	3 (2 × Ser 192, 1 × Gln 196)	8 (1 × Tyr 150, 1 × Glu 153, 1 × Phe 157, 1 × Lys 199, 1 × Ala 291, 1 × Lys 195, 1 × His 242, 1 × Arg 257)
Dihydroxybenzoic acid	2,3-dihydroxybenzoic acid Mol formula: $C_7H_6O_4$ PubChem ID: 19	−6.2	3 (1 × Arg 222, 1 × Ser 287, 1 × Arg 257)	9 (1 × Leu 260, 1 × Ile 290, 1 × Leu 238, 1 × Ala 291, 1 × Arg 257, 1 × Leu 219, 1 × Ile 264, 1 × Ala 261, 1 × Tyr 150)

* Van der Waals, polar, pi–pi interactions, carbon–hydrogen bonds, pi–sigma, pi–alky, etc.

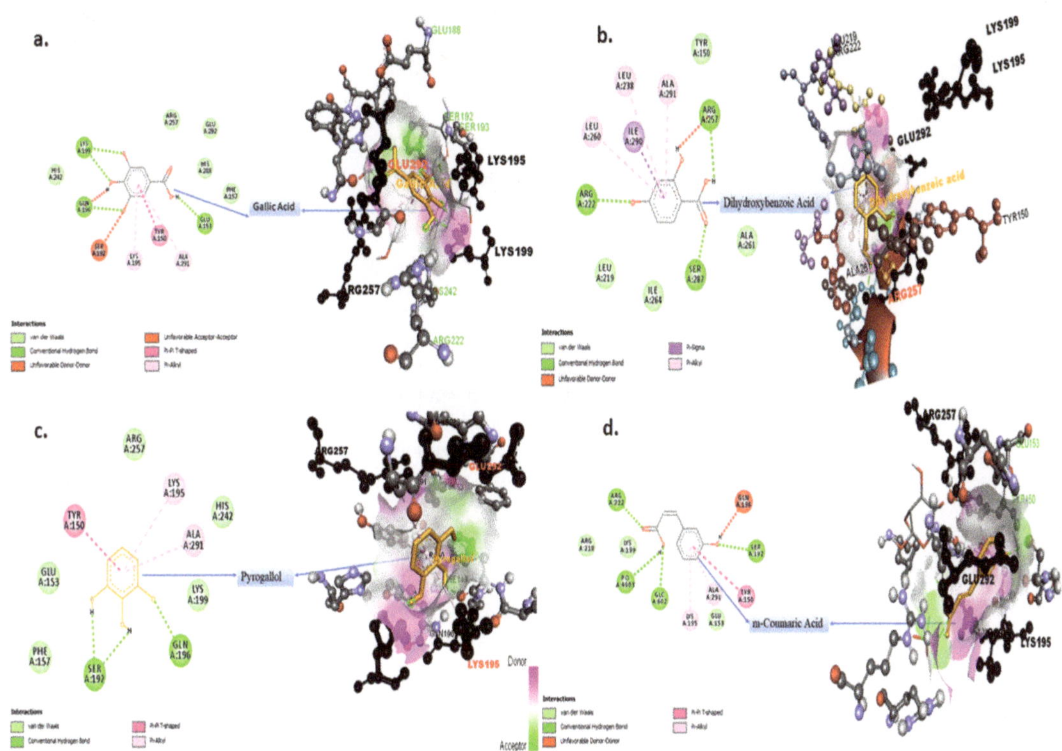

Figure 10. 2D and 3D interaction models of ligands with HSA. Ligands used are (**a**) gallic acid, (**b**) dihydroxy benzoic acid, (**c**) pyrogallol, and (**d**) m-coumaric acid.

3. Discussion

The core cause of many lethal diseases is thought to be connected to metabolic disturbances and inflammatory changes. Diabetes mellitus, characterized by hyperglycemia, is a severe health concern that affects people all over the world [21]. Long-term hyperglycemia has been linked to diabetes complications and biomolecule glycation [22–24].

Nutrition looks at how people might use their food choices to reduce their disease risk and manage their illnesses. If a person's diet lacks the right nutritional balance, they are more prone to develop a variety of health problems. When a person consumes excess or very little amounts of a nutrient, it might cause sickness. A balanced diet, according to growing data, may help you avoid problems including heart disease, cancer, osteoporosis, and type 2 diabetes [25].

Garlic has been reported to have potential therapeutic properties due to its biochemical constituents. Garlic extract can be used against glycation and AGE-related health complications linked with chronic diseases in diabetic patients due to its broad therapeutic potential [26,27]. Compounds such as quercetin, pyrogallol, caffeic acid, gallic acid, m-coumaric acid, and their derivatives are among the beneficial chemicals found in garlics. A high quantity of quercetin has been found in garlic, which is a potent antioxidant compound [28]. The antioxidant capabilities of garlic were well studied in the present investigation through various in vitro methods.

Biochemical analysis of this study showed the presence of phenolic compounds and flavonoids in a respectable amount in garlic extract [28]. According to the current findings, methanolic extract of garlic has a considerable potential to scavenge free radicals, as estimated using DPPH. Garlic also showed high reducing activity that increased with higher concentrations of the extract. It is generally known that the breakdown of H_2O_2 produces hydroxyl free radicals in the blood. The strong antioxidant properties are attributed to the polyphenolic compounds and polysaccharides. The polysaccharides and polyphenolic compounds in garlic might prevent free radical formation, making it an effective antioxidant.

Lowering chronic inflammation may help to postpone, prevent, and possibly treat a variety of chronic illnesses, including cancer [29]. Besides, many ROS, free radicals including NO, superoxide (O_2^-), and their reaction product peroxynitrite ($ONOO^-$) are produced in excessive quantities during the host's response to infections and inflammatory circumstances [16]. Furthermore, oxidative stress and inflammation are connected in glycation pathways. Identifying alternatives to non-steroidal anti-inflammatory drugs and developing innovative, effective, and safe anti-inflammatory medicines has long been a key priority. External stress as well as several compounds may cause protein denaturation that leads to the loss of structural integrity of proteins and thus loss of their functions [30,31]. Our results suggest that garlic extract can inhibit the heat-induced denaturation of HSA.

Several diseases such as familial amyloidosis, Alzheimer's, pancreatic islet amyloidosis, etc. are caused by protein aggregation in the circulation and in organs [32]. These aggregates undergo further reactions and form amyloid fibrils that contain cross beta structures. In glycation reactions, reducing carbohydrates non-enzymatically and covalently binds to lysine and arginine groups of proteins as well as with the N terminus of polypeptides [33]. Increased levels of these aggregates may cause neurological degeneration. It has been evident that protein glycation can induce aggregation. Intense browning was observed in the polyacrylamide gel electrophoresis glycated samples of albumin [1,33].

Glycation-induced microenvironment structural alterations were observed through spectral studies that included UV spectra and AGE-specific fluorescence. These alterations were inhibited when garlic extract was present in the reaction mixture during the incubation of the glycation reaction. However, a significant inhibition in microenvironment alterations was observed at higher concentrations of the extract. Consequently, extracts lent protection from structural alterations to the protein.

Further in-depth analysis of structural alterations in glycated proteins, including the inhibitory effect of garlic extract on these changes, was investigated using CD. CD analysis of glycated HSA showed protein destabilization and reduction in α-helix structure [34]. It has been reported that changes in CD spectra of glycated HSA are based on glucose concentration. It has also been observed that upon HSA glycation there was a partial denaturation with alterations in structural integrity at different glucose concentrations (1 mg/mL and 5 mg/mL) [35]. Some authors showed that glycated albumin was more favored in a β-sheet conformation structure [36]. These previous studies showed coherence with our CD results of native and glycated HSA. Moreover, CD experimental analysis revealed that garlic extract provided protection of protein secondary structure alterations and inhibited the conversion of α-helix to β-pleated sheet structure. There are possibilities of the interaction of the metabolites or compounds present in garlic extract (Table 3) with the glycation sites on the HSA molecule, causing inhibition in secondary structure alterations (conversion of a-helix to β-sheet). These inhibitions in structural alteration of glycated HSA are highly important for their functional integrity of the protein.

Sudlow Site I is made up of three subsites: the SAL subsite, deep-seated at the bottom of the Sudlow Site I and comprised of hydrophobic residues Leu-238 and Ala-291 and the indomethacin (IMD) and 3'-azido-3'-deoxythymidine (ADT) subsites situated near the opening of the Sudlow Site I are rich in positively charged residues, Arg-218, Lys-195, and Glu-292 [20]. In blood plasma, the D-glucose found in blood plasma is a mixture of two anomers—i.e., α-D-glucopyranose and β-D glucopyranose [37]. The mutarotation between the open aldehyde chain form and ring form is quick and dependent upon medium conditions. Thus, glucose can potentially react, as an open aldehyde form or closed ring form, within a short time frame [37]. Wang Yu et al. reported the mechanism of glucose interaction with HSA [20]. Two molecules of glucopyranose were found to interact subsequently at the Sudlow Site I [20]. The first molecule gets bottom deep-seated at the SAL subsite, and this configuration is stabilized by the hydrophobic interactions involving Leu238 and Ala291 (Figure 9a). The second glucopyranose molecule is held at the entrance of the Sudlow Site I marked by IMD and AZT subsites. This region is rich in positively charged residues Lys-195, Arg-218, and Glu-292. The oxygen atom bound to C5 (carbon 5 of the second glucopyranose molecule) is attacked (protonated) by the Lys199. This releases C1 atom, which undergoes a covalent interaction with the Lys195, providing strong stability to the second glucopyranose molecule bound to HSA. The presence of a stable ligand at the SAL subsite of the Sudlow Site I reduces the propensity of protonation of the second glucopyranose molecule by the Lys199. It can be observed that quercetin and gallic acid undergoes a conventional hydrogen bonding interaction with Lys199. In both cases the Lys199 serves as a hydrogen donor; hence, this hydrogen is not available for protonation of C5 of the second glucopyranose molecule. Therefore, no covalent interaction could be established between the second glucopyranose and Lys195. This makes the second glucose molecule, at the entrance of Sudlow Site I, more vulnerable for replacement by a more stable competitor. Since quercetin, gallic acid, catechin, and caffeic acid offer more stable interaction as compared with a non-covalently bound glucose molecule (Table 3), they can offer strong competition.

Quercetin appears to be the most promising inhibitor of glucose interaction since it efficiently prevents the formation of a covalent bond by directly interacting with the Lys199. Interaction of quercetin with the SAL subsite is stabilized by the hydrophobic interaction with Ala291 and Leu238—the same amino acids involved in the stabilization of the first glucopyranose molecule at the SAL subsite. The 3-4 dihydroxyphenyl ring bound to C2 of the deep seated 3,5,7-trihydroxychromen-4-one of quercetin at the SAL subsite (Figure 9c) extends well beyond the Lys199 wall, as described by Wang et al. [20]. This sterically hinders the settlement of the second glucopyranose molecule at the opening of the Sudlow Site I, hence inhibiting the glucose interaction. Gallic acid also prevents covalent bond formation (even more efficiently than quercetin due to two possible hydrogen bond interaction); however, due to the small size of trihydroxy benzoic acid, it remains deep-seated at the SAL subsite and well within the wall defined by the Lys199, thereby incapable of preventing the entry of the second glucopyranose molecule at Sudlow Site I.

Uncontrolled levels of blood glucose together with oxidative stress create conditions with high possibilities for the formation of intermediary metabolites of glycation and AGEs. These AGEs and AGE-related metabolites exert structural and functional alterations of blood proteins, contributing to further complications in patients with hyperglycemia as well patients with other inflammatory diseases. In our study, we proved that the natural products that are present in garlic extract have antioxidant and anti-glycation properties and lend protection against the structural destabilization of proteins such as HSA. As a result, this research will help researchers better grasp the link between glycation and natural products that could be beneficial for disease prevention mechanisms in humans.

4. Materials and Methods

4.1. Materials

Ascorbic acid, trichloroacetic acid, DPPH, Folin–Ciocalteau reagent, ferric chloride, potassium ferricyanide, gallic acid, trypsin, quercetin, and Congo red were purchased from Sigma (Saint Louis, MO, USA). Hydrochloric acid, aluminum chloride, mono sodium dihydrogen phosphate, DMSO, ethanol, methanol, disodium hydrogen phosphate, sodium hydroxide, sodium carbonate, and hydrogen peroxide were purchased from Merck (Darmstadt, Germany). All chemicals and reagents were of the highest analytical grade. All solvents used were HPLC grade.

4.2. Chemical and Biological Studies

4.2.1. Preparation of Aqueous Solution of Garlic Extract

Soft-necked varieties of garlic, i.e., *Allium sativum* var *sativum* were obtained from the local markets in Hail, KSA. Fresh aqueous extract was prepared using a previously published method with slight modifications [26]. The concentration of the aqueous solution was determined based on the total garlic used and the final volume of the extract (100 mL of aqueous extract contained 50 g of garlic, i.e., 500 mg/mL).

4.2.2. Qualitative and Quantitative Analysis for Flavonoids and Phenolics

Qualitative analysis: The presence of flavonoids (alkaline reagent test) and phenolic ($FeCl_3$ test) compounds in the aqueous garlic extract was checked using a protocol published by Alsahli et al. with slight modifications [38].

Quantitative analysis of total phenol and flavonoids content: Total phenol content was estimated using Folin– Ciocalteau reagent as described previously with slight modifications [26]. Gallic acid (0–100 µg/mL) served as a reference point. The total phenolic content in garlic extract was determined from the calibration plot and expressed as mg gallic acid equivalents (GAE). All assays were performed in triplicate. Results are expressed as mg gallic acid equivalent per g dry extract.

$$\text{Total phenolic content (TPC)} = C \times V/M$$

where C is the concentration of gallic acid in mg/mL that was obtained from gallic acid calibration curve, V is the volume of plant extract in mL, and M is the weight of pure plant extract in grams (g).

The aluminum chloride ($AlCl_3$) colorimetric method was used to determine total flavonoid content in garlic extract as described previously [12,39]. The calibration curve was plotted using quercetin (0–250 µg/mL). Garlic extract (50 µg/mL) or standard quercetin solution was added to 2% $AlCl_3$ (500 µL). The resultant solution was incubated for 1 h with occasional stirring. The absorbance of reaction mixture was estimated using a spectrophotometer at 430 nm using ethanol as blank, as 2% $AlCl_3$ solution was prepared in ethanol. Total flavonoid content was determined to be quercetin equivalent (mg/g) (mg QUE/g).

$$\text{Total flavonoid content (TFC)} = Z \times V/m$$

where Z is concentration of quercetin (mg/mL); V is volume (mL) of sample used in the extraction; m is weight of pure dried sample used (g).

4.2.3. Antioxidant and Free Radical Scavenging Activities of Extract

Antioxidant activity—reducing power: The ferric reducing antioxidant power (FRAP) technique was used to measure in vitro antioxidant activity [37]. An amount of 1 mL ascorbic acid (0–100 µg/mL) or garlic extract (0–100 µg/mL) was combined with 2.5 mL in phosphate buffer (0.1 M, pH 6.6) and 1% potassium ferricyanide (2.5 mL). After 20 min at 50 °C, 2.5 mL of trichloroacetic acid (10%) was added to the test tubes to terminate reaction. The mixtures were subsequently centrifuged at 3000 rpm for 10 min, causing the formation of supernatant. Finally, freshly made ferric chloride (0.5 mL, 0.5%) solution was mixed with the supernatant (2.5 mL) and distilled water mixture (2.5 mL). The absorbances of various samples were measured at 700 nm.

$$\text{Percentage free radical scavenging activity} = [(X_{control} - X_{sample})/X_{control}] \times 100$$

$$X_{control} = \text{absorbance of control sample.}$$

$$X_{sample} = \text{absorbance of sample in the presence of extract.}$$

Analysis of free radical scavenging activity by DPPH method: Antioxidant activity of aqueous garlic extracts was estimated using 1,1 difenyl-2-picryl-hydrazyl (DPPH) as published before [12]. Dry powder was obtained from the aqueous extract of garlic. This was dissolved in methanol. One milliliter of 0.3 mM DPPH in methanol was added to the extract solution test sample (2.5 mL) of varying concentrations (0–100 μg/mL) and incubated in the dark for 30 min at room temperature. All samples were read at 517 nm with methanol used as a blank. Inhibitory effect of DPPH was calculated according to the following equation:

$$\text{Percentage of free radical scavenging activity} = [(Ac - As)/Ac] \times 100$$

where, Ac = absorbance of control, and As = absorbance in presence of extract.

4.3. Modification of HSA by Glucose

HSA glycation was carried out as described previously with minor changes [1,2]. Solutions of HSA (3 mg/mL) with and without glucose were incubated for 10 weeks under identical experimental conditions. To analyze the anti-glycation activity of garlic extract, HSA with glucose was also incubated with varying concentrations of garlic extract (0–100 μg/mL) and incubated under similar conditions. Post-incubation, all samples were dialyzed against PBS. Samples were stored at −20 °C. Protein concentrations were determined by NanoDrop™ 2000/2000c spectrophotometer (Thermo Scientific, USA).

4.4. Protein Denaturation Inhibition by Garlic Extract

The protein denaturation inhibition function of garlic extract was studied as described by Sakat et al. [40] and Pandey et al. [41]. A portion of 1% aqueous solution (500 μL) of HSA was added separately to 100 μL of varying concentrations (0–100 μg/mL) of garlic extract. First, the mixtures were incubated at 37 °C for 20 min and then heated for 20 min at 51 °C. Subsequently, the samples were cooled, and turbidity was measured at 660 nm. All samples were run in triplicate.

The percentage inhibition of protein denaturation was calculated using the following equation:

$$\text{Percentage Inhibition} = [(Ac - As)/Ac] \times 100$$

where Ac = absorbance of control, and As = absorbance in presence of extract.

4.5. Measurement of Browning in Glycated Samples

The glycation process is a non-enzymatic event that can produce HSA denaturation. Glycation promotes the browning of proteins by glucose in diabetes. As a result, browning intensity can be used as a preliminary screen for glycation. The intensity of browning of glycated materials was measured by absorbance at 420 nm using a 1 cm path length cell after dilution with distilled water [12]. Experiments were performed in triplicate.

$$\text{Percentage protection from browning} = [(A_{control} - A_{sample})/A_{control}] \times 100$$

where, $A_{control}$ = absorbance of HSA and glucose system, and A_{sample} = absorbance of HSA and glucose system incubated with extract.

4.6. Effect of Garlic Extract on Protein Aggregation Index

Glycation is a major driver of biomolecular structural changes, particularly proteins, resulting in the development of aggregates [42]. Garlic extract was tested for its ability to protect against protein aggregation caused by glycation by determining glycated sample absorbance, treated with glucose, either in the absence or presence of extract (0.78–100 μg/mL). Aggregation index was determined for each sample using their absorbance at 340 and 280 nm [12].

$$\text{Percentage of protein aggregation index} = [A_{340}/(A_{280} - A_{340})] \times 100$$

where, A_{340} = absorbance at 340 nm, and A_{280} = absorbance at 280 nm.

4.7. Spectral Studies

UV absorption measurements were taken with a Perkin Elmer Lambda 35 spectrophotometer (Waltham, MA, USA) with two beams. In the absence/presence of glucose and garlic extract, the UV spectra of HSA (0.2 mg/mL) were measured at a wavelength range of 240–500 nm. Each sample's absorbance intensity was measured at 280 nm [2,42].

AGE-specific fluorescence study: Fluorescence measurements were performed with a Shimadzu spectrofluorometer (model RF-5301PC, Tokyo, Japan). Excitation (350 nm) and emission (400–480 nm) wavelengths were used to determine the generation of fluorescent AGE products. Slit width of 3 nm was used for both excitation and emission [38].

4.8. Circular Dichroism

The CD of native and glycated samples of HSA (2.2 μM), either with or without garlic extracts (0.78–100 μg/mL), was recorded on a Jasco J 810 spectropolarimeter (Tokyo, Japan). The spectropolarimeter has a temperature-controlled sample cell holder attached to a NESLAB model RYE 110 water bath (Tokyo, Japan). [1]. Path length of cuvettes used was 1–10 mm. Each spectrum was the average of three scans. CD spectra were recorded over a wavelength range of 200 to 280 nm and sensitivity of at 5 mm/millidegree (mdeg). Sodium phosphate buffer (20 mM, pH 7.4) was used to prepare all protein solutions. Presence of secondary structural elements was estimated as relative percentage by Chen and Yang equation [19].

4.9. Structural Details of Potential Natural Product Ligands/Inhibitors of HSA Glycation

Natural products contain several phenolic compounds and their metabolites that might be involved in inhibiting the glycation reactions. Some of the competing ligands/inhibitors are given in Table 3, with the HSA molecule characterized using molecular docking studies. All the ligands' .sdf files were derived from PubChem ID (Table 3). The cyclic form of glucose bound to HSA in the crystal structure was obtained from protein data bank (PDB ID: 4iw2) and served as control. The HSA target structure was retrieved from the same PDB ID (4iw2) by removing the ligands. SAL subsite of Sudlow Site I served as the active site.

4.10. Molecular Docking and Scoring of Ligand Poses

Molecular docking of ligands on the active site (defined above) was performed by Autodock-Vina [43]. A standard docking protocol was implemented. The binding efficiency of the glucopyranose and other ligands was also determined by the prodigy-ligand webserver [44,45].

4.11. Statistical Analysis

The experimental results were calculated in triplicate and expressed as mean $\pm$ standard error. Statistical analyses were performed using OriginPro 8.5 followed by t test. Significance was represented by p values < 0.05.

5. Conclusions

Previous researchers have demonstrated that there is a complex network of interdependencies and correlations among qualities of various natural products, which may work together to provide the overall therapeutic capabilities of different medicinal plants. The results show that the inhibition of heat-induced denaturation of protein and prevention of glycation and AGEs formation increased with an increase in garlic extract concentration. Furthermore, garlic extract exhibited significant levels of protection in protein structural stabilization against glycation. Our research significantly supports the numerous therapeutic properties of garlic and addresses the question of the interdependence of various biological activities and their antioxidant capacity. Thus, it is suggested to include garlic into any existing preventative and treatment approach for glycation-induced health complications in diabetic patients. For this purpose, a well-designed research strategy involving animal and clinical study is highly recommended for the verification of the therapeutic advantages of garlic extract.

Author Contributions: Conceptualization, M.W.A.K., A.A.O. and S.S.; methodology, M.W.A.K., A.A.O., S.S., W.A.K., R.S. and S.R.V.; validation, N.A., A.K.D.A. and W.A.K.; formal analysis, M.W.A.K., A.A.O., S.S., R.S. and S.R.V.; investigation, M.W.A.K., A.A.O., S.S. and W.A.K.; resources, M.W.A.K., A.A.O., S.S., S.A.A.-Z., E.M.A. and W.A.K.; data curation, M.W.A.K. and A.A.O.; writing—original draft preparation, M.W.A.K. and S.S.; writing—review and editing, N.A., E.M.A., S.A.A.-Z. and A.K.D.A.; supervision, M.W.A.K.; funding acquisition, M.W.A.K. All authors have read and agreed to the published version of the manuscript.

Funding: This research has been funded by the Scientific Research Deanship at the University of Ha'il–Saudi Arabia through project number RG-20 120.

Institutional Review Board Statement: This study was reviewed and approved by the Research Ethics Committee (REC) at the University of Hail dated 27/11/2020 and approved by university president letter number Nr. 20455/5/42 dated 16/04/1442 H.

Informed Consent Statement: Not applicable.

Data Availability Statement: The data presented in this study are available on request from the corresponding author.

Acknowledgments: Authors are highly thankful to the Molecular Diagnostic and Personalised Therapeutics Unit, University of Ha'il, Ha'il 2440, Saudi Arabia, for providing resources and support for the completion of this research.

Conflicts of Interest: The authors declare no conflict of interest.

Sample Availability: Samples of the compounds are not available from the authors.

References

1. Khan, M.W.A.; Rasheed, Z.; Khan, W.A.; Ali, R. Biochemical, biophysical, and thermodynamic analysis of in vitro glycated human serum albumin. *Biochemistry* **2007**, *72*, 146–152.
2. Khan, M.W.; Al Otaibi, A.; Al-Zahrani, S.A.; Alshammari, E.M.; Haque, A.; Alouffi, S.; Khan, W.A.; Khan, S.N. Experimental and theoretical insight into resistance to glycation of bovine serum albumin. *J. Mol. Struct.* **2021**, *1230*, 129645. [CrossRef]
3. Sirangelo, I.; Iannuzzi, C. Understanding the role of protein glycation in the amyloid aggregation process. *Int. J. Mol. Sci.* **2021**, *22*, 6609. [CrossRef]
4. Khanam, A.; Ahmad, S.; Husain, A.; Rehman, S.; Farooqui, A.; Yusuf, M.A. Glycation and Antioxidants: Hand in the glove of antiglycation and natural antioxidants. *Curr. Protein Pept. Sci.* **2020**, *21*, 899–915. [CrossRef] [PubMed]
5. Singh, V.P.; Bali, A.; Singh, N.; Jaggi, A.S. Advanced glycation end products and diabetic complications. *Korean J. Physiol. Pharmacol.* **2014**, *18*, 1–14. [CrossRef]
6. Anwar, S.; Khan, S.; Almatroudi, A.; Khan, A.A.; Alsahli, M.A.; Almatroodi, S.A.; Rahmani, A.H. A review on mechanism of inhibition of advanced glycation end products formation by plant derived polyphenolic compounds. *Mol. Biol. Rep.* **2021**, *48*, 787–805. [CrossRef] [PubMed]
7. Vlassara, H.; Uribarri, J. Advanced glycation end products (AGE) and diabetes: Cause, effect, or both? *Curr. Diab. Rep.* **2014**, *14*, 453. [CrossRef] [PubMed]
8. Nowotny, K.; Jung, T.; Höhn, A.; Weber, D.; Grune, T. Advanced glycation end products and oxidative stress in type 2 diabetes mellitus. *Biomolecules* **2015**, *5*, 194–222. [CrossRef] [PubMed]
9. de Groot, L.; Hinkema, H.; Westra, J.; Smit, A.J.; Kallenberg, C.G.; Bijl, M.; Posthumus, M.D. Advanced glycation endproducts are increased in rheumatoid arthritis patients with controlled disease. *Arthritis Res. Ther.* **2011**, *13*, R205. [CrossRef]
10. Fishman, S.L.; Sonmez, H.; Basman, C.; Singh, V.; Poretsky, L. The role of advanced glycation end-products in the development of coronary artery disease in patients with and without diabetes mellitus: A review. *Mol. Med.* **2018**, *24*, 59. [CrossRef]
11. Indyk, D.; Bronowicka-Szydełko, A.; Gamian, A.; Kuzan, A. Advanced glycation end products and their receptors in serum of patients with type 2 diabetes. *Sci. Rep.* **2021**, *11*, 13264. [CrossRef] [PubMed]
12. Anwar, S.; Almatroudi, A.; Allemailem, K.S.; Jacob Joseph, R.; Khan, A.A.; Rahmani, A.H. Protective effects of ginger extract against glycation and oxidative stress-induced health complications: An in vitro study. *Processes* **2020**, *8*, 468. [CrossRef]
13. Mittal, M.; Siddiqui, M.R.; Tran, K.; Reddy, S.P.; Malik, A.B. Reactive oxygen species in inflammation and tissue injury. *Antioxid. Redox Signal.* **2014**, *20*, 1126–1167. [CrossRef]
14. Anwar, S.; Almatoudi, A.; AlSahli, M.A.; Khan, M.A.; Khan, A.A.; Rahmani, A. Natural products: Implication in cancer prevention and treatment through modulating various biological activities. *Anticancer Agents Med. Chem.* **2020**, *20*, 2025–2040. [CrossRef]

15. Koch, H.P.; Lawson, L.D. *Garlic: The Science and Therapeutic Application of Allium Sativum L. and Related Species*, 2nd ed.; Williams & Wilkins: Baltimore, MD, USA, 1996.

16. Murray, M.T. *The Healing Power of Herbs: The Enlightened Person's Guide to the Wonders of Medicinal Plants*, 2nd ed.; Prima: Rocklin, CA, USA, 1995.

17. Colín-González, A.L.; Santana, R.A.; Silva-Islas, C.A.; Chánez-Cárdenas, M.E.; Santamaría, A.; Maldonado, P.D. The antioxidant mechanisms underlying the aged garlic extract- and S-allylcysteine-induced protection. *Oxid. Med. Cell. Longev.* **2012**, *2012*, 907162. [CrossRef] [PubMed]

18. Marrelli, M.; Amodeo, V.; Statti, G.; Conforti, F. Biological properties and bioactive components of *Allium cepa L.*: Focus on potential benefits in the treatment of obesity and related comorbidities. *Molecules* **2018**, *24*, 119. [CrossRef]

19. Chen, Y.H.; Yang, J.T. A new approach to the calculation of secondary structures of globular proteins by optical rotatory dispersion and circular dichroism. *Biochem. Biophys. Res. Commun.* **1971**, *44*, 1285–1291. [CrossRef]

20. Wang, Y.; Yu, H.; Shi, X.; Luo, Z.; Lin, D.; Huang, M. Structural mechanism of ring-opening reaction of glucose by human serum albumin. *J. Biol. Chem.* **2013**, *288*, 15980–15987. [CrossRef]

21. Safari, M.R.; Azizi, O.; Heidary, S.S.; Kheiripour, N.; Ravan, A.P. Antiglycation and antioxidant activity of four Iranian medical plant extracts. *J. Pharmacopunct.* **2018**, *21*, 82–89. [CrossRef]

22. Khan, M.W.A.; Banga, K.; Mashal, S.N.; Khan, W.A. Detection of autoantibodies against reactive oxygen species modified glutamic acid decarboxylase-65 in type 1 diabetes associated complications. *BMC Immunol.* **2011**, *12*, 19. [CrossRef]

23. Khan, M.W.A.; Qadrie, Z.L.; Khan, W.A. Antibodies against gluco-oxidative modified HSA-detected in diabetes associated complications. *Int. Arch. Allergy Immunol.* **2010**, *153*, 207–214. [CrossRef] [PubMed]

24. Khan, M.W.A.; Sherwani, S.; Khan, W.A.; Moinuddin; Ali, R. Characterization of hydroxyl radical modified GAD$_{65}$: A potential autoantigen in type 1 diabetes. *Autoimmunity* **2009**, *42*, 150–158. [CrossRef] [PubMed]

25. Tuso, P.J.; Ismail, M.H.; Ha, B.P.; Bartolotto, C. Nutritional update for physicians: Plant-based diets. *Perm. J.* **2013**, *17*, 61–66. [CrossRef] [PubMed]

26. Elosta, A.; Slevin, M.; Rahman, K.; Ahmed, N. Aged garlic has more potent antiglycation and antioxidant properties compared to fresh garlic extract in vitro. *Sci. Rep.* **2017**, *7*, 39613. [CrossRef] [PubMed]

27. Khan, M.W.A.; Otaibi, A.A.; Alhumaid, A.F.M.; Alsukaibi, A.K.D.; Alshamari, A.K.; Alshammari, E.M.; Al-Zahrani, S.A.; Almudyani, A.Y.M.; Sherwani, S. Garlic Extract: Inhibition of Biochemical and Biophysical Changes in Glycated HSA. *Appl. Sci.* **2021**, *11*, 11028. [CrossRef]

28. Nagella, P.; Thiruvengadam, M.; Ahmad, A.; Yoon, J.Y.; Chung, I.M. Comosition of polyphenols and antioxidant activity of garlic bulbs collected from different locations of Korea. *Asian J. Chem.* **2014**, *26*, 897–902. [CrossRef]

29. Furman, D.; Campisi, J.; Verdin, E.; Carrera-Bastos, P.; Targ, S.; Franceschi, C.; Ferrucci, L.; Gilroy, D.W.; Fasano, A.; Miller, G.W.; et al. Chronic inflammation in the etiology of disease across the life span. *Nat. Med.* **2019**, *25*, 1822–1832. [CrossRef]

30. Angel, G.R.; Vimala, B.; Nambisan, B. Antioxidant and anti-inflammatory activities of proteins isolated from eight Curcuma species. *J. Phytopharm.* **2013**, *4*, 96–105.

31. Ruiz-Ruiz, J.C.; Moguel-Ordoñez, Y.B.; Segura-Campos, M.R. Biological activity of Stevia rebaudiana Bertoni and their relationship to health. *Crit. Rev. Food Sci. Nutr.* **2017**, *57*, 2680–2690. [CrossRef]

32. Raimundo, A.F.; Ferreira, S.; Martins, I.C.; Menezes, R. Islet Amyloid Polypeptide: A partner in crime with Aβ in the pathology of Alzheimer's disease. *Front. Mol. Neurosci.* **2020**, *13*, 35. [CrossRef]

33. Bouma, B.; Kroon-Batenburg, L.M.; Wu, Y.P.; Brünjes, B.; Posthuma, G.; Kranenburg, O.; de Groot, P.G.; Voest, E.E.; Gebbink, M.F. Glycation induces formation of amyloid cross-beta structure in albumin. *J. Biol. Chem.* **2003**, *278*, 41810–41819. [CrossRef]

34. Monacelli, F.; Storace, D.; D'Arrigo, C.; Sanguineti, R.; Borghi, R.; Pacini, D.; Furfaro, A.L.; Pronzato, M.A.; Odetti, P.; Traverso, N. Structural alterations of human serum albumin caused by glycative and oxidative stressors revealed by circular dichroism analysis. *Int. J. Mol. Sci.* **2013**, *14*, 10694–10709. [CrossRef]

35. Barzegar, A.; Moosavi-Movahedi, A.A.; Sattarahmady, N.; Hosseinpour-Faizi, M.A.; Aminbakhsh, M.; Ahmad, F.; Saboury, A.A.; Ganjali, M.R.; Norouzi, P. Spectroscopic studies of the effects of glycation of human serum albumin on l-trp binding. *Protein Pept. Lett.* **2007**, *14*, 13–18. [CrossRef] [PubMed]

36. GhoshMoulick, R.; Bhattacharya, J.; Roy, S.; Basak, S.; Dasgupta, A.K. Compensatory secondary structure alterations in protein glycation. *Biochim. Biophys. Acta.* **2007**, *1774*, 233–242. [CrossRef] [PubMed]

37. Miwa, I.; Maeda, K.; Okuda, J. Anomeric compositions of D-glucose in tissues and blood of rat. *Experientia* **1978**, *34*, 167–169. [CrossRef] [PubMed]

38. Alsahli, M.A.; Anwar, S.; Alzahrani, F.M.; Almatroudi, A.; Alfheeaid, H.; Khan, A.A.; Allemailem, K.S.; Almatroodi, S.A.; Rahmani, A.H. Health promoting effect of Phyllanthus emblica and Azadiractha indica against advanced glycation end products formation. *Appl. Sci.* **2021**, *11*, 8819. [CrossRef]

39. Ordonez, A.A.L.; Gomez, J.D.; Vattuone, M.A.; Lsla, M.I. Antioxidant activities of Sechium edule (Jacq.) Swartz extracts. *Food Chem.* **2006**, *97*, 452–458. [CrossRef]

40. Sakat, S.S.; Juvekar, A.R.; Gambhire, M.N. In vitro antioxidant and anti-inflammatory activity of methanol extract of Oxalis corniculata linn. *Int. J. Pharm. Pharm. Sci.* **2010**, *2*, 146–155.

41. Pandey, A.K.; Kashyap, P.P.; Kaur, C.D. Anti-inflammatory activity of novel Schiff bases by in vitro models. *Bangladesh J. Pharmacol.* **2017**, *12*, 41–43. [CrossRef]

42. Anwar, S.; Younus, H. Inhibitory effect of alliin from Allium sativum on the glycation of superoxidedismutase. *Int. J. Biol. Macromol.* **2017**, *103*, 182–193. [CrossRef]

43. Trott, O.; Olson, A.J. AutoDock Vina: Improving the speed and accuracy of docking with a new scoring function, efficient optimization, and multithreading. *J. Comput. Chem.* **2010**, *31*, 455–461. [CrossRef] [PubMed]

44. Vangone, A.; Schaarschmidt, J.; Koukos, P.; Geng, C.; Citro, N.; Trellet, M.E.; Xue, L.C.; Bonvin, A.M. Large-scale prediction of binding affinity in protein–small ligand complexes: The PRODIGY-LIG web server. *Bioinformatics* **2019**, *35*, 1585–1587. [CrossRef] [PubMed]

45. Kurkcuoglu, Z.; Koukos, P.I.; Citro, N.; Trellet, M.E.; Rodrigues, J.P.G.L.M.; Moreira, I.S.; Roel-Touris, J.; Melquiond, A.S.J.; Geng, C.; Schaarschmidt, J.; et al. Performance of HADDOCK and a simple contact-based protein-ligand binding affinity predictor in the D3R Grand Challenge 2. *J. Comp. Aid. Mol. Des.* **2018**, *32*, 175–185. [CrossRef] [PubMed]

*molecules*

MDPI

Article

Protective Role of Quercetin in Carbon Tetrachloride Induced Toxicity in Rat Brain: Biochemical, Spectrophotometric Assays and Computational Approach

Seema Zargar [1,*] and Tanveer A. Wani [2]

[1] Department of Biochemistry, College of Science, King Saud University, Riyadh 11495, Saudi Arabia
[2] Department of Pharmaceutical Chemistry, College of Pharmacy, King Saud University, Riyadh 11451, Saudi Arabia; twani@ksu.edu.sa
* Correspondence: szargar@ksu.edu.sa

Abstract: Carbon tetrachloride (CCL4) induces oxidative stress by free radical toxicities, inflammation, and neurotoxicity. Quercetin (Q), on the other hand, has a role as anti-inflammatory, antioxidant, antibacterial, and free radical-scavenging. This study explored protection given by quercetin against CCL4 induced neurotoxicity in rats at given concentrations. Male Wistar rats were divided into four groups Group C: control group; Group CCL4: given a single oral dose of 1 mL/kg bw CCL4; Group Q: given a single i.p injection of 100 mg/kg bw quercetin; and Group Q + CCL4: given a single i.p injection of 100 mg/kg bw quercetin before two hours of a single oral dose of 1 mL/kg bw CCL4. The results from brain-to-body weight ratio, morphology, lipid peroxidation, brain urea, ascorbic acid, reduced glutathione, sodium, and enzyme alterations (aspartate aminotransferase (AST), alanine aminotransferase (ALT), catalase, and superoxide dismutase) suggested alterations by CCL4 and a significant reversal of these parameters by quercetin. In silico analysis of quercetin with various proteins was conducted to understand the molecular mechanism of its protection. The results identified by BzScore4 D showed moderate binding between quercetin and the following receptors: glucocorticoids, estrogen beta, and androgens and weak binding between quercetin and the following proteins: estrogen alpha, Peroxisome proliferator-activated receptors (PPARγ), Herg k+ channel, Liver x, mineralocorticoid, progesterone, Thyroid α, and Thyroid β. Three-dimensional/four-dimensional visualization of binding modes of quercetin with glucocorticoids, estrogen beta, and androgen receptors was performed. Based on the results, a possible mechanism is hypothesized for quercetin protection against CCL4 toxicity in the rat brain.

Keywords: quercetin; CCL4; neurotoxicity; VirtualToxLab; oxidative stress markers

Citation: Zargar, S.; Wani, T.A. Protective Role of Quercetin in Carbon Tetrachloride Induced Toxicity in Rat Brain: Biochemical, Spectrophotometric Assays and Computational Approach. *Molecules* **2021**, *26*, 7526. https://doi.org/10.3390/molecules26247526

Academic Editor: Stefano Castellani

Received: 18 November 2021
Accepted: 9 December 2021
Published: 12 December 2021

1. Introduction

Exposure to toxic chemicals, environmental changes, and drugs can cause harmful effects and injuries through the metabolic production of reactive oxygen species (ROS). The description of oxidative stress is an imbalance between ROS production and antioxidant defense, which causes cell damage at high levels. The ROS and their pathophysiological effects depend on the concentration, type, and specific production site. When ROS are at a high level, they react with DNA, proteins and cell membrane, and other molecules, causing cellular damage and producing other more reactive radicals [1]. In addition, the formation of ROS leads to DNA strand breaks and oxidative DNA damage that induce changes in mRNA expression of DNA damage responsive genes [2].

Carbon tetrachloride (CCL4), a colorless, transparent, heavy, and non-flammable industrial liquid, is widely used to induce free radical toxicity in various experimental animal tissues such as kidneys, heart, liver, lung, testis, brain, and blood [3]. Exposure to CCL4 initiates a complex process for resistance to toxicity and production of free radicals to metabolize CCL4, leading to further oxidative stress, which participates in the initiation and progression of brain injury [4]. The resulting oxidative stress leads to DNA fragmentation and produces a significant interconnected change of cellular metabolism and destruction of the cells by lipid peroxidation [4]. Acute and harmful tissue injuries are induced by CCL4 metabolites, reactive metabolic trichloromethyl radicals (CCl3), and peroxy trichloromethyl radicals (OOCCl3). The single hepatotoxic dose produces more intense free-radical stress in the brain than in the liver [5]. These free radicals can covalently bind to macromolecules such as lipids, proteins, and nucleic acids present in the brain [6]. Previous studies showed that CCL4-induction caused an observed reduction in p53, a tumor suppressor gene expression [7]. The antioxidant mechanism prevents cells in the G phase of the cell cycle and gives additional time for DNA repair, while severe DNA damage triggers apoptosis, a life-threatening condition [7].

Flavonoids are polyphenolic compounds that play an essential role in free radical detoxification. These polyphenolic compounds are found in vegetables, fruits, and medicinal plants. Quercetin in plants exists as either a free (aglycone) or conjugated with carbohydrates (quercetin glycosides) and alcohols (quercetin methyl ethers). It was reported that quercetin is an anti-inflammatory, antioxidant, antibacterial, radical-scavenging, antiviral, gastroprotective, and immune-modulator, and is used to treat cardiovascular diseases and obesity [8,9]. Quercetin is abundantly present in apples, berries, onions, capers, broccoli, tea, and red wine. Quercetin is reported to protect against CCL4 induced hepatotoxicity by inhibiting Toll-like receptor 2 (TLR2), TLR4 activation, and mitogen-activated protein kinase (MAP Kinase) phosphorylation. These, in turn, inactivate nuclear factor kappa B (NF-kB) and the inflammatory cytokines in livers of the CCL4-treated animals. Quercetin is reported to protect against brain injury in mice through TLR2/4 and MAPK/NF-kB pathway [10]. A high concentration of quercetin metabolites is present in the brain after several hours of quercetin administration [11].

An intraperitoneal dose of quercetin 10 mg/kg body weight in rats before two hours of acrylamide assault resulted in diminutive acrylamide mediated neurotoxicity. Quercetin treatment leads to decreased dopamine, interferon-γ, and 8-hydroxyguanosine levels and the restoration of serotonin levels [12]. In addition, quercetin can increase the body's antioxidant activity by regulating glutathione (GSH) levels. The GSH is a central component of a natural defense mechanism of the body against oxidative stress. The superoxide dismutase (SOD) quickly captures O^{2-} and transforms it into hydrogen peroxide H_2O_2. This enzyme further catalyzes the decomposition of H_2O_2 to the non-toxic H_2O. This reaction requires GSH as a hydrogen donor. Quercetin in several studies was found to induce reduced glutathione (GSH) synthesis. One study reported that the p53 penetrates in the mechanism of cell response to quercetin through modulation of glutathione-related enzyme expression [13,14]. ROS and reactive nitrogen species (RNS) are produced continuously in the body by oxidative metabolism, mitochondrial bioenergetics, and immune function that can cause potential biological damage [15]. As a flawed anti-oxidative system favors the accumulation of free radicals due to a decrease in the activity of antioxidant enzymes, quercetin may find application in the prevention of neurological disorders due to its neuroprotective effects. The present study evaluated the toxicity of CCL4 in rat brains and investigated whether quercetin can protect against the damage caused by it.

The VirtualToxLab can predict the toxic potential of the tested compound, e.g., endocrine and metabolic disruption, some aspects of carcinogenicity, and cardiotoxicity by simulation and quantification of their interactions towards a series of proteins suspected to trigger mentioned adverse effects [16]. This tool follows an automated protocol and calculates the binding affinity of the investigated compound to selected proteins. It is beneficial to understand that interaction mechanism at the molecular level and estimate the toxic potential of the studied drugs. The VirtualToxLab™ was used to study the interaction between quercetin and various proteins to understand the molecular basis of the protective potential of quercetin against carbon tetrachloride toxicity on rat brains.

2. Materials and Methods

2.1. Chemicals

All the reagents and chemicals, including quercetin and CCL4, were obtained from Sigma Chemical C., St Louis, MO, USA. All the kits for enzymatic analysis were purchased from United Diagnostics Industry (Dammam, Saudi Arabia).

2.2. Animals

Male Wistar rats 8–12 weeks of age, weighing 80–90 g (n = 24), were obtained from the Animal House Facility of King Saud University, Riyadh, Saudi Arabia. The animal ethics committee approved the study of King Saud University (approval no. KSU-SE-21-05). Rats were housed at 23–25 °C and 55–60% ambient humidity on a 12:12 h light/dark cycle and were fed a regular diet with fresh drinking water daily.

2.3. Experimental Design

Group C: control group; Group CCL4: exposure to CCL4; Group Q: exposure to quercetin only; Group Q + CCL4: treated with both quercetin and CCL4. Each group consisted of six animals. Sunflower oil (vehicle) was used to dissolve CCL4 and then administered to specific groups of animals. To the control group, only the sunflower oil was administered. Due to CCL4's non-polar nature and high volatility, it was necessary to dissolve it in sunflower oil (3:1) to maintain a consistent, effective dose. Neurotoxicity in mice occurs with a single dose of CCL4 (1 mL/kg bw) [5]. To the CCL4 Group, the CCL4 dose was administered by oral gavage. To the quercetin-only group (Group Q), quercetin was given as a single dose of i.p injection of quercetin (100 mg/kg bw) [17–19]. In the combination group, quercetin (100 mg/kg bw) was administered 2 h before the assault by CCL4 (1 mL/kg bw). The Cmax of quercetin is 2–3 h; therefore, quercetin was administered to obtain protective concentration levels of quercetin in systemic circulation before the CCL4 assault [10]. The animals were sacrificed by carbon dioxide asphyxiation after 24 h of their respective treatments bw) [17–20]. The brains of the studied groups were removed and processed immediately for biochemical assays. The harvested brains were weighed immediately for determining the organ/body weight ratio. The harvested brains were homogenized in 1X PBS, and slices from the cortex were cut and stored in formalin for histopathology.

The samples were processed as follows: fixing the specimens in a 10% neutral buffered formalin solution (Loba Chemie, Colaba, India) block preparation in paraffin (Leica Biosystems, Wetzlar, Germany) cutting sections of 5–6 μm thick, and the sections stained with hematoxylin-eosin stain (Leica Biosystems, Wetzlar, Germany). The sections were photographed and analyzed using an electron microscope (Leica Biosystems, Wetzlar, Germany) by an expert pathologist who was not informed about the sample assigned to the experimental groups.

2.4. Brain Enzymes

Alanine aminotransferase (ALT) and aspartate aminotransferase (AST) were measured with UV-kinetic diagnostic kits according to the kit protocol of the manufacturer (United Diagnostics Industry (UDI, Dammam, Saudi Arabia). The method was based on the oxidation of NADH, and the rate of decrease in absorbance at 340 nm is proportional to the ALT activity of the sample. ALT and AST reagents were reconstituted with 3 mL of distilled water, then 1000 μLof reconstituted reagent were pre-warmed at 37 °C for 2 min followed by mixing with 100 μLof the sample. The mixture was allowed to stand for 60 s for temperature equilibrium. ALT absorbance was measured every 60 s within 3 min at 340 nm against distilled water, and AST absorbance was measured every 60 s within a 2-min interval at 340 nm against distilled water. Eventually, ΔA/min was determined. Each unit of AST and ALT enzyme activity was defined as micromoles of NADH decomposed per minute using a molar absorbance of $6.22 \times 10^3 \times M^{-1} Cm^{-1}$.

Superoxide dismutase was determined by the method of Nishikimi et al. [21]. Briefly, the reaction mixture contained 0.1 mL of sodium pyrophosphate (0.1 mM), 0.1 mL of NBT (0.3 mM), 0.1 mL of NADH (0.47 mM), 0.05 mL of PMS (0.93 µM), and 0.1 mL of enzyme (homogenized brain tissue) in a total volume of 1 mL. The rate of change of absorbance was measured at 560 nm. Values were expressed as units of enzyme min^{-1} mg protein. Catalase activity was estimated in the whole homogenized brain tissue by the method of Aebie, 1984 [22]. The reaction mixture in a total volume of 3 mL contained 0.4 M sodium phosphate buffer pH 7.2, 1.2 mL of H_2O_2, and a suitably diluted enzyme. The reaction was started by adding H_2O_2 and reading the change in absorbance at 240 nm for two minutes. One unit of CAT activity was defined as micromoles of H_2O_2 decomposed per min using molar absorbance of H_2O_2 (43.6 M^{-1} Cm^{-1}).

2.5. Measurement of Brain Sodium

Measurement of brain sodium was done with a sodium estimation kit procured from United Diagnostics Industry (UDI, Dammam, KSA), as per the kit protocol procedures. Na_+/K_+-ATPase is an enzyme found in the cell's membrane. It performs several functions in cell physiology. The assay activates the B-galactosidase enzyme by the sodium present in the sample and the consequent enzymatic transformation of o-nitrophenyl-β, D-glactopyranoside (o-NPG) into o-nitrophenol, and galactose. The o-nitrophenol formed was kinetically measured at 405 nm against distilled water every 20 s for two minutes as per the kit protocol. One unit of sodium level was defined as micromoles of o-NPG (o-nitrophenyl-β, D-glactopyranoside) decomposed per minute using molar absorbance of 18.75×10^3 M^{-1} Cm^{-1}.

2.6. Measurement of Brain Urea

Brain Urea was calculated using GLDH—UV-kinetic diagnostic kit with the SEMI MICRO METHOD from United Diagnostics Industry (UDI, Dammam, Saudi Arabia). Urea is a primary end product of protein nitrogen metabolism. The urea reagent was reconstituted with 3 mL of distilled water, then 1000 µL of reconstituted reagent was pre-warmed at 37 °C for 2 min followed by mixing with 10 µL of the sample. Absorbance was measured every 30 s for 90 s at 340 nm against distilled water, and eventually, ΔA/min was determined. One urea unit was defined as micromoles of NADH decomposed per minute using molar absorbance of $6.22 \times 10^3 \times$ M^{-1} Cm^{-1}.

2.7. Lipid Peroxidation

Lipid peroxidation was done by the method of Utley et al. [23]. Briefly, 1 mL of homogenized brain tissue was incubated in a metabolic shaker at 37 °C for one hour, 1.5 mL of 20% TCA was added, which was then centrifuged at 600 g for 10 min. Next, 1 mL of supernatant was added to 1 mL of freshly prepared Thiobarbituric acid (0.67%). The reaction was kept in the water bath for 10 min. Upon cooling, absorbance was read at 535 using a reagent blank. Values were expressed as nanomoles of malondialdehyde formed hour^{-1} mg protein^{-1}.

2.8. Ascorbic Acid (AsA)

Ascorbic acid was determined by the method of Jagota and Dani [24]. First, 0.2 mL of homogenized brain tissue was treated with 0.8 mL of 10% TCA. After vigorous shaking, tubes were kept in an ice-cold bath for 5 min and centrifuged at 1200 g for 5 min. Next, 0.2 to 0.5 mL of the supernatant were diluted to 2 mL with distilled water, and 0.2 mL of Folin reagent (0.2 M) was added. After 10 min, the absorbance was read at 760 nm against a reagent blank. The amount of ascorbic acid was calculated from the standard graph. Values were expressed as µg of ascorbic acid µg protein^{-1}.

2.9. Reduced Glutathione (GSH)

The estimation was carried out by the method of Beutler et al. [25]. Briefly, 0.4 mL of homogenized brain tissue was mixed with 3.6 mL of double-distilled water and followed by treatment with 0.6 mL of precipitating reagent (containing 1.67 g of glacial metaphosphoric acid, 0.2 g of EDTA, and 30.0 g of NaCl and made up to 100 mL with double distilled water).

The above reaction mixture was centrifuged at 600 g for 10 min. To 0.3 mL of supernatant, 2 mL of Na_2HPO_4 (0.3 M) and 0.25 mL of 5,5' dithio-bis-2-nitrobenzoic acid (0.4% in 1% sodium citrate) were added, and the volume was made up to 3 mL with DDW. OD was read at 412 nm against blank. Values were expressed as µg of reduced glutathione µg protein^{-1}.

2.10. In Silico Testing of Quercetin Toxicity and Molecular Docking

The VirtualToxLab™ is an online platform to estimate the toxicity of drugs, which requires the test compound to be submitted in pdb format. Therefore, the quercetin structure was downloaded from PubChem (PubChem CID: 5280343) in sdf format and was converted to the pdb format using the discovery studio software. The interaction of quercetin with the glucocorticoid, estrogen α, estrogen β, androgen, aryl hydrocarbon, thyroid α, thyroid β, mineralocorticoid, progesterone, hERG, liver X, and PPARγ was evaluated. In addition to these proteins, the enzymes cytochrome P450 1A2, 2C9, 2D6, and 3A4 were assessed for their interactions with quercetin. Flexible molecular docking was conducted for the quercetin and the moderately bound proteins in the VirtualToxLab running on an automated protocol. The low-energy poses are sampled into a dataset. The binding affinities between the quercetin and the target proteins were quantified using the dataset as input for Boltzmann scoring (Software BzScore4D) [16]. The binding mechanisms with these biomolecules were used to assess the protective mechanisms of quercetin to CCL4 toxicity in the brain.

3. Results

CCL4 is lipophilic and passes freely through all biological membranes, including the brain's myelin sheath. A marked effect on brain tissue histology was observed with the cortex showing swelling, vacuolar degeneration, and karyopyknosis. All these changes are morphological characteristics of apoptosis. Figure 1a is the normal histology of the cerebral cortex of the brain. Extensive vacuolization was seen in myelin within the white matter, and a few vacuoles were also seen in the gray matter of rats treated with CCL4. CCL4-treated neurons had large nuclei turning from basophilic to pyknotic revealing apoptosis (Figure 1b). However, the normal histology was observed in quercetin-treated rats (Figure 1c). Figure 1d shows a complete reversal of altered histology in rats treated with quercetin two hours before CCL4.

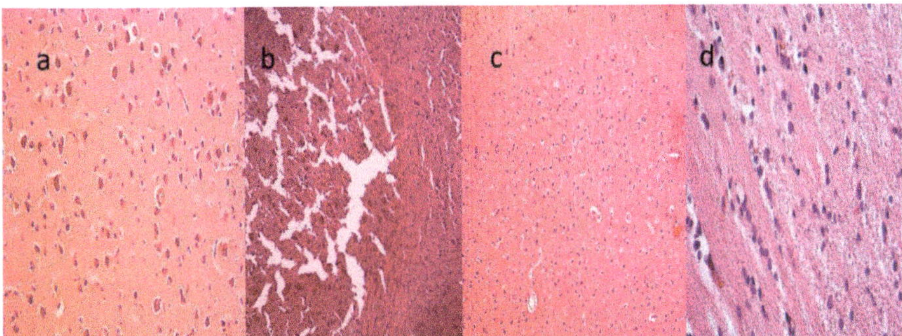

Figure 1. Cortex sections of normal and treated groups at a scale bar of 100 µm. (**a**) Normal histological appearance of brain tissues with neurocytes having well-defined nuclei. (**b**) CCL4-treated brain cortex section with widespread intracellular vacuolization and infiltration of inflammatory cells (aster). Neurocytes have dark eosinophilic cytoplasm, with cells having heterochromatic nuclei. (**c**) Quercetin-treated brain tissue has fewer vacuoles and inflammation. (**d**) Q + CCL4-treated brain section with mild vacuolization and mild infiltration of inflammatory cells (*n* = 3).

The effect of CCL4 on the brain-to-body weight ratio is presented in Figure 2a. The brain-to-body weight ratio decreased significantly ($p < 0.0001$) in the CCL4 treatment group, showing this group's metabolic or growth disorders. Treatments with quercetin and combination groups completely reversed these groups' metabolic or growth disorders and showed non-significant differences with control group growth patterns.

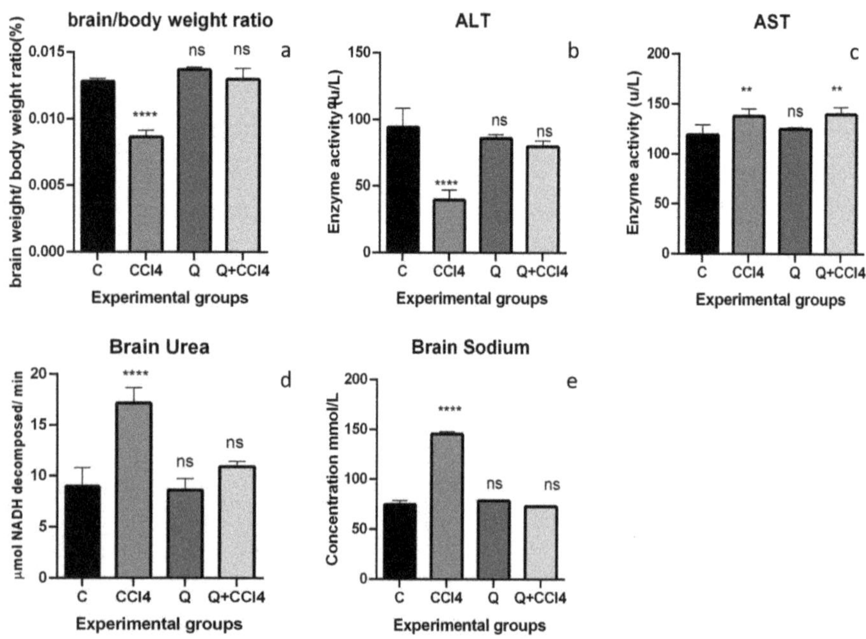

Figure 2. (a–e). Effect of CCL4 (1 mg/kg) and Q (100 mg/kg b.w) on the brain-to-body weight ratio, ALT, AST, brain urea, and brain sodium ($n = 6$). Data were analyzed by (one-way ANOVA), and Tukey's test was used for multiple comparisons. Treated groups are compared to the control group. **** $p < 0.0001$, ** $p < 0.01$, ns is non-significant.

Figure 2b shows significantly decreased levels of ALT in group CCL4-treated rats ($p < 0.0001$) when compared to Group C (control). Quercetin treatment alone (Group Q) caused a non-significant decrease in the ALT levels ($p > 0.05$) compared to controls. Pretreatment of quercetin in Group IV reversed the decreased ALT levels and showed significant protection when compared with Group CCL4. The decreased ALT levels in the group Q + CCL4 were non-significant ($p > 0.05$) compared to the control group.

Figure 2c shows significantly increased levels of AST in Group CCL4 rats ($p < 0.0001$) when compared to control, indicating the induction of brain damage as AST catalytic activities in the CSF are linked to high risk. Quercetin treatment alone was relatively similar to control in the AST levels ($p > 0.05$) compared to control. However, pretreatment of quercetin in the Q + CCL4 group could not reverse the increased levels of AST.

Figure 2d shows significantly increased levels of urea in Group CCL4 rats ($p < 0.0001$) when compared to control, indicating brain damage and increased metabolic activity. Quercetin treatment alone caused a non-significant increase in urea levels compared to controls. However, pretreatment of Quercetin in the Q + CCL4 group decreased the increased urea levels significantly.

Figure 2e shows significantly increased sodium levels in the group Q rats ($p < 0.0001$) compared to controls, which is another indication of brain damage due to the voltage potential. Quercetin treatment alone was relatively similar to control; non-significant change in the sodium levels ($p > 0.05$) compared to controls. Pretreatment of quercetin in the Group Q + CCL4 reversed the increased sodium levels and showed significant protection compared with the group CCL4. The increase of sodium levels in group Q + CCL4 was significantly decreased ($p < 0.001$) compared to the control group, relatively similar to controls.

Table 1 shows the levels of biochemical oxidative stress biomarkers lipid hydroperoxides, ascorbic acid, and reduced glutathione catalase and superoxide dismutase in control and experimental rats (Table 1). The levels of lipid hydroperoxides and AsA were significantly ($p < 0.05$) increased in CCL4- treated rats compared to control and quercetin groups. The treatment of rats with quercetin resulted in significant recovery of these free radicals generated in brain tissues of group Q + CCL4. Exposure with CCL4 significantly decreased the reduced glutathione ($p < 0.05$) that was reversed considerably in group Q + CCL4 ($p < 0.05$). Catalase activity was reduced substantially with CCL4 treatment that was significantly recovered by treatment with quercetin in group Q + CCL4 (Table 1).

Table 1. Effect of quercetin on CCL4 induced oxidative stress in the brains of control and experimental rats.

Treatment Groups	Control	CCL4 (1 mg/kg)	Q (100 mg/kg)	Q + CCL4
Lipid Peroxidation (mmoles of malonaldehyde formed/hour/mg protein)	1.98 ± 0.44 [b,d]	8.33 ± 0.82 [a,c,d]	0.94 ± 0.15 [b,c]	0.83 ± 0.35 [a,b,c]
Ascorbic acid (μg of ascorbic acid/μg protein)	0.73 ± 0.15 [b,d]	3.23 ± 1.82 [a,c,d]	0.45 ± 0.06 [b,d]	1.70 ± 0.14 [a,b,c]
Reduced glutathione (μg of GSH/mg protein)	5.30 ± 0.80 [b]	2.53 ± 1.47 [a,c,d]	5.55 ± 3.47 [b]	6.90 ± 0.41 [b]
Catalase mmol H_2O_2/min/mg protein	3.05 ± 0.55 [b,c]	0.26 ± 0.30 [a,c,d]	2.30 ± 0.45 [a,b,d]	2.43 ± 0.85 [b,c]
Superoxide dismutase Units/mg protein	135.29 ± 44.79 [b]	63.11 ± 22.48 [a,d]	131.26 ± 7.29 [d]	126.10 ± 25.73 [b,c]

Quercetin (100 mg/kg b.w) was administered 2 h before CCL4 assault. Data are representative of mean $\pm$ SD of three independent experiments, each group containing six mice. [a] significant ($p < 0.05$) when compared to control; [b] significant ($p < 0.05$) when compared to CCL4 (1 mg/kg) group; [c] significant ($p < 0.05$) when compared to Q (100 mg/kg) group; [d] significant ($p < 0.05$) when compared to Q + CCL4 group.

The results from the VirtualToxLab™ are presented in Table 2. Quercetin was found to have a strong binding affinity to the androgens and had moderate binding to glucocorticoid and estrogen beta receptors. In addition, the quercetin was found to have weak binding to Estrogen receptor α (ERα), hERG, Peroxisome Proliferator-Activated receptor γ (PPARγ) and Thyroid receptor β (TRβ). Furthermore, quercetin did not bind to Aryl hydrocarbon receptor (AhR), Thyroid receptor α, Mineralocorticoid receptor (MR), progesterone receptor (PR), hERG, and Liver X receptor (LXR), nor to any of the cytochrome P450 enzymes. Figure 3 shows real-time 3D/4D visualization of binding modes of quercetin with glucocorticoids, estrogen beta, and androgen receptors in concomitance with binding results from BzScore4D. Based on the results, a possible mechanism of quercetin protection against CCL4 toxicity in the rat brain was hypothesized.

Table 2. Binding of quercetin to various proteins.

Target	Binding Type	Binding Affinity (VirtualToxLab)
Androgens	moderate binding	948 nM
Aryl hydrocarbon	negligible	>100 μM
CYP1A2	negligible	>100 μM
CYP2C9	negligible	>100 μM
CYP2D6	negligible	>100 μM
CYP3A4	negligible	>100 μM
Estrogen α,	weak binding	3.54 μM

Table 2. *Cont.*

Estrogen β	moderate binding	448 nM
Target	**Binding Type**	**Binding Affinity (VirtualToxLab)**
Glucocorticoid	moderate binding	574 nM
Herg k⁺ channel	weak binding	4.87 μM
Liver x	weak binding	72.7 μM
Mineralocorticoid	weak binding	14.7 μM
Pparγ	weak binding	4.53 μM
Progesterone	weak binding	37.2 μM
Thyroid α,	weak binding	15.3 μM
Thyroid β	weak binding	5.85 μM

Overall toxic potential was found to be 0.418 [16].

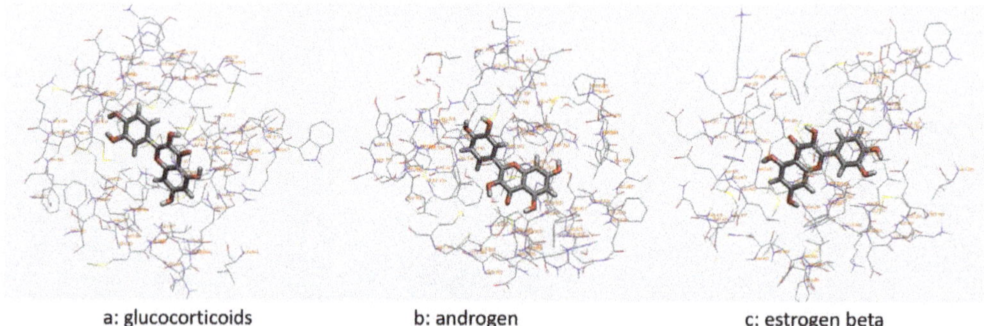

Figure 3. Docking conformation of quercetin with different targets. (**a**) Predicted bonded interactions (blue dashed lines) between quercetin and glucocorticoid; (**b**) binding interaction between quercetin and androgen; (**c**) binding interaction of quercetin and estrogen alpha. The ligand is based on atom type and the protein-based on amino acid residue type coloring.

4. Discussion

CCL4 is a lipophilic colorless liquid and readily crosses cell membranes, including the blood–brain barrier. All body tissues rapidly take up CCL4, but the toxicity on the brain remains poorly understood. Some studies reported that adverse events of CCL4 in rodents are mainly based on systemic effects in the liver (centrilobular necrosis) and some impact on the kidney in inducing free radical toxicity and some tissues injuries [26]. However, in this study, we evaluated all the parameters in the brain tissue itself. In this study, the elevation of ROS was reported in brain cells on exposures to a single dose of CCL4 that is scavenged by antioxidant quercetin during normal cell metabolism. Histopathological analysis on brain tissues indicated a CCL4 induced karyopyknosis, apoptosis, and swelling. The toxic effect of CCL4 is believed to be due to trichloromethyl radicals produced during oxidative stress, as per previous literature [27]. The results of our study are in corroboration with Ritesh et al., 2015, who reported that a single hepatotoxic dose of CCL4 is equally neurotoxic to rats as many consecutive doses [5]. Organ and body weights showed significantly decreased growth changes in the brain-to-body weight ratios. The brain/bw ratio is one of the most sensitive parameters measured for detecting the effect of exposure to toxins on growth and development [28]. CCL4 significantly altered AST and ALT activity in brain tissue, and no modified ALT levels were observed in the prophylactically quercetin-treated group. CCL4 treatment causes brain tissue damage followed by the release of AST molecules into the extracellular spaces of brain tissue. The function of this enzyme is the reversible transport of amines from aspartate to α-ketoglutarate [29]. In concomitance to this study,

Kelbich et al. reported an increase in AST catalytic activities in the cerebrospinal fluids of cerebral ischemia patients [30].

Furthermore, the increased AST activity could be connected with increased transport of NADH from the cytosol to mitochondria. In contrast, the increased ALT activity would represent more transformation of pyruvate to alanine due to increased glycolysis and hence increased pyruvate [31]. Increased brain urea by CCL4 was reversed by quercetin. The 'reverse urea effect causes brain edema', i.e., the significant urea gradient between blood and brain causes an inflow of water into the brains of induced animals [32]. For the first time, this study has evaluated brain sodium levels with CCL4 exposure. Brain sodium is increased with sympathoexcitatory and pressor responses [33], which further destroys brain potential and leads to brain inflammation. In addition, increased sodium levels hyperactivate Na/K channels that trigger excitotoxic neuron death [34]. In concomitance to sodium levels, the Herg k^+ channel (Pottasium channel) was found to show weak binding with quercetin in silico analysis. Hence, it can be predicted that quercetin protected via the action of Na^+/K^+ pump. However, the action of protection by quercetin with respect to Na^+/K^+ pump must be verified in future studies. Glutathione eliminates reactive oxygen species (ROS) produced in oxidative stress [35]. GSH, a ubiquitous intracellular cytosolic tripeptide at millimolar concentrations, is the primary non-enzymatic biomarker of redox homeostasis. The reduced GSH in brains after the exposure to CCL4 could result from increased GSH-peroxidase activity in the exposed rat. The decreased GSH was further associated with ascorbic acid and lipid peroxidation [19,36,37]

Additionally, glutathione depletion induces glycogenolysis-dependent ascorbic acid synthesis in murine hepatocytes in vitro [38]. The significant decreased levels of antioxidant enzymes such as CAT and SOD were reversed to normal by quercetin supplementation (Table 1). Based on our results, we conclude that quercetin shows significant prophylactic effects against CCL4 with respect to the studied parameters. Thus, it was indicated that quercetin prevents alterations in oxidative stress parameters and neurotransmitters parameters [39]. Our results highlight the importance of understanding the potential prophylactic effects of quercetin against neurotoxicity.

Based on the binding affinities of quercetin to various proteins, the normalized binding affinity of quercetin to the studied proteins was found to be 0.418 (average). It should be noted that the toxicity values can be overestimated since binding to a particular protein may or may not lead to any adverse event. The software used for the flexible docking in VirtualToxLab is Alignator and Cheetah. All the binding modes between the quercetin and the target proteins were identified during the study. The interaction between quercetin and the target proteins, which showed moderate binding, is given in Figure 3. The binding affinity are defined between >100 μM (not binding) to <1.0 nM (strong binding). The overall binding potential ranged from 0.0 (benign) to 1.0 (extreme) [16].

Moderate binding between quercetin and glucocorticoids, estrogen beta, and androgens was evaluated. Previous studies suggested that oxidative stress increases glucocorticoid (GC) hormones; hence, the hippocampus, which has a high concentration of GC receptors, is especially vulnerable to elevated levels of GCs. The GCs have been suggested to endanger hippocampal neurons by exacerbating the excitotoxic glutamate-calcium-reactive oxygen species (ROS) cascade [40–42]. Our binding results suggest quercetin binding to glucocorticoids as a preventing mechanism for preventing ROS cascade generated by CCL4 neurotoxicity.

Previous studies demonstrated that estrogen receptor β (ERβ) signaling alleviates systemic inflammation in animal models, and suggested that ERβ-selective agonists may deactivate microglia and suppress T cell activity via the downregulation of nuclear factor k-light-chain-enhancer of activated B cells (NF-kB) [43]. The estrogen receptor agonists play an essential role in protecting the central nervous system against neuroinflammation and neurodegeneration. Quercetin aglycone and its glucuronide possess estrogenic activity, and quercetin is also classified as a phytoestrogen [8,44]. Thus, the binding of quercetin to estrogen alpha and beta might have caused the protection against the stress induced by carbon tetrachloride [8]. Future studies are needed to verify this mechanism for protection. Another possible mechanism for the protection of quercetin can be the suppression of stress-induced plasma corticosterone and adenocorticosterone levels. The DNA binding activity of the glucocorticoid receptor is modulated on binding to quercetin and is one of the possible reasons for suppressing plasma corticosterone and adrenocorticotropic hormone levels [45,46]. The effect of androgens on the cerebral vasculature is a complex mechanism. They have both protective and detrimental effects, depending on several factors, such as age, drug dose, and state of disease. Chronically elevated androgens are pro-angiogenic, promote vasoconstriction, and influence blood–brain barrier permeability. In addition to these, androgens have been shown to affect the cerebral vasculature [44] directly. This study found moderate binding with androgens; hence, we propose that elevated androgen levels by CCL4, which could otherwise be harmful to the brain, were probably prevented by quercetin by binding with androgen receptors. Future studies are needed to verify our hypothesis.

Author Contributions: Conceptualization: S.Z.; Methodology: S.Z.; Software: T.A.W.; Formal analysis: T.A.W.; S.Z.; Investigation: T.A.W.; S.Z.; Resources: S.Z.; Writing: S.Z.; T.A.W.; Review & Editing: S.Z.; Project administration: T.A.W. All authors have read and agreed to the published version of the manuscript.

Funding: Research Groups, Deanship of Scientific Research, King Saud University, Grant Number: RG-1438-042.

Institutional Review Board Statement: The study was conducted according to the guidelines of the Declaration of Helsinki, and approved by the Institutional Review Board of King Saud University approval no. KSU-SE-21-05 dated 06-12-2020.

Informed Consent Statement: Not applicable.

Data Availability Statement: Data will be available on request to corresponding author.

Acknowledgments: This work was supported by the Deanship of Scientific Research, King Saud University; Research group No. RG-1438-042.

Conflicts of Interest: The authors declare no conflict of interest.

References

1. Usui, T.; Foster, S.S.; Petrini, J.H. Maintenance of the DNA-damage checkpoint requires DNA-damage-induced mediator protein oligomerization. *Mol. Cell* **2009**, *33*, 147–159. [CrossRef]
2. Petković, J.; Žegura, B.; Stevanović, M.; Drnovšek, N.; Uskoković, D.; Novak, S.; Filipič, M. DNA damage and alterations in expression of DNA damage responsive genes induced by TiO_2 nanoparticles in human hepatoma HepG2 cells. *Nanotoxicology* **2011**, *5*, 341–353. [CrossRef]
3. Dong, S.; Chen, Q.-L.; Song, Y.-N.; Sun, Y.; Wei, B.; Li, X.-Y.; Hu, Y.-Y.; Liu, P.; Su, S.-B. Mechanisms of CCl4-induced liver fibrosis with combined transcriptomic and proteomic analysis. *J. Toxicol. Sci.* **2016**, *41*, 561–572. [CrossRef]
4. Alkreathy, H.M.; Khan, R.A.; Khan, M.R.; Sahreen, S. CCl 4 induced genotoxicity and DNA oxidative damages in rats: Hepatoprotective effect of Sonchus arvensis. *BMC Complementary Altern. Med.* **2014**, *14*, 452. [CrossRef] [PubMed]
5. Ritesh, K.; Suganya, A.; Dileepkumar, H.; Rajashekar, Y.; Shivanandappa, T. A single acute hepatotoxic dose of CCl4 causes oxidative stress in the rat brain. *Toxicol. Rep.* **2015**, *2*, 891–895. [CrossRef] [PubMed]
6. Hafez, M.M.; Al-Shabanah, O.A.; Al-Harbi, N.O.; Al-Harbi, M.M.; Al-Rejaie, S.S.; Alsurayea, S.M.; Sayed-Ahmed, M.M. Association between paraoxonases gene expression and oxidative stress in hepatotoxicity induced by CCl4. *Oxidative Med. Cell. Longev.* **2014**, *2014*, 893212. [CrossRef]
7. Khan, R.A.; Khan, M.R.; Sahreen, S. CCl 4-induced hepatotoxicity: Protective effect of rutin on p53, CYP2E1 and the antioxidative status in rat. *BMC Complementary Altern. Med.* **2012**, *12*, 178. [CrossRef] [PubMed]
8. Costa, L.G.; Garrick, J.M.; Roquè, P.J.; Pellacani, C. Mechanisms of neuroprotection by quercetin: Counteracting oxidative stress and more. *Oxidative Med. Cell. Longev.* **2016**, *2016*, 2986796. [CrossRef]
9. Massi, A.; Bortolini, O.; Ragno, D.; Bernardi, T.; Sacchetti, G.; Tacchini, M.; De Risi, C. Research progress in the modification of quercetin leading to anticancer agents. *Molecules* **2017**, *22*, 1270. [CrossRef]
10. Ma, J.-Q.; Li, Z.; Xie, W.-R.; Liu, C.-M.; Liu, S.-S. Quercetin protects mouse liver against CCl4-induced inflammation by the TLR2/4 and MAPK/NF-κB pathway. *Int. Immunopharmacol.* **2015**, *28*, 531–539. [CrossRef]

11. Paulke, A.; Eckert, G.P.; Schubert-Zsilavecz, M.; Wurglics, M. Isoquercitrin provides better bioavailability than quercetin: Comparison of quercetin metabolites in body tissue and brain sections after six days administration of isoquercitrin and quercetin. *Die Pharm. Int. J. Pharm. Sci.* **2012**, *67*, 991–996.

12. Zargar, S.; Siddiqi, N.J.; Ansar, S.; Alsulaimani, M.S.; El Ansary, A.K. Therapeutic role of quercetin on oxidative damage induced by acrylamide in rat brain. *Pharm. Biol.* **2016**, *54*, 1763–1767. [CrossRef]

13. Granado-Serrano, A.B.; Martín, M.A.; Bravo, L.; Goya, L.; Ramos, S. Quercetin modulates Nrf2 and glutathione-related defenses in HepG2 cells: Involvement of p38. *Chem.-Biol. Interact.* **2012**, *195*, 154–164. [CrossRef] [PubMed]

14. Xu, D.; Hu, M.-J.; Wang, Y.-Q.; Cui, Y.-L. Antioxidant activities of quercetin and its complexes for medicinal application. *Molecules* **2019**, *24*, 1123. [CrossRef]

15. Gulec, S.; Ozdol, C.; Vurgun, K.; Selcuk, T.; Turhan, S.; Duzen, V.; Temizhan, A.; Ozturk, S.; Ozdemir, A.O.; Erol, C. The effect of high-dose aspirin pre-treatment on the incidence of myonecrosis following elective coronary stenting. *Atherosclerosis* **2008**, *197*, 171–176. [CrossRef]

16. Vedani, A.; Dobler, M.; Hu, Z.; Smieško, M. OpenVirtualToxLab—A platform for generating and exchanging in silico toxicity data. *Toxicol. Lett.* **2015**, *232*, 519–532. [CrossRef] [PubMed]

17. Siddiqi, N.J.; Zargar, S. Effect of quercetin on cadmium fluoride-induced alterations in hydroxyproline/collagen content in mice liver. *Connect. Tissue Res.* **2014**, *55*, 234–238. [CrossRef]

18. Ansar, S.; Siddiqi, N.J.; Zargar, S.; Ganaie, M.A.; Abudawood, M. Hepatoprotective effect of Quercetin supplementation against Acrylamide-induced DNA damage in wistar rats. *BMC Complementary Altern. Med.* **2016**, *16*, 327. [CrossRef] [PubMed]

19. Zargar, S.; Siddiqi, N.J.; Al Daihan, S.K.; Wani, T.A. Protective effects of quercetin on cadmium fluoride induced oxidative stress at different intervals of time in mouse liver. *Acta Biochim. Pol.* **2015**, *62*, 207–213. [CrossRef]

20. Zargar, S.; Siddiqi, N.J.; Khan, T.H.; Elredah, I.E. Effect of cadmium fluoride and quercetin on in vivo activity of indoleamine 2,3-dioxygenase in mice liver and kidney. *Fluoride* **2014**, *47*, 31–42.

21. Nishikimi, M.; Rao, N.A.; Yagi, K. The occurrence of superoxide anion in the reaction of reduced phenazine methosulfate and molecular oxygen. *Biochem. Biophys. Res. Commun.* **1972**, *46*, 849–854. [CrossRef]

22. Aebi, H. [13] Catalase in vitro. *Methods Enzymol.* **1984**, *105*, 121–126. [PubMed]

23. Utley, H.G.; Bernheim, F.; Hochstein, P. Effect of sulfhydryl reagents on peroxidation in microsomes. *Arch. Biochem. Biophys.* **1967**, *118*, 29–32. [CrossRef]

24. Jagota, S.; Dani, H. A new colorimetric technique for the estimation of vitamin C using Folin phenol reagent. *Anal. Biochem.* **1982**, *127*, 178–182. [CrossRef]

25. Beutler, E. Improved method for the determination of blood glutathione. *J. Lab. Clin. Med.* **1963**, *61*, 882–888.

26. Masuda, Y. Learning toxicology from carbon tetrachloride-induced hepatotoxicity. *Yakugaku Zasshi J. Pharm. Soc. Jpn.* **2006**, *126*, 885–899. [CrossRef]

27. Shah, G.H.; Patel, B.G.; Shah, G.B. Development of carbon tetrachloride-induced chronic hepatotoxicity model in rats and its application in evaluation of hepatoprotective activity of silymarin. *Asian J. Pharm. Clin. Res.* **2017**, *10*, 274–278. [CrossRef]

28. Cragg, B.; Rees, S. Increased body: Brain weight ratio in developing rats after low exposure to organic lead. *Exp. Neurol.* **1984**, *86*, 113–121. [CrossRef]

29. Graefe, G.; Karlson, P. *Kurzes Lehrbuch der Biochemie für Mediziner und Naturwissenschaftler*; Geleitwort von A. Butenandt, 8., völlig neubearbeitete Auflage, Georg Thieme Verlag, Stuttgart, 1972. 405 Seiten, 82 Abb. in 105 Einzeldarstellungen, 21 Tabellen, 1 Falttafel, PVC-kartoniert DM 29, 80; Wiley Online Library: Hoboken, NJ, USA, 1973.

30. Kelbich, P.; Radovnický, T.; Selke-Krulichová, I.; Lodin, J.; Matuchová, I.; Sameš, M.; Procházka, J.; Krejsek, J.; Hanuljaková, E.; Hejčl, A. Can Aspartate Aminotransferase in the Cerebrospinal Fluid Be a Reliable Predictive Parameter? *Brain Sci.* **2020**, *10*, 698. [CrossRef]

31. Netopilová, M.; Haugvicová, R.; Kubová, H.; Dršata, J.; Mareš, P. Influence of convulsants on rat brain activities of alanine aminotransferase and aspartate aminotransferase. *Neurochem. Res.* **2001**, *26*, 1285–1291. [CrossRef]

32. Trinh-Trang-Tan, M.-M.; Cartron, J.-P.; Bankir, L. Molecular basis for the dialysis disequilibrium syndrome: Altered aquaporin and urea transporter expression in the brain. *Nephrol. Dial. Transplant.* **2005**, *20*, 1984–1988. [CrossRef]

33. Huang, B.S.; Leenen, F. Sympathoexcitatory and pressor responses to increased brain sodium and ouabain are mediated via brain ANG II. *Am. J. Physiol. -Heart Circ. Physiol.* **1996**, *270*, H275–H280. [CrossRef] [PubMed]

34. Zou, S.; Chisholm, R.; Tauskela, J.S.; Mealing, G.A.; Johnston, L.J.; Morris, C.E. Force spectroscopy measurements show that cortical neurons exposed to excitotoxic agonists stiffen before showing evidence of bleb damage. *PLoS ONE* **2013**, *8*, e73499. [CrossRef]

35. Lushchak, V.I. Glutathione homeostasis and functions: Potential targets for medical interventions. *J. Amino Acids* **2012**, *2012*, 736837. [CrossRef] [PubMed]

36. Bánhegyi, G.; Csala, M.; Braun, L.; Garzó, T.; Mandl, J. Ascorbate synthesis-dependent glutathione consumption in mouse liver. *FEBS Lett.* **1996**, *381*, 39–41. [CrossRef]

37. Mårtensson, J.; Meister, A. Glutathione deficiency increases hepatic ascorbic acid synthesis in adult mice. *Proc. Natl. Acad. Sci. USA* **1992**, *89*, 11566–11568. [CrossRef]

38. Braun, L.; Csala, M.; Poussu, A.; Garzó, T.; Mandl, J.; Bánhegyi, G. Glutathione depletion induces glycogenolysis dependent ascorbate synthesis in isolated murine hepatocytes. *FEBS Lett.* **1996**, *388*, 173–176. [CrossRef]

39. Abdalla, F.H.; Schmatz, R.; Cardoso, A.M.; Carvalho, F.B.; Baldissarelli, J.; de Oliveira, J.S.; Rosa, M.M.; Nunes, M.A.G.; Rubin, M.A.; da Cruz, I.B. Quercetin protects the impairment of memory and anxiogenic-like behavior in rats exposed to cadmium: Possible involvement of the acetylcholinesterase and Na+, K+-ATPase activities. *Physiol. Behav.* **2014**, *135*, 152–167. [CrossRef]
40. You, J.-M.; Yun, S.-J.; Nam, K.N.; Kang, C.; Won, R.; Lee, E.H. Mechanism of glucocorticoid-induced oxidative stress in rat hippocampal slice cultures. *Can. J. Physiol. Pharmacol.* **2009**, *87*, 440–447. [CrossRef]
41. Uno, H.; Eisele, S.; Sakai, A.; Shelton, S.; Baker, E.; Dejesus, O.; Holden, J. Neurotoxicity of glucocorticoids in the primate brain. *Horm Behav.* **1994**, *28*, 336–348. [CrossRef] [PubMed]
42. Barton, L.; BUTTERBACH-BAHL, K.; Kiese, R.; Murphy, D.V. Nitrous oxide fluxes from a grain–legume crop (narrow-leafed lupin) grown in a semiarid climate. *Global Change Biol.* **2011**, *17*, 1153–1166. [CrossRef]
43. Xiao, L.; Luo, Y.; Tai, R.; Zhang, N. Estrogen receptor β suppresses inflammation and the progression of prostate cancer. *Mol. Med. Rep.* **2019**, *19*, 3555–3563. [CrossRef]
44. Ruotolo, R.; Calani, L.; Brighenti, F.; Crozier, A.; Ottonello, S.; Del Rio, D. Glucuronidation does not suppress the estrogenic activity of quercetin in yeast and human breast cancer cell model systems. *Arch. Biochem. Biophys.* **2014**, *559*, 62–67. [CrossRef] [PubMed]
45. Kawabata, K.; Kawai, Y.; Terao, J. Suppressive effect of quercetin on acute stress-induced hypothalamic-pituitary-adrenal axis response in Wistar rats. *J. Nutr. Biochem.* **2010**, *21*, 374–380. [CrossRef] [PubMed]
46. Gélinas, S.; Martinoli, M.G. Neuroprotective effect of estradiol and phytoestrogens on MPP+-induced cytotoxicity in neuronal PC12 cells. *J. Neurosci. Res.* **2002**, *70*, 90–96. [CrossRef] [PubMed]

Article

Biochemical and Biophysical Characterisation of the Hepatitis E Virus Guanine-7-Methyltransferase

Preeti Hooda [1,†], Mohd Ishtikhar [1,†], Shweta Saraswat [1], Pooja Bhatia [1], Deepali Mishra [1], Aditya Trivedi [1], Rajkumar Kulandaisamy [2], Soumya Aggarwal [3], Manoj Munde [3], Nemat Ali [4], Abdullah F. AlAsmari [4], Mohd A. Rauf [5], Krishna K. Inampudi [2,*] and Deepak Sehgal [1,*]

[1] Virology Lab, Department of Life Sciences, Shiv Nadar University, Greater Noida 201314, India; ph236@snu.edu.in (P.H.); iftikharbiochem@gmail.com (M.I.); shweta.saraswat@nii.ac.in (S.S.); pb671@snu.edu.in (P.B.); mishratani2012@gmail.com (D.M.); at662@snu.edu.in (A.T.)
[2] Department of Biophysics, All India Institute of Medical Sciences (AIIMS), New Delhi 110029, India; danielrajkumar@aiims.ac.in
[3] School of Physical Sciences, Jawaharlal Nehru University (JNU), New Delhi 110067, India; soumya22_sps@jnu.ac.in (S.A.); mmunde@mail.jnu.ac.in (M.M.)
[4] Department of Pharmacology and Toxicology, College of Pharmacy, King Saud University, P.O. Box 55760, Riyadh 11451, Saudi Arabia; nali1@ksu.edu.sa (N.A.); afalasmari@ksu.edu.sa (A.F.A.)
[5] Department of Pharmaceutical Sciences, Wayne State University, Detroit, MI 48201, USA; ahmarrauf2@gmail.com
* Correspondence: krishna.inampudi@aiims.edu (K.K.I.); deepak.sehgal@snu.edu.in (D.S.); Tel.: +91-9640600447 (K.K.I.); +91-9910031591 (D.S.)
† These authors contributed equally to this work.

Citation: Hooda, P.; Ishtikhar, M.; Saraswat, S.; Bhatia, P.; Mishra, D.; Trivedi, A.; Kulandaisamy, R.; Aggarwal, S.; Munde, M.; Ali, N.; et al. Biochemical and Biophysical Characterisation of the Hepatitis E Virus Guanine-7-Methyltransferase. *Molecules* **2022**, *27*, 1505. https://doi.org/10.3390/molecules27051505

Academic Editor: Luigi A. Agrofoglio

Received: 28 December 2021
Accepted: 18 February 2022
Published: 23 February 2022

Publisher's Note: MDPI stays neutral with regard to jurisdictional claims in published maps and institutional affiliations.

Abstract: Hepatitis E virus (HEV) is an understudied pathogen that causes infection through fecal contaminated drinking water and is prominently found in South Asian countries. The virus affects ~20 million people annually, leading to ~60,000 infections per year. The positive-stranded RNA genome of the HEV genotype 1 has four conserved open reading frames (ORFs), of which ORF1 encodes a polyprotein of 180 kDa in size, which is processed into four non-structural enzymes: methyltransferase (MTase), papain-like cysteine protease, RNA-dependent RNA polymerase, and RNA helicase. MTase is known to methylate guanosine triphosphate at the 5′-end of viral RNA, thereby preventing its degradation by host nucleases. In the present study, we cloned, expressed, and purified MTase spanning 33–353 amino acids of HEV genotype 1. The activity of the purified enzyme and the conformational changes were established through biochemical and biophysical studies. The binding affinity of MTase with magnesium ions (Mg^{2+}) was studied by isothermal calorimetry (ITC), microscale thermophoresis (MST), far-UV CD analysis and, fluorescence quenching. In summary, a short stretch of nucleotides has been cloned, coding for the HEV MTase of 37 kDa, which binds Mg^{2+} and modulate its activity. The chelation of magnesium reversed the changes, confirming its role in enzyme activity.

Keywords: hepatitis E virus; methyltransferase; fluorescence quenching; protein–ligand interaction; protein stability; enzyme assay

1. Introduction

Hepatitis E virus (HEV) is inherently hepatotropic, causing acute hepatitis and chronic infection in immunocompromised patients, as well as leading to extrahepatic manifestations in some patients [1–4]. Globally, hepatitis E accounts for an estimated mortality rate of ~3.3% in the infected population and causes fulminant hepatitis failure in 25–30% of infected pregnant women [5]. HEV has a genome of ~7.2 kb plus-stranded RNA with a 5′-methylguanine (m^7G) cap accorded by guanylyltransferase (GTase) and methyltransferase (MTase). The RNA capping is essential for the viruses to evade the host immune system and produce other viral proteins by protecting the viral mRNA from nucleases.

In the case of HEV, the $5'$ m^{7}G cap has been demonstrated to play an additional role, increasing infectivity in non-human primates and cultured hepatoma cells [6,7]. Despite being an important enzyme, few studies have examined the functional and structural aspects of the HEV MTase. Magden et al. 2001 expressed a 110 kDa protein from the 1–979 amino acid region of HEV cDNA and demonstrated both the activities of MTase and GTase [8]. Before that, using computational homology modelling, Koonin et al. derived that the MTase domain lay in the region of aa 56–240 of the HEV genome [9]. Through bioinformatic studies, Emerson et al. predicted that the MTase lay in the region of aa 33–353 of HEV genotype 1 [6]. This fragment was further preferred since it formed a section of the MTase domain of aa 1–979 expressed by Magden and overlapping the region of amino acids 56–240, computationally predicted to be MTase by Koonin. Howver, expressing the protein containing this predicted fragment and validating the MTase activity is an unexplored area.

Other viruses in which capping is essential for MTase activity include the Dengue virus [10], coronaviruses [11], and flaviviruses [12]. MTase is an integral enzyme required for capping, which is dependent on magnesium (Mg^{2+}) as a cofactor for its activity [13]. The dependence of MTase on divalent cations for its activity has been observed in the case of many viruses, such as the Zika virus [14] and respiratory syncytial virus (RSV) [15]. Many viruses, such as the Chlorella virus and hepatitis C virus (HCV), coronaviruses, and flaviviruses, also need Mg^{2+} for the activity of viral enzymes [16–20]. Alteration of the activity may be due to conformational changes resulting from the binding of magnesium to the enzyme. Following this hypothesis, the role of Mg^{2+} in the MTase activity of coronaviruses has been attributed to the conformational changes in the nsp10/nsp16 enzyme complex for $2'$O-methylation [21]. The binding is postulated to induce the change in the structural, conformational, and interactional properties of MTase. As seen in the Dengue virus, Mg^{2+} stabilises the RNA cap by coordinating with the inverted triphosphate moiety from the solvent-exposed side of the RNA cap [22].

Briefly, in the present study, we have reported and expressed the HEV MTase from the predicted domain of the 33–353 amino acid fragment of HEV genotype 1. The molecular weight of the enzyme with MTase activity was found to be 37 kDa. The enzyme activity was shown to be associated with Mg^{2+} using biochemical and biophysical studies.

Our work suggested that HEV MTase requires Mg^{2+} for its activity, and future studies should help to establish the direct relationship between RNA capping and host cellular metal ions. Further, the identification of drug-like inhibitors that are structurally compatible with the domains responsible for the enzyme activity could be performed. Using this information, the co-crystallisation of MTase with magnesium can be performed, to elucidate the exact binding pocket of Mg^{2+} and hence design structurally compatible inhibitors.

2. Results

2.1. Expression and Purification of MTase

For studying the biochemical and biophysical characteristics of the enzyme MTase, the gene for MTase was cloned into the pET-28a vector and validated by restriction digestion of a ~969 bp fragment (Figure 1A) and confirmed by DNA sequencing (data not shown). The recombinant MTase construct was transformed into BL21-DE3 cells to express the protein, as demonstrated using SDS-PAGE and Western blotting with proper controls (Figure 1B,C; Lanes 2 and 3). The protein was solubilized using different detergents, but achieved maximum solubility with 0.5% NLS (Figure 1B,C; Lane 4). The recombinant protein of 37 kDa, with a His-tag, was purified using Ni-NTA chromatography and confirmed to be MTase, as shown by SDS-PAGE and western blotting using epitope-specific antibodies (Figure 1B,C; Lane 5). In conclusion, the protein was purified and confirmed to be MTase through enzyme assay using GTP as the substrate.

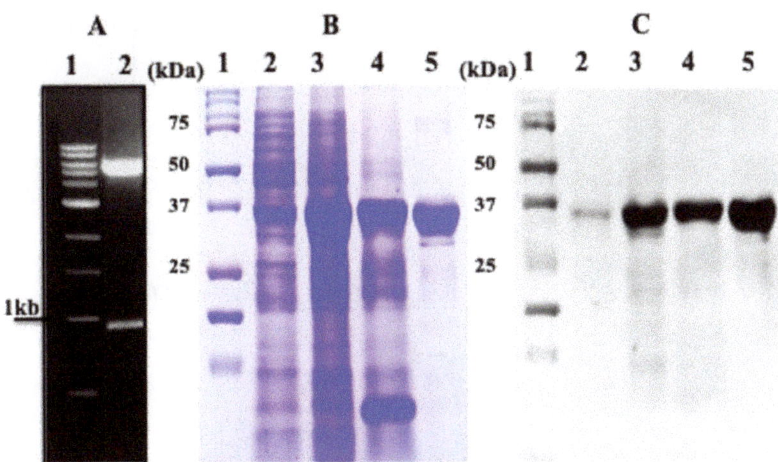

Figure 1. Cloning, expression, and purification of MTase: (A) The cloned MTase gene (969 bp) was confirmed by restriction digestion using NdeI and XhoI (Panel A). **(B)** Coomassie-stained gel representing the expressed and purified fractions of HEV MTase. Lane 1, marker; lane 2, uninduced cell lysate; lane 3, induced cell lysate; lane 4, solubilised fraction; lane 5, Ni-NTA purified fraction. **(C)** Western blotting analysis using anti-HEV MTase primary antibody. Lane 1, marker; lane 2, uninduced cell lysate; lane 3, induced cell lysate; lane 4, solubilised fraction; lane 5, Ni-NTA purified fraction.

2.2. MTase Activity of 37 kDa Protein

2.2.1. Enzyme Kinetics

MTase kinetics were studied as described in Materials and Methods to determine the nature of the enzyme and its activity. The activity of the purified protein was determined through different parameters, including various ranges of temperature, pH, and time. A linear increase in the enzyme activity was observed upon the increase in MTase concentration from 0–5 µM (Figure 2A). Further, enzyme activity was studied in the pH range from 6 to 12, leading to a bell-shaped curve with an optimal pH of 8.0 (Figure 2B). Additionally, the optimal temperature of the enzyme activity was found to be 37 °C (Figure 2C), above which the enzyme became denatured. Further, through the time-course assay, the optimal enzyme activity was found at 150 min, after which it plateaued (Figure 2D). For all future studies, the concentration of the enzyme used was 0.675 µM, while the concentration of GTP used was 0.4 mM. The graphs were plotted in terms of S-adenosyl homocysteine (SAH) concentration, determined using the standard curve. All the reactions were performed in triplicate, and the readings were subtracted from those of the no enzyme control. Each data point in the graph represents the mean value, and the error bars indicate the standard deviation.

2.2.2. MTase Activity in the Presence of Guanosine Triphosphate (GTP) as a Substrate

The MTase kinetics were performed in the presence of GTP as its substrate while keeping the concentrations of S-adenosyl methionine (SAM) (methyl donor) and the enzyme constant at 1 µM and 0.675 µM, respectively. The concentration of the GTP substrate (methyl acceptor) ranged from 0 to 10 mM. The Michaelis–Menten equation was used to calculate the k_m value for GTP, which was found to be 0.387 mM (Figure 3A). The k_m value was also determined using a Lineweaver–Burke plot produced by GraphPad Prism 9. The straight-line equation for the plot, $Y = 0.0003207 \times X + 0.0008274$, was used to calculate the k_m and V_{max} values (Figure 3B). The calculated k_m and V_{max} values from the Lineweaver–Burke plot were found to be 0.387 mM and 120,496, respectively.

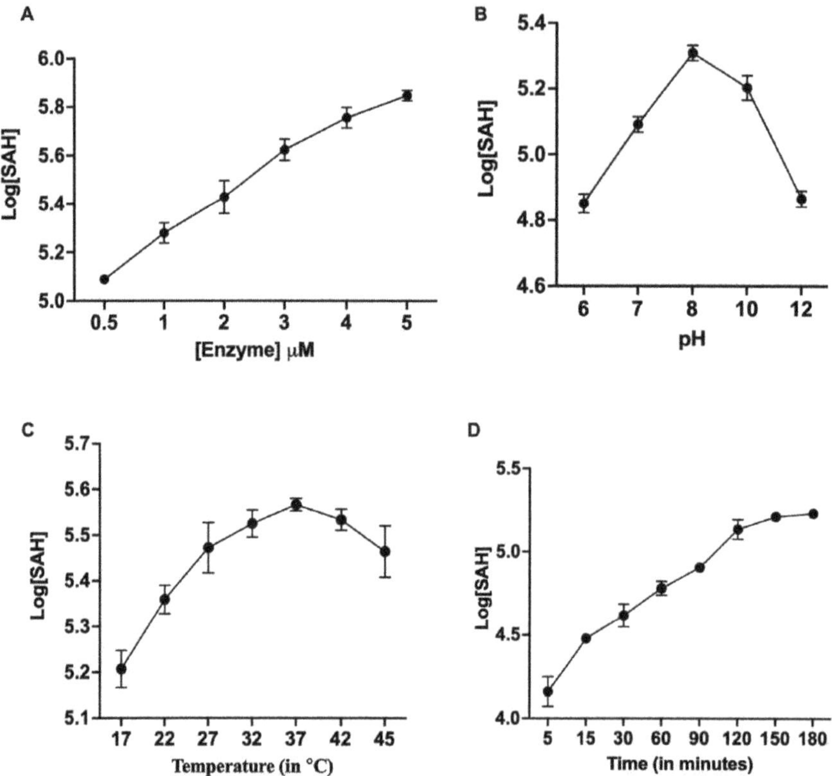

Figure 2. Effect of various parameters on MTase activity: The MTase activity was optimised by adjusting various parameters. (**A**) MTase activity at varying enzyme concentrations. (**B**) Enzyme activity at different pH values. (**C**) Enzyme activity at different temperatures. (**D**) Enzyme activity at various time points. The observations are made from experiments performed in triplicate. Each data point on the graph represents the mean value, and the error bars indicate the standard deviation.

2.2.3. Substrate Specificity of MTase

The substrate specificity of the enzyme was studied using different nucleotides, cytidine triphosphate (CTP), uridine triphosphate (UTP), cyclic adenosine monophosphate (cAMP), guanosine diphosphate (GDP), GTP, and deoxy guanosine triphosphate (dGTP), as well as the capped analogues, 7-methyl guanosine diphosphate (m^7-GDP) and 7-methyl-guanosine triphosphate (m^7-GTP). The reaction was performed, in which 0.675 μM of MTase was tested for enzyme activity in the presence of 0.5 mM of different nucleotide triphosphates (NTPs) as cap analogues, as mentioned above. The concentration of SAM, the methyl donor, was kept constant at 1 μM in the reactions. The activity assay demonstrated that GDP is a better substrate than GTP, while minimal activity was observed for CTP, UTP, cAMP, and dGTP (Figure 4). The cap analogues, m^7-GDP and m^7-GTP, showed minimal activity. These results showed the specificity of MTase activity, indicating that there must be a free position on the substrate to transfer the methyl group. The minimal activity for m^7-GDP and m^7-GTP was observed since the N-7 position of the guanosine is already methylated [8] (Figure 4).

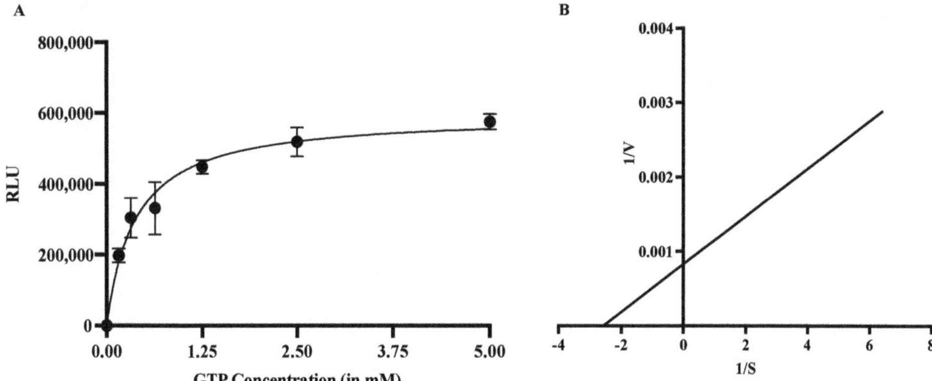

Figure 3. Kinetic studies of MTase: (**A**) The reaction mixture contained 1 μM SAM and different concentrations of GTP. Each data point represents the mean value, and the error bars indicate the standard deviation. The K_m value of the substrate, GTP, was calculated using the Michaelis–Menten equation and found to be approximately 0.38 mM. (**B**) The Lineweaver–Burke plot was produced by GraphPad Prism 9. A straight-line equation was determined using K_m and V_{max} ($Y = 0.0003207 \times X + 0.0008274$). The calculated K_m and V_{max} values using the Lineweaver–Burke plot were 0.387 mM and 120,496, respectively.

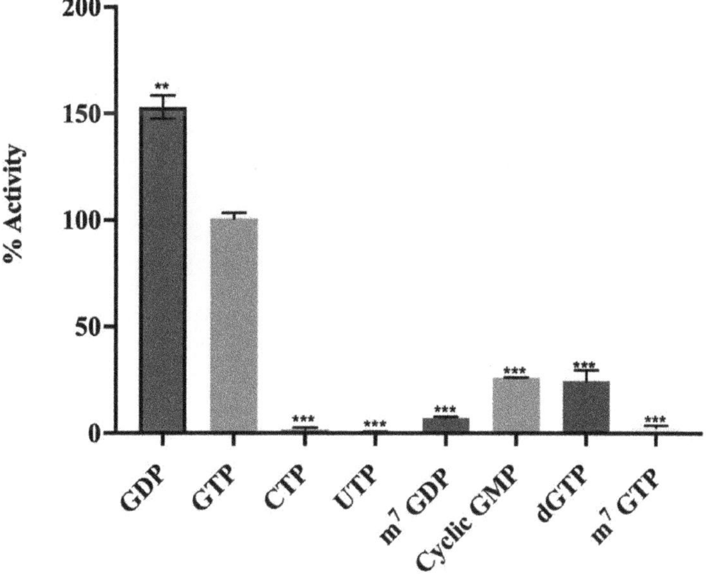

Figure 4. Effect of different substrates on HEV MTase activity: The graph represents the activity of MTase using different substrate analogues. The bars represent various nucleotide analogues, as indicated in the figure. The values have been normalised to the percentage activity of the enzyme in the presence of GTP. The enzyme assay was performed in triplicate; the graph represents the mean value while the error bar indicates the standard deviation. The statistical significance of the data was determined using Student's *t*-test. ** *p*-value < 0.05, *** *p*-value < 0.0005.

2.2.4. Effect of Magnesium (Mg^{2+}) on MTase Activity

Although divalent cations are crucial for the activity of many viral enzymes, no study has been observed the role of divalent cations on MTase activity. Hence, MTase activity in response to varying concentrations of Mg^{2+} was studied. This resulted in the linear increase in MTase activity for up to 2 mM of magnesium, after which it became saturated (Figure 5A). The effect of another important cation, Ca^{2+}, was checked for its activity of the enzyme, but no effect on MTase activity was seen (data not shown). To further validate the impact of Mg^{2+}, the reaction was performed in the presence of EDTA, which is supposed to be a chelating agent for the magnesium. When increasing the concentration of EDTA from 0 to 8 mM, the enzyme activity decreased significantly, establishing that the presence of magnesium is essential for the MTase activity (Figure 5B).

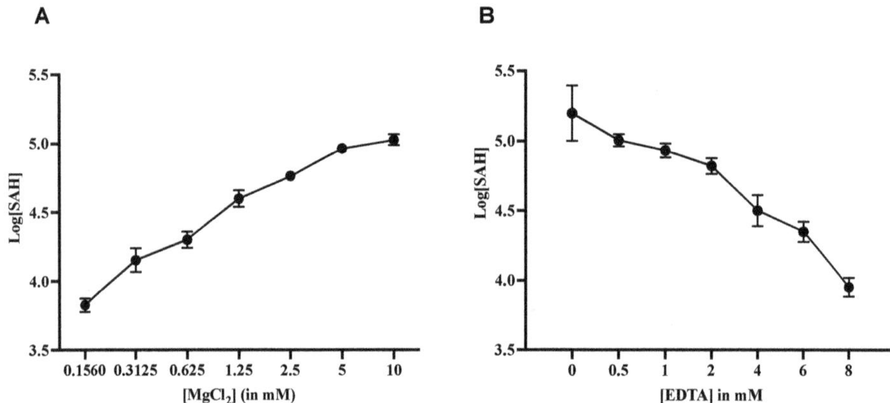

Figure 5. Effect of magnesium on MTase activity: (**A**) The graph represents MTase activity when increasing the concentration of $MgCl_2$ up to 10 mM. (**B**) The effect of EDTA on the enzyme activity. The graphs represent the mean value of three different readings and the error bars indicate the standard deviation.

2.3. Circular Dichroism (CD) Analysis

Far-UV CD experiments were performed to investigate the changes in secondary structural content of MTase at different pH values or Mg^{2+} concentrations [23]. The obtained MTase spectra showed two negative peaks around 208 nm and 222 nm. A positive peak at 195 nm (not shown in the figure due to high HT values at this region) reflects a characteristic α-helical spectrum of MTase (Figure 6). In addition, the negative peak at 218 indicates the presence of the β-sheets. Overall, at pH 8.0, it contains 27% α-helix, 23% β-sheet, and the rest of the content as random coils, loops, turns, etc. Hence, we report that 50% of the MTase protein is comprised of α-helical and β-sheet structures. Similar percentage of α-helices and β-sheets for MTase were obtained by secondary structure prediction methods (data not shown). The secondary structure of MTase was calculated using online DichroWeb software (http://dichroweb.cryst.bbk.ac.uk/html/home.shtml) (accessed on 27 December 2021) based on the K2D model and analysed by the method of Chen et al. [24].

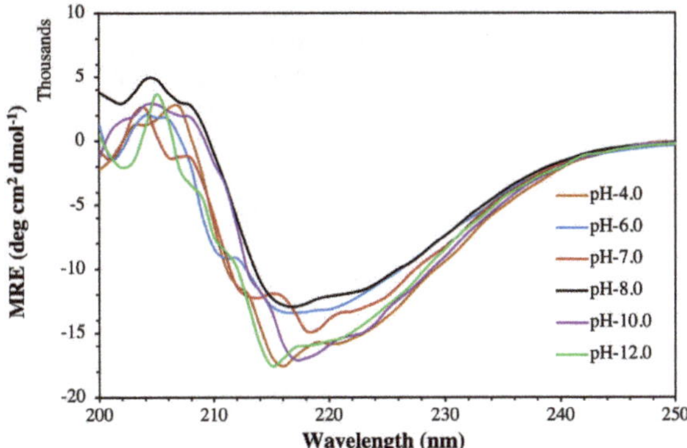

Figure 6. Far-UV CD spectra of MTase: To determine the change in the secondary structure of MTase in the presence of N-lauryl sarcosine sodium salt and NaCl at pH 4.0, 6.0, 7.0, 8.0, 10.0, and 12.0 at 25 °C.

2.4. Structural and Thermal Stability of MTase in the Presence of Mg^{2+}

The structural stability of MTase was investigated in the presence of Mg^{2+} at different pH conditions. The experimental results revealed a minimum change in the secondary structure of MTase in the presence of Mg^{2+} at pH 4.0 and pH 8.0 compared to pH 6.0, pH 7.0, pH 10.0, and pH 12.0 (Figure 7). As per our observations, Mg^{2+} is responsible for the change in the secondary structure at different pH conditions. However, it has been found that shifting the pH values towards acidic (pH 4.0) or alkaline (pH 12.0) is responsible for the increment in the α-helix content of MTase. This observation confirmed that the MTase secondary structure in the presence of 0.2 mM Mg^{2+} is more stable at pH 4.0 and pH 8.0 compared to other pH values. Therefore, we selected pH 8.0 for further studies, which is closer to physiological pH. This will help provide a comparative analysis for protein activity and drug discovery.

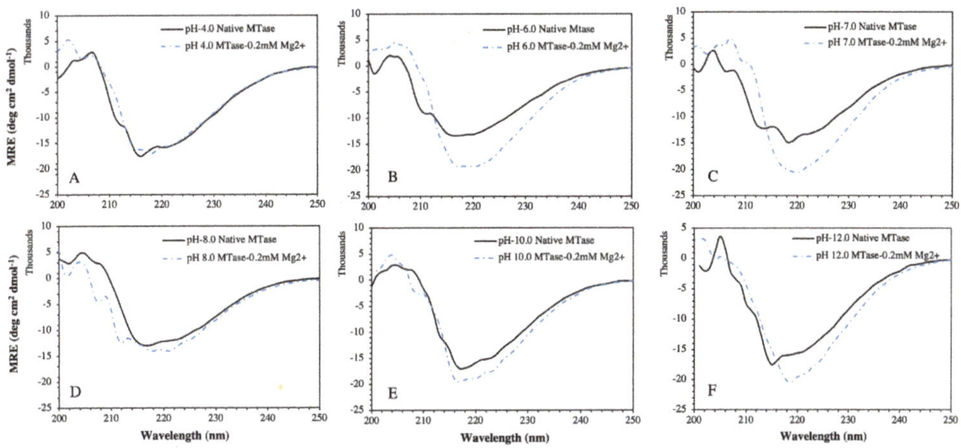

Figure 7. Far-UV CD spectra of MTase: CD spectra at (**A**) pH 4.0, (**B**) pH 6.0, (**C**) pH 7.0, (**D**) pH 8.0, (**E**) pH 10.0, and (**F**) pH 12.0 under standard temperature in the absence and presence of 0.2 mM Mg^{2+}; all respective buffer solutions contain N-lauryl sarcosine sodium salt and NaCl to provide structural stability at respective pH.

Further, we have determined the effect of Mg^{2+} on MTase secondary structural changes by varying the Mg^{2+} concentrations (Figure 8A). We found that 0.2 mM Mg^{2+} effectively changed the secondary structure of MTase compared to 1.0 mM of Mg^{2+}. Furthermore, the far-UV CD spectra revealed that lower Mg^{2+} concentrations induced the formation of β-sheets that was proportionally not supported at higher concentrations. Moreover, we determined the thermal stability at similar conditions. The obtained experimental denaturation graphs (Figure 8B) at 0.2 mM Mg^{2+} showed that the metal ions effectively induced the thermal stability of MTase compared to 1.0 mM (Table 1). Far UV-CD spectra and enzymatic assays also supported these results.

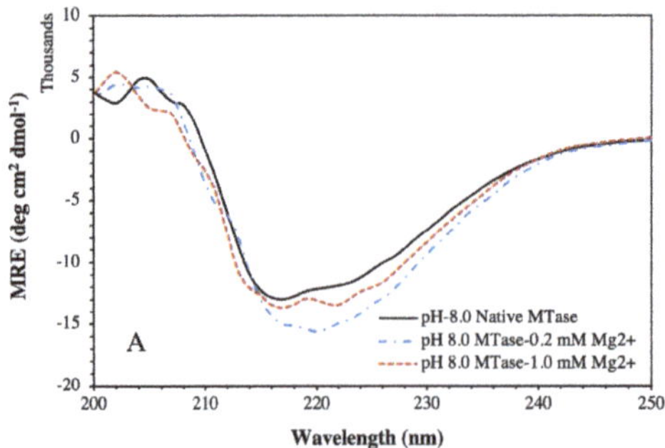

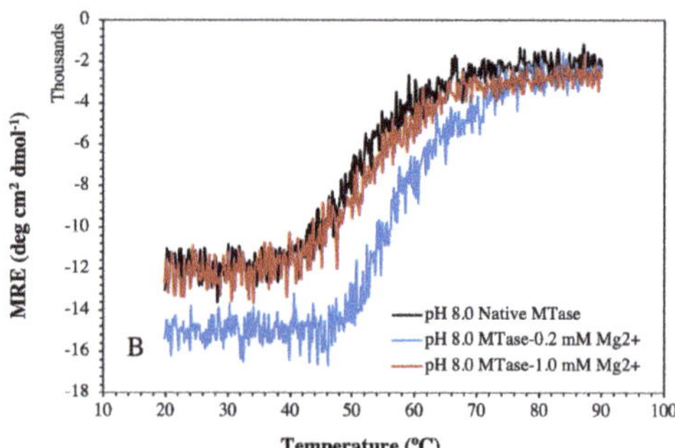

Figure 8. Far-UV CD spectra and thermal denaturation of MTase: (A) Far-UV Cd spectra at standard temperature (25 °C). **(B)** Thermal denaturation spectra of MTase at 222 nm in absence and presence of 0.2 mM and 1.0 mM Mg^{2+}. Tris buffer of pH 8.0 containing N-lauryl sarcosine sodium salt and NaCl was used in each experimental sample under minimum concentrations.

Table 1. Percentage change in secondary structure content and thermal denaturation values of MTase in the presence and absence of Mg^{2+}. This revealed changes in the structural conformational and stability of MTase at pH 8.0.

Sample System	% α-Helices	% β-Sheets	% Random Coils and Other Sec. Structures	T_m (°C)
Native MTase	27.08	23.03	48.09	58.83
MTase + 0.2 mM Mg^{2+}	22.64	36.96	40.40	61.57
MTase + 1.0 mM Mg^{2+}	25.93	24.87	49.20	59.73

We calculated the thermal unfolding of MTase by the two-state folding–unfolding model and the related Equations (2) and (3), which are used to resolve the temperature mid-point (T_m) by fitting the ellipticity. The results showed that secondary structural contents of MTase were transformed during the thermal denaturation process. The loss of protein function and structure were directly related to the decrease in the ellipticity at 222 nm with temperature. The obtained T_m values of native MTase were calculated at around ~58.8 °C, which changed significantly compared to the presence of Mg^{2+} (Figure 8B, Table 1). The thermal denaturation of MTase and waning of hydrophobic and other non-covalent interactions might be responsible.

2.5. Determination of Binding Affinity and Mechanism of Mg^{2+} Ion with MTase by Fluorescence Quenching Method

We also performed fluorescence quenching experiments to determine the binding affinity of Mg^{2+} with MTase. For this, we titrated Mg^{2+} against MTase (5 μM) at 25 °C. Figure 9A shows that MTase possesses a sharp fluorescence emission peak around 340 nm when excited at 280 nm. The quenching of MTase fluorescence occurs by increasing Mg^{2+} to its saturation level, at a millimolar concentration. It is possible that Mg^{2+} intercalates with MTase at a site close to tryptophan or other aromatic amino acid residues. This region is predominantly responsible for the change in the emission peak at around 340 nm after excitation at 280 nm [25]. Therefore, an ongoing reduction in the emission spectral intensity of MTase has been found, without remarkable variation in the wavelength of maximal fluorescence emission (λ_{max}) until the final quenching concentration [26].

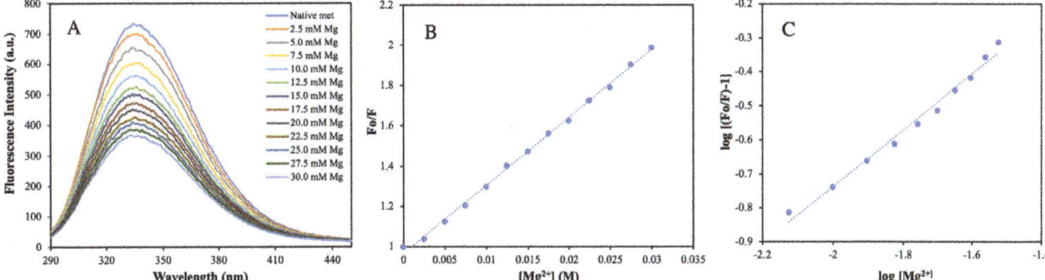

Figure 9. (**A**) Fluorescence quenching measurement: fluorescence emission spectra of MTase (5 μM) in the presence of Mg^{2+}. (**B**) Stern–Volmer plot for the MTase–Mg^{2+} interaction. (**C**) Binding parameter measurements: plot of log [(Fo/F) − 1] vs. log [Mg^{2+}] for the determination of binding constants and binding stoichiometry for the MTase–Mg^{2+} interaction at room temperature and pH 8.0.

To determine the binding affinity of MTase, we performed the titration of Mg^{2+} against MTase at 25 °C. However, the k_q value for the MTase–Mg^{2+} system was ten times higher than the highest scatter collision quenching constant of innumerable quenchers with polymers (2×10^{10} $M^{-1}s^{-1}$) [27]. This reflects that quenching is not commenced by dynamic diffusion, but arises by creating powerful complex formation between MTase and Mg^{2+}.

The intrinsic fluorescence intensity (FI) of aromatic amino acids decreases continuously by increasing the metal ion concentration. For example, the emission spectra become saturated at 30 mM Mg^{2+}, as shown in Figure 9A. The decrease in FI upon adding ions was analysed using the Stern–Volmer equation. It is a fact that the slopes in Figure 9B indicate that the binding of the ligand to the protein is responsible for quenching [28]. Based on the Stern–Volmer plot, the K_q value of Mg^{2+} is 5.81×10^9, reflecting lower scatter collision quenching value initiated by the dynamic diffusion of molecules.

The binding constant and related binding stoichiometry of Mg^{2+} were calculated as per log [(Fo/F) − 1] plotted against log [Mg^{2+}], as presented in Figure 9C. The slope of these plots reveals that binding stoichiometry (n) and corresponding intercept value give the information about binding constant (K_b), which was calculated from Equation (5), with computed values are reflected in Table 2.

Table 2. Binding parameters for the MTase and metal (Mg^{2+}) ion complex at standard temperature (25 °C).

Systems	K_{sv} (M^{-1})	$K_q \times 10^{-9}$ ($M^{-1}S^{-1}$)	N	K_b (M^{-1})	ΔG (kcal mol^{-1})
HEV MTase–Mg^{2+}	33.61± 0.02	5.81 ± 0.02	0.96	8.11 ± 0.59	−1.239 ± 0.010

Therefore, our CD and fluorescence data imply that under slightly basic conditions (at pH 8.0), the structural and conformational alteration in MTase was minimal compared to other pH and more feasible for the study of interaction with other inhibitors in the presence of Mg^{2+}. These consequences occur because of the pH-induced alteration in the vicinity of the metal-binding site of the MTase that is responsible for the change in the mode and mechanism of protein, and finally, the interaction of MTase with Mg^{2+} [29].

2.6. Isothermal Titration Calorimetry (ITC) Analysis

The thermodynamic parameters of Mg^{2+} upon binding to MTase were studied by ITC. The injected heat signals for Mg^{2+} binding with MTase and the integrated heat of the reaction for each injection are displayed in Figure 10. The calculated output results of the ITC data are shown in Table 3, which demonstrate that there are two binding sites available for Mg^{2+} in MTase. The first binding (6.7×10^4 M^{-1}) site is much stronger, which is driven by a small negative enthalpy (-0.08 ± 0.06 kcal/mol) and a large positive entropy (6.49 ± 0.10 kcal/mol). The favourable entropy contributes to the solvation entropy because of the loss of water from the binding interface, indicating the presence of electrostatic interaction during the complex formation. Meanwhile, the secondary binding site is weaker (3.0×10^2 M^{-1}) and endothermic. Based on the thermodynamic analysis, it is suggested that the protein–Mg^{2+} complexation is mainly driven by electrostatic interaction.

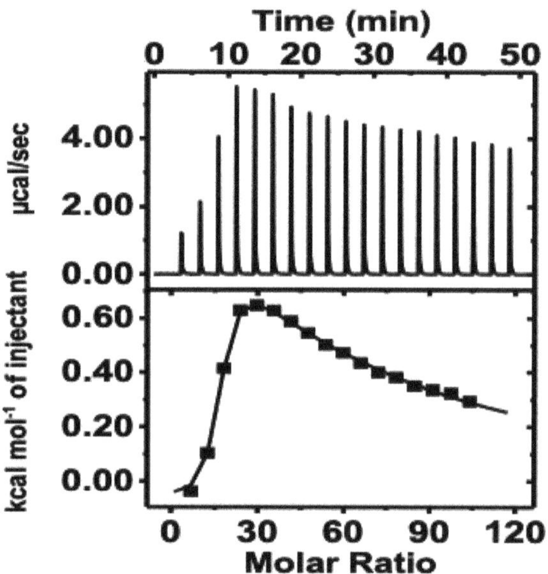

Figure 10. ITC thermogram: ITC thermogram showing the binding of protein with Mg^{2+}. The experiment was performed at 25 °C; top panel, detected heat signals; bottom panel, integrated heat of reaction for each titration.

Table 3. Thermodynamic parameters obtained from ITC thermogram.

Systems	K_1 (M^{-1})	ΔH_1 (kcal/mol)	$T\Delta S_1$ (kcal/mol)	ΔG_1 (kcal/mol)	K_2 (M^{-1})	ΔH_2 (kcal/mol)	$T\Delta S_2$ (kcal/mol)	ΔG_2 (kcal/mol)
HEV MTase–Mg^{2+}	6.7×10^4 $\pm 1.15 \times 10^4$	-0.08 ± 0.06	6.49 ± 0.10	-6.57 ± 0.16	300 ± 55	4.00 ± 0.36	7.39 ± 0.44	-3.39 ± 0.09

2.7. Binding Study of Magnesium with MTase

Further, the binding affinity of magnesium and MTase was determined using microscale thermophoresis (MST), a tool to study the interaction of the biomolecules. During the study, the MTase concentration was set at 50 nM, and the concentration of Mg^{2+} was varied from 5 mM to 300 nM in 16 different dilutions. The graph was plotted using the concentration of Mg^{2+} on the X-axis in log [M], while the Y-axis displayed the normalised fluorescence. The sample fluorescence is recorded during an MST experiment, starting with 3 s at ambient temperature to monitor steady-state fluorescence, followed by IR laser activation for a defined MST-on time. The MST analysis was performed using MO Control and MO Affinity Analysis Software (Monolith NT.115, NanoTemper Technologies, München, Germany). The calculated dissociation constant (K_D) between MTase and Mg^{2+} is approximately 15 µM (Figure 11).

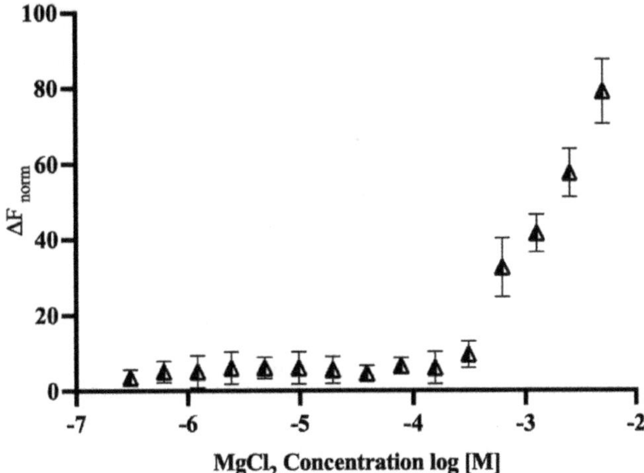

Figure 11. Microscale thermophoresis: Dose–response curve for the binding interaction between RED-Tris-NTA-labelled MTase and Mg^{2+}. The concentration of RED-Tris-NTA-labelled MTase is constant, while the concentration of Mg^{2+} varies between 5 mM and 300 nM. The K_D value for $MgCl_2$ with MTase is 15 μM. The *Y*-axis on the graph represents the fluorescence change, and the *X*-axis on the graph represents the concentration of Mg^{2+}. The graph describes the values of three different experiments.

3. Discussion

The plus-stranded viruses have a $5'$-capped genome catalysed by MTase [30,31] and found to be essential for their infectivity and replication [6,7]. In line with this, the $5'$ non-coding region (NCR) of HEV RNA has been demonstrated to have an m^7G-cap that is indispensable for its life cycle [6,32]. Similar results have been reported in alphavirus nsP1, tobacco mosaic virus P126, brome mosaic virus replicase protein 1a, and bamboo mosaic virus nonstructural protein [8]. Despite being an important enzyme that may act as a drug target, not many structural or functional studies have been conducted on MTase. Previously, the molecular weight of the active enzyme has been demonstrated to be 110 kDa, encoding amino acids 1 to 979 of the HEV genome [8]. In another study, using a computational approach, the MTase region has been predicted to be from 56 to 240 amino acids [9]. Emerson et al.'s computational predictions revealed that a region of 33–353 amino acids on the HEV pSK-HEV2 genome could exhibit MTase activity [6]. Therefore, we expressed this region to translate a protein of 37 kDa in size, as confirmed by western blotting and MALDI-TOF (Supplementary data). A recent study also demonstrated that a ~37 kDa MTase enzyme was processed from the HEV-ORF1 polyprotein when Huh7 cells were transduced with BacMam-HEV [33]. In our previous study, the digestion of ORF1 polyprotein, using cysteine protease, yielded a ~37 kDa protein, detected by MTase epitope-specific antibodies [34]. The enzyme was thus expressed and found to be active, as determined using the luminescence-based assay. The activity was altered by various cap analogues, as seen in an earlier study by Magden et al. [8]. The enzyme activity was also confirmed using enzyme kinetics and binding studies.

Another objective of this study is the metal dependency of MTase activity and several viral MTases, as other enzymes use divalent cations for their activity [17–20], which prompted us to study the effect of Ca^{2+} and Mg^{2+} on MTase activity. While there was no considerable effect on the enzyme activity was observed in the presence of Ca^{2+} (data not shown), significant MTase activity was observed in the case of Mg^{2+}. The results confirmed that activity was decreased when Mg^{2+} was depleted by the chelating agent, EDTA. In eukaryotes, mRNA capping also requires Mg^{2+} for the catalysis of lysine–GMP intermediate formation [35]. The metal ions are known to stabilise the random coil regions of enzymes by forming metal-binding pockets and protecting the unstructured part from the protease activity; they act as electron donors at the catalytic centre to expand the biochemical palette and regulate a wide range of functions.

In this work, we performed biophysical studies to explore the importance of Mg^{2+}, which affects the stability of MTase at different ranges of pH and temperature. The far-UV-CD spectra (190–250 nm) clearly showed that MTase contains both α-helices (27%), and β-sheets (23%), but different pH environments induced changes in the secondary structure components. The similar secondary structural components were seen in most SAM-dependent MTases with a Rossmann-like fold. Further, the CD experiments at different Mg^{2+} concentrations revealed that the secondary structure was proportional to Mg^{2+} concentration, which is evident in metal-binding proteins. Thermal denaturation experiments in the presence of Mg^{2+} were also performed to understand the effect of metal ions on MTase folding by increasing the temperature. Analysis of the plot revealed the reduction in the percentage of the secondary structure with increased temperature, indicating the absence of any intermediate unfolding states during the thermal unfolding pathway of MTase. The obtained T_m values of MTase were calculated using native conditions that significantly increased in the presence of Mg^{2+}. The waning of hydrophobic and polar interactions might be responsible for the denaturation of MTase with Mg^{2+}; without it, MTase is very unstable and is precipitated. The binding affinity of Mg^{2+} determined using the fluorescence quenching experiment identified its strong association with MTase. The decrease in the emission intensity at 340 nm is reflected when increasing concentration of the Mg^{2+}, which are accountable for the quenching of fluorescence intensity from the aromatic amino acids closely associated with Mg^{2+} (Figure 9). The ITC experiments revealed that MTase has two Mg^{2+} binding sites. The first one is stronger, and entropy plays a significant role in the binding (Table 2); the second is weaker and endothermic. Therefore, ITC experiments suggested that the MTase–Mg^{2+} complexation is mainly driven by electrostatic interaction. MST analysis was also performed to determine the Mg^{2+} binding with MTase, and these results suggested that the Mg^{2+} ion has an affinity towards MTase.

4. Conclusions

Briefly, we expressed MTase and determined its activity in its shortest functional form, 33–353 amino acids of HEV polyprotein, which was previously predicted by computational studies. The active MTase was determined to be 37 kDa in size by MALDI-TOF. The activity of the enzyme was confirmed using enzymatic assay, with SAM as a methyl donor and GTP as a methyl acceptor. GDP showed the highest activity and GTP showed the second highest as a methyl acceptor in the MTase activity assay. The activity of the enzyme was found to be increased in the presence of Mg^{2+}, which is a feature of many RNA capping enzymes. Similarly, the activity of MTase was decreased when it was chelated with EDTA. The circular dichroism, fluorescence quenching, and thermal denaturation studies provided the MTase with structural stability in the presence of Mg^{2+}. The binding affinity of Mg^{2+} with MTase was determined by ITC and MST experiments. Overall, our study has shown the indispensable role of Mg^{2+} in MTase activity and stability. Further, this work established the optimal experimental conditions that would be helpful for the screening of inhibitor libraries against HEV MTase to identify potential inhibitors.

5. Materials and Methods

5.1. Expression and Purification of MTase

The histidine-tagged MTase (MTase) gene of approximately 960 bases coding for 33–353 amino acids of genotype 1 (GenBank accession no. AF444002.1) was cloned in pET 28a (+) vector (GenScript, New Jersey, USA) and confirmed by restriction digestion using NdeI and XhoI enzymes. The positive pET28a (+)-MTase construct was transformed into BL21 (DE3) *E. coli* cells to express the protein. A single positive clone carrying the MTase gene was grown in LB broth, and the logarithmic phase culture was induced with 1 mM IPTG at 37 °C for 3 h. The induced culture was centrifuged at 5000× *g* for 20 min at 4 °C, and the pellet was suspended in lysis buffer (10 mM Tris–Cl pH 8.0, 1 mM EDTA, 100 mM

NaCl, 5 mM DTT, and 500 μg/mL lysozyme) and sonicated for 5 min (10 s on, 20 s off). The cell lysate was centrifuged for 15,000× g at 4 °C for 60 min. The pellet was further resuspended in 20 mL IB (inclusion bodies) wash buffer (10 mM Tris–Cl pH 8.0, 1 mM EDTA, 100 mM NaCl, 1% Triton X-100) by continuous stirring for 2 h and centrifuged at 15,000× g for 45 min at 4 °C. The pellet obtained in IBs was solubilised in buffer (10 mM Tris–HCl pH 8.0, 100 mM NaCl, and 0.5% NLS) (N-lauryl sarcosine, Sigma Aldrich, Burlington, MA, USA) by continuous stirring for overnight at 4 °C. The solubilised pellet was centrifuged at 15,000× g for 45 min at 4 °C and subsequently filtered through a 0.45 μM syringe filter (Millipore). The solubilised protein was purified using metal (Ni-NTA) affinity chromatography using AKTA start (GE Healthcare, Chicago, IL, USA). The protein was eluted in 10 mM Tris–Cl, pH 8.0, 100 mM NaCl, 0.01% NLS, and 200 mM imidazole. The purified MTase was further characterised by Western blot analysis, as described previously [33,34]. The membrane was probed with MTase epitope-specific primary antibody followed by HRP-conjugated goat anti-rabbit secondary antibody (Invitrogen, Waltham, MA, USA). Signal was detected using electrochemiluminescence (Bio-Rad Western ECL substrate, Hercules, CA, USA).

5.2. Functional Aspects of MTase

The functional analysis of MTase was determined using bioluminescence-based assay with GTP as a substrate. The reaction was performed in white-bottomed 96-well plate (Tarsons) in the presence of 20 mM Tris buffer of pH 8.0, 50 mM NaCl, 1 mM EDTA, 3 mM MgCl$_2$, 0.1 mg/mL BSA, 1 mM DTT, and 1 μM SAM. SAM acts as a methyl donor, and GTP acts as the methyl acceptor. In principle, MTase transfers a methyl group from SAM to GTP and converts it into SAH and m^{7}GTP. Then MTase-Glo™ reagent (Promega, Madison, WI, USA) changes SAH to ADP, which is turned into ATP by MTase-Glo™ detection solution (Promega, USA). The luminescence was measured on a Tecan Plate reader. The enzymatic activity was determined under various reaction conditions, such as pH (6–12), temperature (17–45 °C), time (5–120 min), and various enzyme (0 to 5 μM) and substrate concentrations. The substrate-specificity assay was performed to test its ability to incorporate methyl groups on different nucleoside triphosphates, ATP, CTP, UTP guanine nucleotides (GMP and GTP), and cap analogues (m^{7}GDP, m^{7}GTP, and dGTP) at 0.5 mM. The standard reaction was carried out at 37 °C for 120 min in a 20 μL solution that contained 0.675 μM MTase, 0.4 mM GTP, 20 mM Tris buffer, pH 8.0, 50 mM NaCl, 1 mM EDTA, 3 mM MgCl$_2$, 0.1 mg/mL BSA, 1 mM DTT, and 1 μM SAM. The reaction was stopped by adding 0.5% TFA (trifluoroacetic acid) (Sigma) and incubated for 10 min at room temperature, followed by incubation with MTase Glo reagent at room temperature for 30 min. Further, MTase detection reagent was added to the reaction mixture, and luminescence was detected using Tecan Plate Reader after 30 min. Similarly, the MTase activity was determined by increasing the concentrations of Mg^{2+} and EDTA. A standard curve was generated using a serial dilution of SAH ranging from 0 to 1000 nM to correlate luminescence with SAH concentration. The luminescence was plotted on Y-axis against the SAH concentration on X-axis, and the straight-line equation ($Y = 19.71 \times X + 227.60$) was calculated using linear regression in GraphPad Prism 9.0. The graphs were plotted as a function of the concentration of SAH produced.

5.3. Secondary Structure Analysis Using CD Spectroscopy

Circular dichroism (CD) spectra were performed on a Jasco J-1500 model spectropolarimeter. For the instrument calibration, (+)-10-camphor sulfonic acid was used. All CD experiments were performed at standard temperature (25 °C), which was thermostatically controlled by the Jasco Peltier PTC-423S/15 attached to the cell holder with a precision of ±0.1 °C. The change in the secondary structure of MTase under native conditions and the complex form with magnesium (Mg^{2+}) ions were observed to be in the range of 200–250 nm by using a 0.1 cm cell path length. The HT voltage of the scans was kept below 600 V, and the reference signal spectrum was subtracted for each scan. The scan speed of 100 mm/min and the response time of 1 s was set for each scan measurement; each spectrum was an average of three scans. Furthermore, the secondary structure content of MTase was calculated using online DichroWeb software and Chen et al. method [24]. The spectra were smoothed by the Savitzky–Golay method with 15 convolution widths. The results were expressed as mean residual ellipticity (MRE) in deg. cm^2 $dmol^{-1}$, which is defined as:

$$\text{MRE} = \frac{\theta_{obs}(mdeg)}{10 \times n \times C \times l} \tag{1}$$

where θ_{obs} is the observed ellipticity in degrees, C is the molar fraction, n is the number of amino acid residues $(321 - 1 = 320)$, and l is the length of the light path in centimeters.

5.4. Thermal Stability of MTase with Mg^{2+}

A thermal denaturation study of MTase was performed on a specific wavelength (222 nm) using far-UV CD experiments, further analysed based upon the two-state unfolding model. In the case of a single-step unfolding process, $N \rightleftharpoons U$, where N refers to the native state and U to the unfolded state, and K_u for the equilibrium constant with f_u and f_n are the molar fractions of U and N, respectively.

$$K_u = \frac{f_u}{f_n} \tag{2}$$

$$f_d = \frac{(Y_{obs} - Y_n)}{(Y_u - Y_n)} \tag{3}$$

where Y_{obs}, Y_n, and Y_u reflect the observed property, property of the native state, and property of the unfolded state, respectively.

5.5. Fluorescence Quenching Measurements

Mg^{2+} binding with MTase was performed by fluorescence measurements and was performed on Fluorolog TCSPC Horiba FL-1057 spectrofluorometer attached to a temperature-controlled water bath with an accuracy of ± 0.1 °C. The change in fluorescence intensity of MTase was observed at 340 nm and further analysed by using the Stern–Volmer equation [36]

$$\frac{F_o}{F} = K_{sv}[Q] + 1 \tag{4}$$

where F_o and F are the MTase fluorescence intensities in the absence and presence of Mg^{2+} (quencher), and K_{SV} is the Stern–Volmer quenching constant was calculated from the equation

$$K_{sv} = k_q \cdot \tau_o \tag{5}$$

where k_q is the bimolecular rate constant of the protein–ligand reaction process and τ_o is the average integral fluorescence lifetime of Trp, which is $\sim 5.78 \times 10^{-9}$ s [37]. Therefore, binding constants and binding stoichiometry were calculated [38,39].

$$\log\left(\frac{F_o}{F} - 1\right) = \log K_b + n \log[Q] \tag{6}$$

where K_b is the binding constant and n is binding stoichiometry.

5.6. Binding Affinity Using Isothermal Titration Calorimetry (ITC)

The binding affinity of MTase and Mg^{2+} at 25 °C was determined using isothermal titration calorimetry (ITC). MTase and Mg^{2+} were dissolved in 50 mM phosphate buffer, 50 mM NaCl, pH 7.4, and gently degassed. Forty microliters of water was added to a sample cell containing 280 µL of MTase. There was a total of 20 injections, with each infusion of 2 µL of 17 mM Mg^{2+} ion titrated into the sample cell containing 30 µM of MTase simultaneously. Intervals of 150 s separated each injection to allow the signal to return to baseline. A constant stirring speed of 750 rpm was maintained to ensure proper mixing after each infusion. Control experiments were performed under similar conditions by titrating Mg^{2+} into the buffer and were subtracted to correct for the heat of dilution. Thermodynamic parameters were obtained by fitting the data to a two-set site model using Origin software (7.0, OriginLab Corporation, Northampton, MA, USA).

5.7. MicroScale Thermophoresis Measurements

Further, MST experiments were conducted to find the affinity of Mg^{2+} to MTase by following the manufacturer's protocol {Monolith NT.115 Series, NanoTemper Technologies). The process involved fluorescent labelling of MTase using a His-tag labelling kit (NanoTemper Technologies). Briefly, 50 nM of the enzyme was incubated with the labelling buffer at RT for 30 min. For Mg^{2+}, the two-fold serial dilutions were made starting from 5 mM to 300 nM in 16 steps. The assay was performed in buffer (20 mM HEPES, 100 mM NaCl, 0.08% NLS, and 5% glycerol) that provides a good stability strength for protein and Mg^{2+}. The mixture was loaded into glass capillaries and the MST analysis was performed using Monolith NT.115 (NanoTemper Technologies) at 25 °C.

5.8. Statistical Analysis

All the statistical analyses were performed using GraphPad software, version 9.0.0 (San Diego, CA, USA). The statistical significance of the data was determined using Student's *t*-test (unpaired).

Supplementary Materials: The following supporting information can be downloaded online. Figure S1: MALDI TOF analysis of purified HEV-MTase. The major peak corresponds to the molecular weight of the HEV-MTase i.e., 37.07 kDa.

Author Contributions: Conceptualisation by P.H., M.I. and D.S.; methodology by P.H., M.I., P.B., S.S., D.M., M.M. and S.A.; Biophysical validation and software by R.K., N.A. and M.A.R.; validation by M.I., P.H. and K.K.I.; formal analysis by P.H., M.I. and D.S.; data curation by P.H. and M.I.; writing—original and draft preparation by P.H. and M.I.; review and editing by M.I., A.T., N.A., A.F.A. and M.A.R.; supervision by D.S. and K.K.I.; funding acquisition by D.S. All authors have read and agreed to the published version of the manuscript.

Funding: Research work was funded by the Department of Biotechnology (DBT), India Sanction Order No., BT/PR15094/BRB/10/1446/2015 and Researchers Supporting Project number-RSP-2021/335, King Saud University, Riyadh, Saudi Arabia.

Institutional Review Board Statement: Not applicable.

Informed Consent Statement: Not applicable.

Data Availability Statement: Data is available within the article and supplementary file.

Acknowledgments: P.H. is thankful to Shiv Nadar University for financial support in the form of SRF. M.I. is thankful to the Department of Science and Technology (DST), New Delhi, India, for financial aid as Research Associate under project No. E.M.R./2015/001072. Authors are thankful to the Researchers Supporting Project number-RSP-2021/335, King Saud University, Riyadh, Saudi Arabia. The authors are also grateful to the Department of Biophysics, AIIMS, New Delhi and the Department of Chemistry, Shiv Nadar University, Greater Noida, India, for providing the instrumentational facility. The authors thank S.U. Emerson for providing the pSK-HEV-2, cDNA clone of HEV-Genotype 1.

Conflicts of Interest: The authors declare no conflict of interest.

Sample Availability: Samples of the compounds and MTase are available from the Deepak Sehgal's laboratory for the current research work.

Abbreviations

HEV-MTase	Hepatitis E Virus Methyltransferase
ORFs	Open reading frames
FI	Fluorescence Intensity
Mg^{2+}	Magnesium ion
ITC	Isothermal Titration Calorimetry
CD	Circular Dichroism
MST	MicroScale Thermophoresis
T_m	Melting point

References

1. Chandra, V.; Taneja, S.; Kalia, M.; Jameel, S. Molecular biology and pathogenesis of hepatitis E virus. *J. Biosci.* **2008**, *33*, 451–464. [CrossRef] [PubMed]
2. Parvez, M.K. Molecular characterization of hepatitis E virus ORF1 gene supports a papain-like cysteine protease (PCP)-domain activity. *Virus Res.* **2013**, *178*, 553–556. [CrossRef] [PubMed]
3. Kamar, N.; Izopet, J.; Tripon, S.; Bismuth, M.; Hillaire, S.; Dumortier, J.; Radenne, S.; Coilly, A.; Garrigue, V.; D'Alteroche, L.; et al. Ribavirin for chronic hepatitis E virus infection in transplant recipients. *N. Engl. J. Med.* **2014**, *370*, 1111–1120. [CrossRef] [PubMed]
4. Parvez, M.K.; Khan, A.A. Molecular modeling and analysis of hepatitis E virus (HEV) papain-like cysteine protease. *Virus Res.* **2014**, *179*, 220–224. [CrossRef] [PubMed]
5. Navaneethan, U.; Al Mohajer, M.; Shata, M.T. Hepatitis E and pregnancy: Understanding the pathogenesis. *Liver Int.* **2008**, *28*, 1190–1199. [CrossRef]
6. Emerson, S.U.; Zhang, M.; Meng, X.J.; Nguyen, H.; St. Claire, M.; Govindarajan, S.; Huang, Y.K.; Purcell, R.H. Recombinant hepatitis E virus genomes infectious for primates: Importance of capping and discovery of a cis-reactive element. *Proc. Natl. Acad. Sci. USA* **2001**, *98*, 15270–15275. [CrossRef]
7. Emerson, S.U.; Nguyen, H.; Graff, J.; Stephany, D.A.; Brockington, A.; Purcell, R.H. In Vitro Replication of Hepatitis E Virus (HEV) Genomes and of an HEV Replicon Expressing Green Fluorescent Protein. *J. Virol.* **2004**, *78*, 4838–4846. [CrossRef]
8. Magden, J.; Takeda, N.; Li, T.; Auvinen, P.; Ahola, T.; Miyamura, T.; Merits, A.; Kaariainen, L. Virus-Specific mRNA Capping Enzyme Encoded by Hepatitis E Virus. *J. Virol.* **2001**, *75*, 6249–6255. [CrossRef]
9. Koonin, E.V.; Gorbalenya, A.E.; Purdy, M.A.; Rozanov, M.N.; Reyes, G.R.; Bradley, D.W. Computer-assisted assignment of functional domains in the nonstructural polyprotein of hepatitis E virus: Delineation of an additional group of positive-strand RNA plant and animal viruses. *Proc. Natl. Acad. Sci. USA* **1992**, *89*, 8259–8263. [CrossRef]
10. Zhao, Y.; Soh, T.S.; Zheng, J.; Chan, K.W.K.; Phoo, W.W.; Lee, C.C.; Tay, M.Y.F.; Swaminathan, K.; Cornvik, T.C.; Lim, S.P.; et al. A Crystal Structure of the Dengue Virus NS5 Protein Reveals a Novel Inter-domain Interface Essential for Protein Flexibility and Virus Replication. *PLoS Pathog.* **2015**, *11*, e1004682. [CrossRef]
11. Krafcikova, P.; Silhan, J.; Nencka, R.; Boura, E. Structural analysis of the SARS-CoV-2 methyltransferase complex involved in RNA cap creation bound to sinefungin. *Nat. Commun.* **2020**, *11*, 3717. [CrossRef] [PubMed]
12. Dong, H.; Fink, K.; Züst, R.; Lim, S.P.; Qin, C.-F.; Shi, P.-Y. Flavivirus RNA methylation. *J. Gen. Virol.* **2014**, *95*, 763–778. [CrossRef]
13. Venkatesan, S.; Moss, B. Eukaryotic mRNA capping enzyme-guanylate covalent intermediate. *Proc. Natl. Acad. Sci. USA* **1982**, *79*, 340–344. [CrossRef]
14. Coutard, B.; Barral, K.; Lichière, J.; Selisko, B.; Martin, B.; Aouadi, W.; Lombardia, M.O.; Debart, F.; Vasseur, J.-J.; Guillemot, J.C.; et al. Zika Virus Methyltransferase: Structure and Functions for Drug Design Perspectives. *J. Virol.* **2017**, *91*, 1190–1199. [CrossRef]
15. Sutto-Ortiz, P.; Tcherniuk, S.; Ysebaert, N.; Abeywickrema, P.; Noël, M.; Decombe, A.; Debart, F.; Vasseur, J.-J.; Canard, B.; Roymans, D.; et al. The methyltransferase domain of the Respiratory Syncytial Virus L protein catalyzes cap N7 and 2'-O-methylation. *PLoS Pathog.* **2021**, *17*, e1009562. [CrossRef] [PubMed]
16. Martin, B.; Coutard, B.; Guez, T.; Paesen, G.C.; Canard, B.; Debart, F.; Vasseur, J.-J.; Grimes, J.M.; Decroly, E. The methyltransferase domain of the Sudan ebolavirus L protein specifically targets internal adenosines of RNA substrates, in addition to the cap structure. *Nucleic Acids Res.* **2018**, *46*, 7902–7912. [CrossRef] [PubMed]
17. Bouvet, M.; Debarnot, C.; Imbert, I.; Selisko, B.; Snijder, E.J.; Canard, B.; Decroly, E. In Vitro Reconstitution of SARS-Coronavirus mRNA Cap Methylation. *PLoS Pathog.* **2010**, *6*, e1000863. [CrossRef]
18. Issur, M.; Geiss, B.J.; Bougie, I.; Picard-Jean, F.; Despins, S.; Mayette, J.; Hobdey, S.E.; Bisaillon, M. The flavivirus NS5 protein is a true RNA guanylyltransferase that catalyzes a two-step reaction to form the RNA cap structure. *RNA* **2009**, *15*, 2340–2350. [CrossRef]
19. Bougie, I.; Charpentier, S.; Bisaillon, M. Characterization of the Metal Ion Binding Properties of the Hepatitis C Virus RNA Polymerase. *J. Biol. Chem.* **2003**, *278*, 3868–3875. [CrossRef]

20. Decroly, E.; Debarnot, C.; Ferron, F.; Bouvet, M.; Coutard, B.; Imbert, I.; Gluais, L.; Papageorgiou, N.; Sharff, A.; Bricogne, G.; et al. Crystal structure and functional analysis of the SARS-coronavirus RNA cap 2′-O-methyltransferase nsp10/nsp16 complex. *PLoS Pathog.* **2011**, *7*, e1002059. [CrossRef]

21. Viswanathan, T.; Misra, A.; Chan, S.-H.; Qi, S.; Dai, N.; Arya, S.; Martinez-Sobrido, L.; Gupta, Y.K. A metal ion orients SARS-CoV-2 mRNA to ensure accurate 2′-O methylation of its first nucleotide. *Nat. Commun.* **2021**, *12*, 3287. [CrossRef] [PubMed]

22. Zhao, Y.; Soh, T.S.; Lim, S.P.; Chung, K.Y.; Swaminathan, K.; Vasudevan, S.G.; Shi, P.-Y.; Lescar, J.; Luo, D. Molecular basis for specific viral RNA recognition and 2′-O-ribose methylation by the dengue virus nonstructural protein 5 (NS5). *Proc. Natl. Acad. Sci. USA* **2015**, *112*, 14834–14839. [CrossRef] [PubMed]

23. Kelly, S.M.; Jess, T.J.; Price, N.C. How to study proteins by circular dichroism. *Biochim. Biophys. Acta-Proteins Proteom.* **2005**, *1751*, 119–139. [CrossRef] [PubMed]

24. Chen, Y.H.; Yang, J.T.; Martinez, H.M. Determination of the Secondary Structures of Proteins by Circular Dichroism and Optical Rotatory Dispersion. *Biochemistry* **1972**, *11*, 4120–4131. [CrossRef] [PubMed]

25. Marszalek, M.; Konarska, A.; Szajdzinska-Pietek, E.; Wolszczak, M. Interaction of cationic protoberberine alkaloids with human serum albumin. No spectroscopic evidence on binding to Sudlow's site 1. *J. Phys. Chem. B* **2013**, *117*, 15987–15993. [CrossRef]

26. Zolfagharzadeh, M.; Pirouzi, M.; Asoodeh, A.; Saberi, M.R.; Chamani, J. A comparison investigation of DNP-binding effects to HSA and HTF by spectroscopic and molecular modeling techniques. *J. Biomol. Struct. Dyn.* **2014**, *32*, 1936–1952. [CrossRef]

27. Ishtikhar, M.; Chandel, T.I.; Ahmad, A.; Ali, M.S.; Allohadan, H.A.; Atta, A.M.; Khan, R.H. Rosin surfactant QRMAE can be utilized as an amorphous aggregate inducer: A case study of mammalian serum albumin. *PLoS ONE* **2015**, *10*, e0139027. [CrossRef]

28. Ishtikhar, M.; Ali, M.S.; Atta, A.M.; Al-Lohedan, H.A.; Nigam, L.; Subbarao, N.; Hasan Khan, R. Interaction of biocompatible natural rosin-based surfactants with human serum albumin: A biophysical study. *J. Lumin.* **2015**, *167*, 399–407. [CrossRef]

29. Ishtikhar, M.; Rabbani, G.; Khan, R.H. Interaction of 5-fluoro-5′-deoxyuridine with human serum albumin under physiological and non-physiological condition: A biophysical investigation. *Colloids Surf. B Biointerfaces* **2014**, *123*, 469–477. [CrossRef]

30. Scheidel, L.M.; Durbin, R.K.; Stollar, V. SVLM21, a Sindbis virus mutant resistant to methionine deprivation, encodes an altered methyltransferase. *Virology* **1989**, *173*, 408–414. [CrossRef]

31. Ahola, T.; den Boon, J.A.; Ahlquist, P. Helicase and Capping Enzyme Active Site Mutations in Brome Mosaic Virus Protein 1a Cause Defects in Template Recruitment, Negative-Strand RNA Synthesis, and Viral RNA Capping. *J. Virol.* **2000**, *74*, 8803–8811. [CrossRef] [PubMed]

32. Kabrane-Lazizi, Y.; Meng, X.-J.; Purcell, R.H.; Emerson, S.U. Evidence that the Genomic RNA of Hepatitis E Virus Is Capped. *J. Virol.* **1999**, *73*, 8848–8850. [CrossRef] [PubMed]

33. Kumar, M.; Hooda, P.; Khanna, M.; Patel, U.; Sehgal, D. Development of BacMam Induced Hepatitis E Virus Replication Model in Hepatoma Cells to Study the Polyprotein Processing. *Front. Microbiol.* **2020**, *11*, 1347. [CrossRef] [PubMed]

34. Saraswat, S.; Chaudhary, M.; Sehgal, D. Hepatitis E Virus Cysteine Protease Has Papain Like Properties Validated by in silico Modeling and Cell-Free Inhibition Assays. *Front. Cell. Infect. Microbiol.* **2020**, *9*, 478. [CrossRef] [PubMed]

35. Swift, R.V.; Ong, C.D.; Amaro, R.E. Magnesium-Induced Nucleophile Activation in the Guanylyltransferase mRNA Capping Enzyme. *Biochemistry* **2012**, *51*, 10236–10243. [CrossRef] [PubMed]

36. Lakowicz, J.R. *Quenching of Fluorescence BT-Principles of Fluorescence Spectroscopy*; Lakowicz, J.R., Ed.; Springer: Boston, MA, USA, 1983; pp. 257–301, ISBN 978-1-4615-7658-7.

37. Anand, U.; Jash, C.; Mukherjee, S. Spectroscopic probing of the microenvironment in a protein-surfactant assembly. *J. Phys. Chem. B* **2010**, *114*, 15839–15845. [CrossRef]

38. Feng, X.; Bai, C.; Lin, Z.; Wang, N.; Wang, C. The interaction between acridine orange and bovine serum albumin. *Fenxi Huaxue* **1998**, *26*, 154–157. [CrossRef]

39. Lissi, E.; Calderón, C.; Campos, A. Evaluation of the number of binding sites in proteins from their intrinsic fluorescence: Limitations and pitfalls. *Photochem. Photobiol.* **2013**, *89*, 1413–1416. [CrossRef]

molecules

MDPI

Article

Interaction Characterization of a Tyrosine Kinase Inhibitor Erlotinib with a Model Transport Protein in the Presence of Quercetin: A Drug–Protein and Drug–Drug Interaction Investigation Using Multi-Spectroscopic and Computational Approaches

Tanveer A. Wani [1,*], Mohammed M. Alanazi [1], Nawaf A. Alsaif [1], Ahmed H. Bakheit [1], Seema Zargar [2], Ommalhasan Mohammed Alsalami [2] and Azmat Ali Khan [1]

[1] Department of Pharmaceutical Chemistry, College of Pharmacy, King Saud University, P.O. Box 2457, Riyadh 11451, Saudi Arabia; mmalanazi@ksu.edu.sa (M.M.A.); nalsaif@ksu.edu.sa (N.A.A.); abakheit@ksu.edu.sa (A.H.B.); azkhan@ksu.edu.sa (A.A.K.)

[2] Department of Biochemistry, College of Science, King Saud University, P.O. Box 22452, Riyadh 11451, Saudi Arabia; szargar@ksu.edu.sa (S.Z.); oalfageeh@ksu.edu.sa (O.M.A.)

* Correspondence: twani@ksu.edu.sa

Citation: Wani, T.A.; Alanazi, M.M.; Alsaif, N.A.; Bakheit, A.H.; Zargar, S.; Alsalami, O.M.; Khan, A.A. Interaction Characterization of a Tyrosine Kinase Inhibitor Erlotinib with a Model Transport Protein in the Presence of Quercetin: A Drug–Protein and Drug–Drug Interaction Investigation Using Multi-Spectroscopic and Computational Approaches. *Molecules* **2022**, *27*, 1265. https://doi.org/10.3390/molecules27041265

Academic Editors: Brullo Chiara and Thomas Mavromoustakos

Received: 27 December 2021
Accepted: 11 February 2022
Published: 14 February 2022

Publisher's Note: MDPI stays neutral with regard to jurisdictional claims in published maps and institutional affiliations.

Abstract: The interaction between erlotinib (ERL) and bovine serum albumin (BSA) was studied in the presence of quercetin (QUR), a flavonoid with antioxidant properties. Ligands bind to the transport protein BSA resulting in competition between different ligands and displacing a bound ligand, resulting in higher plasma concentrations. Therefore, various spectroscopic experiments were conducted in addition to in silico studies to evaluate the interaction behavior of the BSA-ERL system in the presence and absence of QUR. The quenching curve and binding constants values suggest competition between QUR and ERL to bind to BSA. The binding constant for the BSA-ERL system decreased from 2.07×10^4 to 0.02×10^2 in the presence of QUR. The interaction of ERL with BSA at Site II is ruled out based on the site marker studies. The suggested Site on BSA for interaction with ERL is Site I. Stability of the BSA-ERL system was established with molecular dynamic simulation studies for both Site I and Site III interaction. In addition, the analysis can significantly help evaluate the effect of various quercetin-containing foods and supplements during the ERL-treatment regimen. In vitro binding evaluation provides a cheaper alternative approach to investigate ligand-protein interaction before clinical studies.

Keywords: erlotinib; bovine serum albumin; fluorescence quenching; binding interaction; quercetin; competition

1. Introduction

The most affluent blood protein present in vertebrates is serum albumin (S.A.) and it is mainly responsible for transporting both endogenous and exogenous molecules to their target sites. It is the principal constituent of the plasma proteins ($\approx$60%) [1] and plays a significant role in the pharmacokinetics, pharmacodynamics, and toxicity of drugs [2]. The two most common serum albumin used as model transport proteins are BSA and HSA (human serum albumin). There is almost 75.6% sequence homology and similar ligand binding characteristics between the two. Therefore, BSA is commonly used as an alternative in protein–ligand binding studies because of its ready availability and cost-effectiveness [3,4]. The two most common ligand binding sites on BSA are Sudllow Site I and II positioned in subdomain IIA and IIIA [5–7].

Simultaneous administration of two or more drugs might lead to a competition between the two drugs to bind to a similar site present on the BSA. The conformation of the protein binding pockets can be altered on interaction with ligands, thereby influencing the binding of other ligands to the same binding pockets [8,9]. The therapeutic or adverse effects of the drug depend on the free fraction of the drug available in the plasma, and this property of drugs is taken into account when prescribing two or more drugs in a multidrug therapy due to possible competition at the plasma protein binding level [10,11].

Erlotinib (ERL), a tyrosine kinase inhibitor, has anticancer properties and is approved to treat non-small cell lung cancer. There is a high epidermal growth factor receptor (EGRF) expression in certain cancers, and ERL inhibits EGFR [12]. One of the mechanisms of action suggested for the ERL acts by disabling the phosphorylation ability of EGFR, thus inhibiting the signal transduction cascade leading to malignant cell apoptosis. Even though several mechanisms of action for ERL are proposed, the exact mechanism is yet unknown [13,14]. The binding of ERL to serum proteins from different species such as humans, mice, and rats are 92%, 95%, and 92%, respectively [15].

Fresh fruits and vegetables contain phenolic compounds such as flavonoids and tannins, and one such flavonoid is quercetin [16]. Quercetin (QUR) is also present in various food supplements and has antioxidant properties [17]. Quercetin has chemoprotective and radioprotective properties. In addition, it protects normal cells from the adverse effects of chemotherapy and radiotherapy [18]. The pharmacokinetics of quercetin is variable in humans due to their diet history, genetic polymorphism, and metabolic variation because of gut microbiota [19,20]. Enhancement in the antitumor activity of a breast cancer drug on concurrent use of QUR has been reported [21].

Quercetin is highly metabolized, and the amount of QUR in systemic circulation varies based on inter-individual pharmacokinetics and the source of QUR. In addition, although different organs metabolize quercetin, its transport to these organs is dependent on systemic circulation [20,22]. Thus, quercetin may interfere with the binding of the other concomitantly used drugs, impacting their pharmacokinetics and displacing them from their albumin binding sites. Such interference may lead to a toxic or a subtherapeutic response of the drug in the presence of QUR [23–26]. In addition, the pharmacokinetics of some drugs is affected by concurrent usage of QUR [7,23,25]. For example, a study reported the sudden death of experimental animals on the simultaneous use of QUR and digoxin [26].

The flavonoid QUR binds to Site I, subdomain IIA of BSA. Further, the interaction of QUR with other concomitantly used drugs has also been investigated [27–31]. In addition, possible competition between ERL and QUR to bind to BSA may affect the binding of ERL to BSA and lead to undesired effects of ERL [27–34]. Therefore, the current study investigated the ERL and BSA interaction with multispectroscopic and computational methods. Additionally, the impact of QUR was analyzed on the binding of ERL to BSA.

2. Materials and Methods

2.1. Chemicals

The chemicals used in the study ERL and QUR were procured from Weihua Pharma Co., Limited (Hangzhou, Zhejiang, China) and fatty acid-free BSA from Sigma-Aldrich (St. Louis, MO, USA). The phosphate buffer saline (PBS) pH-7 was prepared afresh to prepare the stock solution. QUR and ERL were dissolved in DMSO (dimethyl sulfoxide) followed by dilution with PBS. Millipore water (type-I) was used to prepare the stock solutions.

2.2. Fluorescence Measurements

The spectrofluorometer FP-8200 (JASCO, Hachioji, Tokyo, Japan) was used to measure the fluorescence spectra. The instrument had a quartz cell, and the slit width was 5 nm. The inner filter effects were corrected for the reabsorption of emitted light using the equation [35]:

$$F_{cor} = F_{obs} \times e^{(A_{ex}+A_{em})/2}$$

F_{cor} and F_{obs} are the corrected and observed fluorescence in the above equation, respectively, and A_{ex} and A_{em} are excitation and emission wavelengths absorbance values, respectively. The BSA spectra were recorded for the binary system, which consisted of BSA-ERL and BSA-QUR, and the ternary system consisted of BSA-QUR-ERL and BSA-ERL-QUR systems. For the measurement of spectra in the binary system, the concentration of BSA was held constant at 1.50 µM and the concentration of ERL was between 0.00–27.5 µM and QUR was between (0.00–35 µM) The spectra were recorded at three temperatures of 298, 303 and 310 K for BSA-ERL system and 298 K for BSA-QUR system. The concentration of BSA was 1.50 µM in the presence of ERL, and the spectra were recorded with ERL 0.00–27.50 µM in the presence of QUR 5.5 µM in the ternary system.

The site marker studies were undertaken to establish the binding Site for ERL with phenylbutazone for Site I and Ibuprofen for Site II as markers. The spectra for BSA were recorded with ERL in the presence of site markers phenylbutazone and ibuprofen.

2.3. U.V. Absorption Measurements

The UV-Vis absorbance spectra were recorded with Shimadzu 1800 (Kyoto, Japan). The spectra for BSA were recorded in the presence of ERL (0.00–15.00 µM) and a fixed concentration of BSA µM.

2.4. Synchronous Fluorescence Spectroscopic (SFS) Studies

The microenvironmental changes in the fluorophore residues were identified with the SFS technique. The changes in the amino acid residues Tyr (tyrosine) and Trp (tryptophan) can be observed at $\Delta\lambda = 15$ nm and $\Delta\lambda = 60$ nm. A shift in the emission spectra indicates changes in the polarity of amino acid residues around the fluorophore molecules.

2.5. Three-Dimensional (3D) Fluorescence Spectroscopy

Structural changes and conformational changes in the protein are identified with 3D fluorescence spectra. In addition, microenvironmental fluctuations in the vicinity of fluorophore residues lead to conformational or structural changes in the protein molecules. The BSA-ERL 3D spectra were matched to the 3D spectra of BSA to identify any such variations in the fluorophore microenvironment.

2.6. Molecular Docking and Molecular Dynamic Simulation (MDS)

The molecular docking was performed with bioinformatics tools MGL Tools [36], Auto Dock Vina [37], Molecular Operating Environment software (MOE), and Discovery Studio visualization tool. The PDB structure for the BSA (PDB ID: 6qs9) obtained from Protein Data Bank was used for the molecular docking [38]. The molecular docking was carried out for the three binding subdomains of BSA, and three runs were carried out for each binding Site. The top-scoring conformations were selected and analyzed for the interaction. MOE default parameters were used for docking analysis. The molecular dynamic simulation was carried out with NAMD 2.13 Suite (http://www.ks.uiuc.edu/Research/namd (accessed on 20 September 2021)) [39]. The ligand structures were immersed in the TIP3P water box. The charges neutralization was carried out by adding Na^+ or Cl^- with their maximum concentration of 0.15 M. The minimization of the system was carried out in MOE. The default MOE parameters were used for the minimization, which included Amber10: EHT force field. Partial Charge on LL toms was calculated using Amber10: EHT force field. The system's energy was minimized to an RMS gradient of 0.1 kcal/mol/A2. As per the MOE manual, an RMS of 0.1 is sufficient for structure minimization for molecular dynamic simulation. The minimized system was gradually heated. The temperature raised from 0 K to 310 K at an incremental level of 50 K for 100 ps and equilibrated at each temperature step followed by an unrestrained run of 20 ns. Constrained light bond with 2 fs time step. Long-range electrostatics were treated via the particle mesh Ewald method. The grid box formed had following parameters and dimensions: solvent molecules—26,824, 1.009 g/cm^3; Space group—P1, Triclinic, 1, 1×; size—(105.8, 83.9, 102.2); cell shape—(90.0, 90.0, 90.0). The parameters RMSD and RMSF were obtained for the Protein BSA and the protein in the presence of ligand (BSA-ERL) systems. The trajectory was visualized and analyzed with VMD 1.9.3 tool.

3. Results and Discussion

3.1. Fluorescence Quenching and Enhancement

The fluorescence spectra for the protein BSA for the binary and the ternary system were recorded at 280 nm (excitation wavelength) and an emission wavelength of 300–500 nm (Figure 1a–c). Both the ligands ERL and QUR decreased the fluorescence intensity of BSA. At increased ligand concentration ERL and QUR, the fluorescence intensity of BSA reduced further. The reduction in fluorescence intensity of BSA in the presence of QUR was higher than in the presence of ERL. A redshift of 8 nm was observed in the emission wavelength of BSA on interaction with ERL, indicating increased polarity and less hydrophilicity in the aromatic amino acid microenvironment [40–42]. A blue shift was also observed in the fluorescence emission spectra of BSA in the presence of QUR, as reported by earlier studies indicating higher hydrophobicity and a decline in polarity in the microenvironment of aromatic amino acid residues present in BSA [43]. The quenching behavior for the BSA-ERL binary system was determined with the help of the Stern Volmer equation [44]:

$$\frac{F_0}{F} = 1 + K_{sv}[Q] = 1 + k_q\tau_0[Q]$$

$$k_q = K_{sv}/\tau_0 \tag{1}$$

where F_0 is BSA's fluorescence intensity, and F is BSA's fluorescence intensity in the presence of quencher. Stern Volmer constant is given as K_{sv}, and (Q) is quencher concentration. The k_q biomolecular quenching constant and τ_0 the lifetime of the fluorophore in the absence of quencher and is valued at 10^{-8} s for biopolymers.

The Stern Volmer plot for the BSA-ERL system is presented in Figure 1d. A decrease in the K_{sv} values was observed with a rise in temperature for the BSA-ERL system (Table 1). Therefore, reducing K_{sv} values with a temperature rise is associated with the static quenching mechanism. In the case of dynamic quenching, there is an increase in the K_{sv} values [15,44]. Thus, a static quenching mechanism for the BSA-ERL system is suggested based on complex formation between BSA and ERL. The BSA-QUR system also follows a static quenching behavior as revealed by earlier studies [43]. The static quenching mechanism for the BSA-ERL system can also be established based on biomolecular quenching constants k_q, which are of the order of 2×10^{10} L·mol^{-1}s^{-1} for dynamic quenching and k_q values higher than the maximum value of 2×10^{10} L·mol^{-1}s^{-1} can be attained for biomolecular quenching constants only during a static quenching. Since the k_q value for the BSA-ERL system given in (Table 1) was high, a static quenching mechanism is suggested for this system in accordance with earlier studies [15,44]. The quenching constant at room temperature for the (BSA-QUR)-ERL ternary system is presented in Table 2. The UV absorbance studies for the BSA-ERL system have reported an increase in the absorbance of BSA at 280 nm.

Table 1. Stern Volmer K_{sv} and bimolecular quenching constant k_q.

System	T (K)	R	$K_{sv} \pm$ SD * (M^{-1})	$k_q \times 10^{12}$ (M^{-1}S^{-1})
BSA-ERL	298	0.9915	19,255.84 $\pm$ 625	1.93
	303	0.9887	18,450.98 $\pm$ 542	1.85
	307	0.9956	18,008.45 $\pm$ 489	1.80
BSA-Quercetin	298	0.9874	460,702 $\pm$ 1875	46.07
(BSA-QUR)-ERL	298	0.9954	26,970 $\pm$ 105	2.70

* standard deviation.

Table 2. Binding parameters binary and ternary systems and thermodynamic parameters for BSA-ERL system.

System	T (K)	$K_b \pm$ SD *	n	ΔG° (kJ·mol^{-1})	ΔH° (kJ·mol^{-1})	ΔS° (J mol^{-1}·K^{-1})
BSA-ERL	298	$2.07 \pm 0.11 \times 10^4$	1.01	-24.40		
	303	$2.83 \pm 0.08 \times 10^3$	0.84	-20.43	-260.80	-793.31
	307	$3.44 \pm 0.04 \times 10^2$	0.73	-17.26		
BSA-QUR	298	$6.33 \pm 0.07 \times 10^6$	-	-	-	-
(BSA-QUR)-ERL	298	$0.20 \pm 0.05 \times 10^2$	-	-	-	-

* standard deviation.

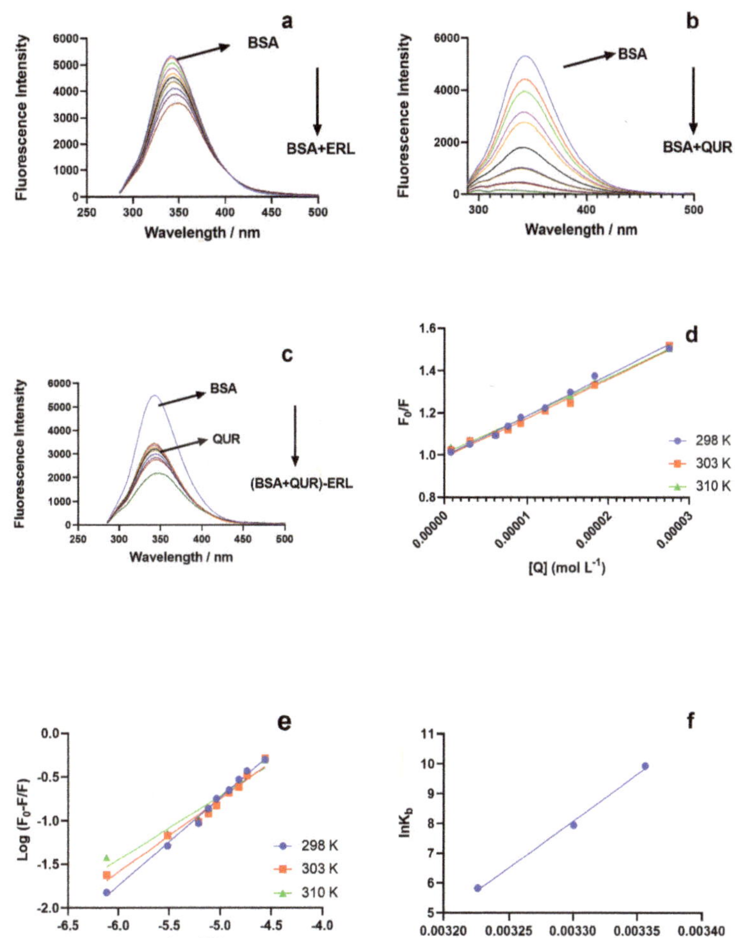

Figure 1. The fluorescence spectra for BSA (1.5 μM) with: (**a**) ERL (0.00–27.5 μM); (**b**) QUR (0.00–35 μM); (**c**) QUR (5.5 μM)-ERL (0.00–27.5 μM) at (λ_{ex} = 280 nm and λ_{em} = 300–500 nm); (**d**) Stern Volmer Plot for BSA(1.5 μM) -ERL (0.00–27.5 μM system at a temperature of 298, 303, 310 K; (**e**) double reciprocal plot [$(F_0 - F)/F$] versus log [Q] for the BSA-ERL system to obtain the binding constant (**b**); (**f**) van't Hoff plots to obtain the for thermodynamic parameters for the BSA-ERL system interaction.

3.2. Binding Constant and Number of Binding Sites

The binding constants and binding stoichiometry are were determined using a double logarithmic regression plot derived from the equation: $\log \frac{(F_0-F)}{F} = \log K_b + n \log[Q]$.

K_b and n represent the binding constant and binding stoichiometry in the above equation. The representative plot for the binding constants at the three studied temperatures of the binary system BSA-ERL is given in Figure 1e. The binding constant was obtained from the intercept of the double log plot, and the binding stoichiometry from its slope is presented in Table 2. Thus, the binding stoichiometry value of ≈1 suggests a single class of binding site was involved in the BSA-ERL interaction. The binding constants for the BSA-ERL system were of the order of 10^4 M^{-1} suggesting a moderate binding [45]. The binding constant at room temperature determined for the BSA-QUR system was of the order of ≈(>10^6). The ternary system (BSA-QUR)-ERL had binding constants of the order of ≈10^2 (Table 2).

The binding constant for the BSA-ERL system was studied at various temperatures since the binding constants are temperature-dependent. Therefore, the thermodynamic processes involved in the BSA-ERL interaction were also investigated using the van't Hoff equation and plot.

$$\ln K_b = -\frac{\Delta H^\circ}{RT} + \frac{\Delta S^\circ}{R}$$

$$\Delta G^\circ = \Delta H^\circ - T\Delta S^\circ$$

In the equation above, ΔH° is enthalpy change, ΔS° is entropy change and ΔG° is Gibbs free energy, R is the universal gas constant, and T is the temperature in kelvins (K).

The thermodynamic parameters ΔH°, ΔS° and ΔG°, were calculated from the van't Hoff plot for $\ln(K_b)$ vs. $1/T$ and are given in Figure 1f.

All the three parameters, enthalpy change, entropy change, and Gibbs free energy, given in Table 2, attained negative values. A negative Gibbs free energy indicates a spontaneous interaction. Furthermore, the BSA-ERL interaction is suggested by van der Waals force and hydrogen bonds based on negative enthalpy and entropy. Since the enthalpy was -260 kJ·mol^{-1} whereas entropy was -0.79 kJ·mol^{-1}, the BSA and ERL interaction indicate the interaction being enthalpy-driven, and the interaction had an unfavorable entropy.

3.3. Comparison of Binary and Ternary System Interactions

In the BSA-ERL binary system, the quenching constants (Table 2) decreased at higher temperatures, suggesting the formation of a complex between BSA and ERL and a static quenching mechanism. Furthermore, the BSA-QUR system quenching mechanism was found to be in accordance with earlier studies [31,46] which reported a static quenching mechanism between them. For the ternary system (BSA-QUR)-ERL, the quenching constant in the presence of QUR for the BSA-ERL system was higher than in its absence. The rise of the quenching constant of the BSA-ERL system can be attributed to the fact that in the presence of QUR, the accessibility of ERL to BSA increases, improving the quenching efficacy of ERL. An earlier study reported a similar phenomenon whereby the quenching efficiency of gliclazide increased in the presence of QUR [43]. The fluorescence quenching plot for the binary system and the ternary system is provided in Figure 2a. In the BSA-ERL system, ERL reduced the fluorescence intensity of BSA by 19%. Whereas in the (BSA-QUR)-ERL system, the fluorescence intensity of BSA reduction was by almost 45%. Therefore, the presence of QUR in the BSA-ERL system further reduced the fluorescence intensity of the BSA-ERL system by 26%. The decrease in the fluorescence intensity of the BSA-ERL system suggests a strong influence of QUR in the BSA-ERL interaction.

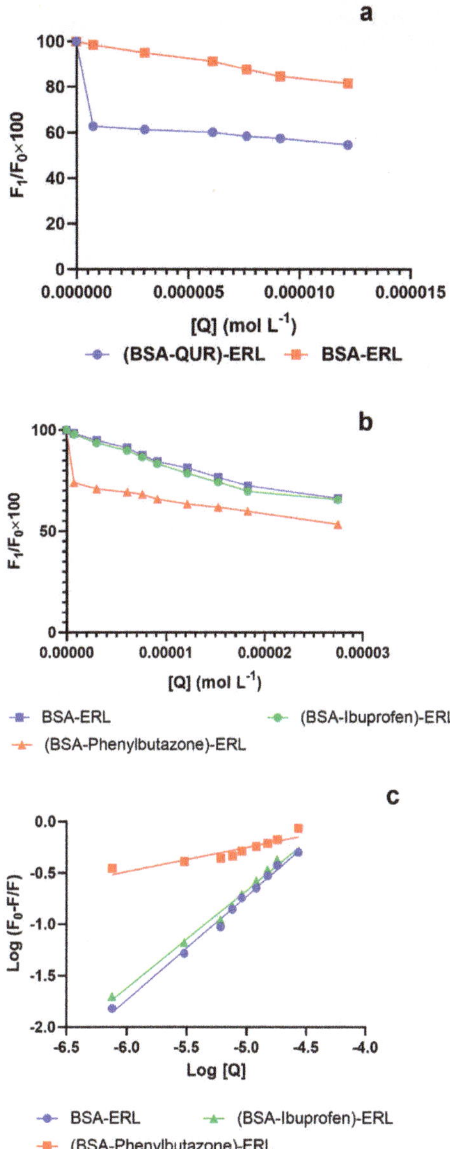

Figure 2. Quenching curve for BSA (1.5 µM)-ERL(0.00–27.5) (**a**); and BSA (1.5 µM)-ERL(0.00–27.5 µM) in presence of QUR (5.5 µM) (**b**); Quenching curve (**c**) and Binding constant for ERL in the presence of site markers phenylbutazone and ibuprofen.

The BSA-ERL binary system's binding constant of $\approx 10^4$ suggests moderate binding between the ERL and BSA. Our results corroborated earlier studies for the interaction between BSA and ERL [15]. The binding constants for the BSA-ERL system lowered with a rise in temperature (Table 2). In the other binding system that consisted of BSA-QUR, it was observed that QUR had a strong binding interaction with BSA. Some studies have reported a strong binding interaction between QUR and serum albumin [43,47]. As a result of this interaction between QUR and BSA, ERL, which is moderately bound to BSA, may not displace the bound QUR from the binding sites present on BSA. Therefore, a ternary system was developed whereby the interaction between BSA and ERL was studied in the presence of QUR. Finally, it was concluded that QUR considerably affected the BSA-ERL system, and the presence of QUR caused a decrease in the binding constant of the BSA-ERL system from 2.0×10^4 to 0.2×10^2. (Table 2).

The binding Site for ERL was identified using phenylbutazone and ibuprofen as site-specific markers for Site I and Site II of BSA, respectively. Quenching curve and binding constants of the BSA-ERL system in the presence of site markers were compared to the binding constant of the BSA-ERL system in the absence of site markers. Ibuprofen presence did not influence the quenching behavior of the BSA-ERL system (Figure 2b), and binding constants in the presence and absence of ibuprofen were similar (Table 2). Thus, these results rule out the binding of ERL to BSA Site II. However, the presence of phenylbutazone in the BSA-ERL system strongly influenced the quenching behavior and the binding constant of the BSA-ERL system (Figure 2c). Hence, it was concluded that ERL binds to Site I of BSA. Furthermore, since QUR also binds to Site I of BSA [43,47], the presence of QUR in the BSA-ERL system can markedly influence the interaction between the BSA-ERL system.

Furthermore, QUR in the BSA-ERL system reduced the binding constants of the system, implying that the free drug fraction of ERL may rise in the systemic circulation in the presence of QUR. A study conducted in non-small cell lung cancer patients taking ERL reported that adherence to the treatment regimen of ERL depends on the severity of side effects in these patients [48]. It also reported that side effects were more severe in patients with higher plasma area under the curve (AUC) for ERL. One of the most common adverse events related to the ERL treatment regimen is skin disorders, including acneiform rash, xeroderma (dry skin), pruritus, and paronychia. Sometimes, these side effects are so severe and necessitate treatment interruption or cessation [49]. One of the reasons attributed to the severity of these side effects for ERL is its plasma concentrations. Therefore, the severity of side effects influences patients' adherence to the treatment regimen. Since QUR may affect the AUC of ERL in plasma, increasing the plasma concentrations of ERL which might lead to unwanted adverse events affecting patients' adherence to the treatment regimen.

3.4. UV Absorption Studies

The two critical aspects that are studied by UV absorption are changes in the protein structure and protein–ligand complex formation [50]. The absorption spectral changes for BSA occur in protein–ligand complex formation and therefore infers a static quenching mechanism, whereas no spectral changes will be recorded in dynamic quenching [51]. Furthermore, the intensity of the BSA absorption peak at 280 nm was higher on interaction with ERL (Figure 3a), suggesting a static quenching mechanism between BSA and ERL.

Spectra for BSA was compared to spectra of the studied ligands ERL and QUR, BSA-ERL complex, and BSA-QUR complex. An increase in absorption spectra was observed for the BSA-ERL system compared to BSA alone. The ERL showed different absorption peaks compared to both BSA and BSA-ERL systems. The higher absorption peak for the BSA-QUR system was observed compared to BSA alone, and the spectra were different from the QUR absorption spectra.

The absorption spectra for the ternary system (BSA-QUR)-ERL was also studied, and it was different from both the binary system spectra (Figure 3b).

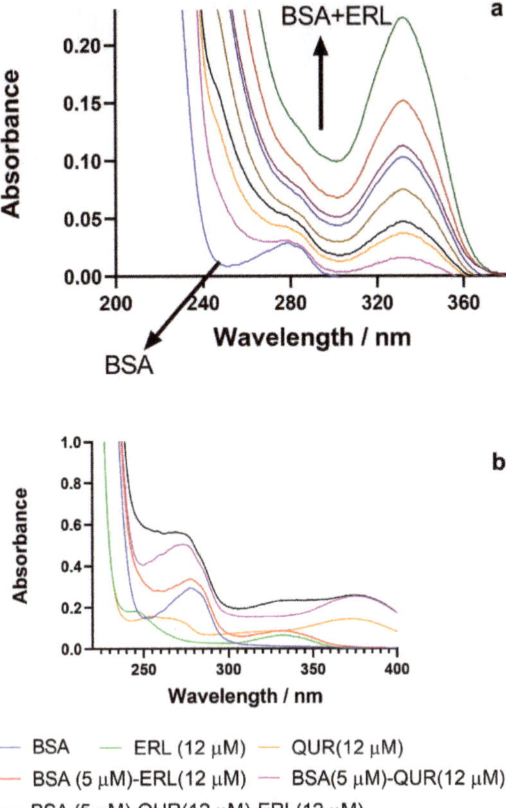

Figure 3. Ultraviolet–visible spectra for BSA (1.5 µM) in presence and absence of ERL (0.00–27.5 µM) (**a**); UV-visible spectra comparison for BSA, ERL, BSA-ERL, QUR, BSA-QUR, and BSA-ERL in the presence of QUR (**b**).

3.5. Synchronous Fluorescence Spectroscopic Studies

The microenvironment alterations in the fluorophore residues can be identified with synchronous fluorescence spectroscopy. The changes in emission wavelength indicate polarity change in the fluorophore residue microenvironment [52]. The synchronous spectra recorded at $\Delta\lambda = 15$ nm and $\Delta\lambda = 60$ nm provide information about the Tyr and Trp residue microenvironment. The spectra for Tyr showed no shift in the emission wavelength, whereas the spectra for Trp showed a slight 1 nm shift. A blue-shift at $\Delta\lambda = 60$ nm suggests micro-environmental modifications in the vicinity of Trp residue. Further, it is recommended that the BSA experienced an increased hydrophobicity on interaction with ERL.

3.6. Three Dimensional (3D) Fluorescence Spectroscopy

The 3D fluorescence spectroscopy provides information about protein structural alteration on interaction with a ligand [41]. The 3D fluorescence spectra were accessed for the protein BSA and the protein–ligand BSA-ERL system (Figure 4a,b). Four peaks observed in the 3D spectra were identified as Peak a and b representing Rayleigh scattering ($\lambda_{ex} = \lambda_{em}$) and IInd order scattering peak ($\lambda_{em} = 2_{ex}$). Higher intensity and scattering were observed for Peak a in the BSA-ERL system compared to BSA since the BSA-ERL complex formed a bigger macromolecule than BSA. The other peaks, Peak I representing

Trp and Tyr fluorophore residues and Peak II (polypeptide backbone structures), were evaluated in the 3D spectra. The intensity of both peaks, Peak I and Peak II, declined in the BSA-ERL protein–ligand system compared to the protein BSA. Hence, it is concluded that the interaction of ERL with BSA altered the microenvironment of the fluorophore residues, which led to these changes in the 3D spectra. Hence it was concluded that structural changes might have occurred in BSA on interaction with ERL.

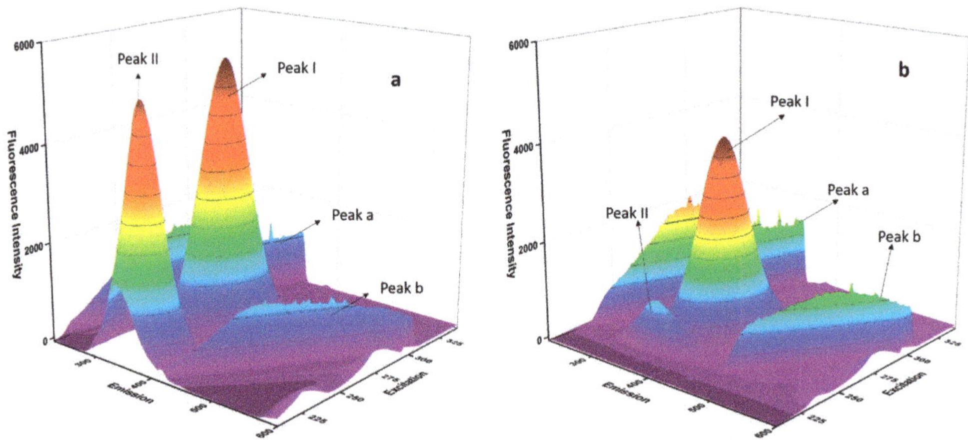

Figure 4. Three-dimensional fluorescence spectra for BSA (**a**) and BSA-ERL system (**b**).

3.7. Molecular Docking and Molecular Dynamic Simulation (MDS)

Molecular docking studies confirmed the experimental results for the BSA-ERL interaction. The site probe experiments with phenylbutazone suggest that the ERL interacted with BSA's subdomain IIA (Site I). However, an earlier study using molecular docking suggested that ERL binds to subdomain IB (Site III) and subdomain IIIA (site II) of BSA [53]. Furthermore, the site probe study with ibuprofen (Site II marker) did not suggest the interaction of ERL at Site II of BSA. Moreover, the quenching curves for the BSA with ERL in the presence or absence of ibuprofen did not change, suggesting no interaction of ERL with site II of BSA.

Molecular docking conformation for the interaction of ERL to Site I and Site III of BSA is given in Figure 5a,c. The binding energy for Site I -33.63 kJ·mol^{-1} was lower than the binding energy -29.91 kJ·mol^{-1} for Site III of BSA. The binding pocket of Site I of BSA (Figure 5b) was surrounded by Glu291, Leu218, Leu233, Arg198, Phe222, Arg256, Arg217, Ser286, Leu237, Leu259, His241, Ile289, Tyr149, Ile263, Ala290, Trp213, Gln195, Ser191, Ala260, Arg194, and the amino acids that surrounded the binding pocket at Site III were Lys 116, Pro117, Leu115, Ile181, Leu122, Glu125, Tyr137, Lys136, Arg185, Ile141, Tyr160, Phe133, Met184. Furthermore, the fluorophore residues Trp213, Tyr149, and Phe222 were found in the vicinity of the binding pocket. Thus, the interaction of ERL with BSA might influence these fluorophore residues. In addition, the hydrogen bond between Ligand C19 and Ser286, 3.41 Å, and pi hydrogen bond between 6-ring and Ala290, 3.79 Å were also observed in the BSA-ERL interaction. The BSA-ERL interaction was also investigated for any changes in the presence of QUR. In this study, three hydrogen bonds, Ligand C15 and Phe133, 2.95 Å, Ligand O3 and Phe133, 2.65 Å, Ligand O4, and Lys136, 3.08 Å, were found. In addition, two pi-H bonds 6-ring and Leu115, 4.16 Å, 6-ring and Leu122, 3.90 Å, were observed.

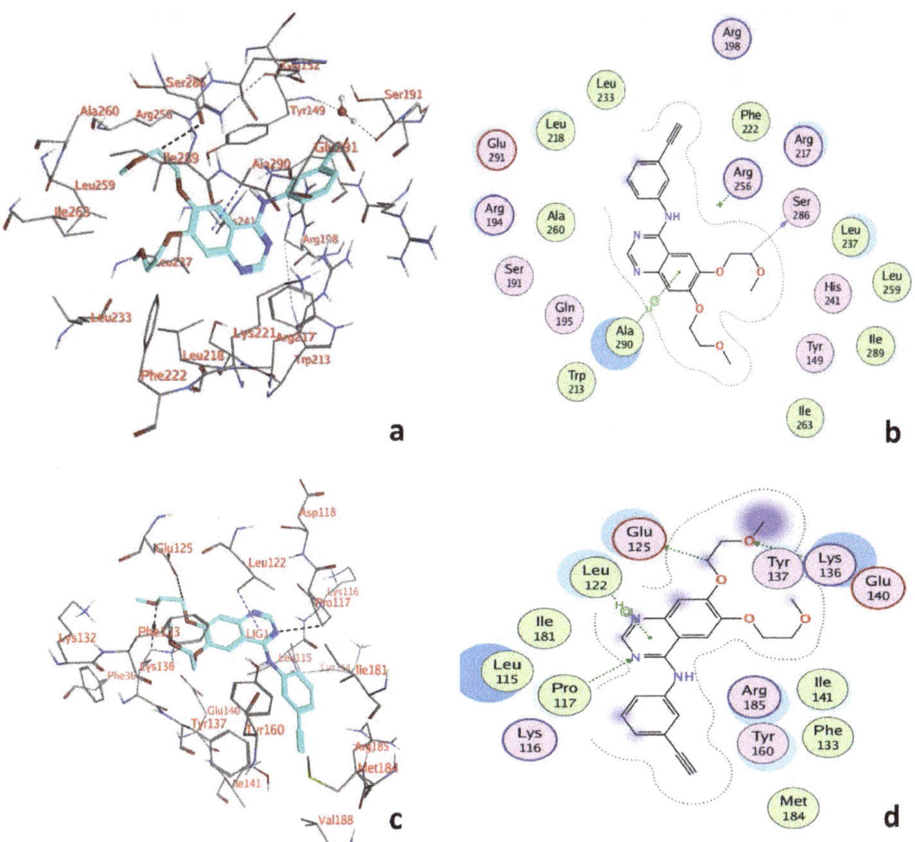

Figure 5. Two-dimensional molecular docking conformation for BSA ERL system at Site I (**a**) and Site III (**c**); three-dimensional docking conformation of BSA-ERL system Site I (**b**) and Site III (**d**).

At Site III (Figure 5d), the BSA and ERL formed three hydrogen bonds and one pi–hydrogen bond. The three hydrogen bonds formed were between Ligand C17 and Glu125, 3.51 Å, Ligand O4 and Lys136, 3.15 Å, Ligand N7, and PRO117, 3.15 Å, and the one pi-hydrogen bond formed was between 6-ring and Leu122, 3.79 Å.

As reported by earlier studies, QUR also binds to site I of serum albumin [47], and thus competition for the same binding Site might occur. Therefore, the molecular docking and the experimental results concluded that the presence of QUR might influence the binding of ERL to BSA.

The conformation stability for the BSA-ERL system was studied with MDS. The MDS study was conducted for Site I and Site III of BSA with ERL. The complex's stability was evaluated based on the root mean square deviation (RMSD) and root mean square fluctuation (RMSF) studies. The RMSD studies for BSA and BSA Site I and Site III with ERL are provided in Figure 6a,b, respectively, whereas the RMSF plots are given in Figure 6c,d, respectively. The most critical deviations from the crystal structure are found at the residue level in the most mobile parts of the protein, i.e., loops, terminal regions, and helix ends. In contrast, the transmembrane segments remain stable in all the simulations.

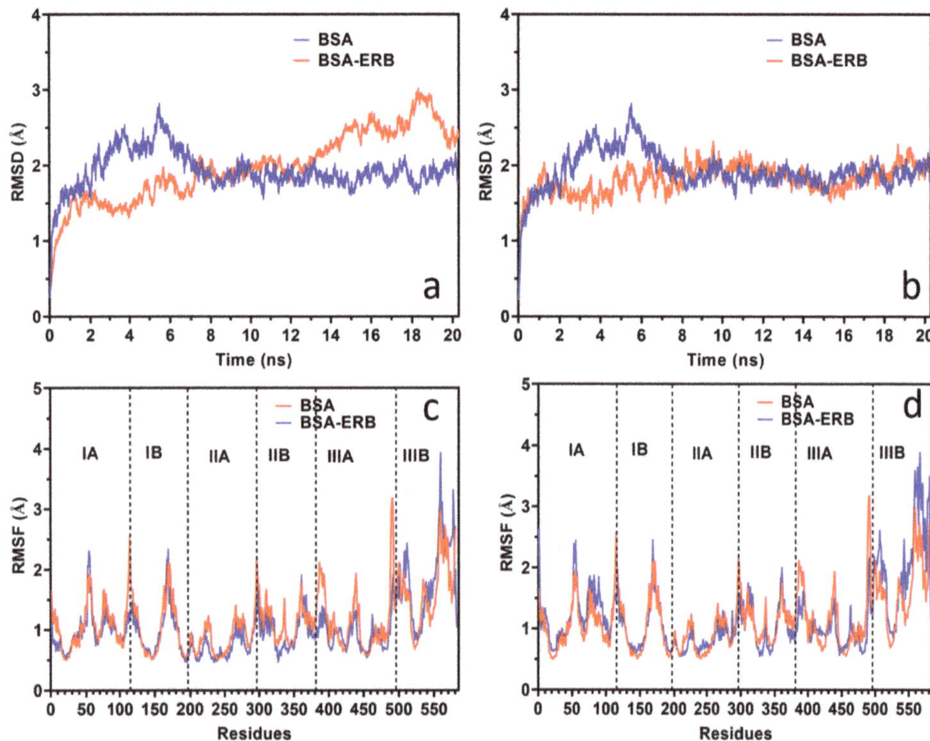

Figure 6. Molecular dynamic simulation RMSD plot for Site I (**a**) and Site III (**b**) and RMSF plot for Site I (**c**) and Site III (**d**).

Figure 6a shows the RMSD per residue for all the simulations reported in the present work for Site I. Three major peaks in subdomain 1 (loops C2) Ala78, Pro110, and Cys168 on the N-terminus were observed.

Further, in the simulation of the complex and comparison to the native protein residue Cys168 and Ala78, the N-terminus seems to be responsible for the higher RMSD values from 14–20 ns. Additional minor peaks can also be observed in the remaining loops 299, 309, and 504 and the C-terminus.

In the simulations between 2 to 8 ns, the resulting RMSD plots seem to be responsible for the higher RMSD values beyond residue Pro110 on the N-terminus for the native protein. No fluctuation was observed in the RMSD or RMSF plots of the BSA-ERL system at either of the binding sites, Site I or Site III. The RMSD averages were 1.98 and 1.82 Å, respectively, and the variation in the RMSD for the BSA-ERL system was between 0.258–3.013 Å for Site I and 0.214–2.318 Å for Site III. Therefore, a stable complex between BSA and ERL is concluded as the RMSD values did not fluctuate too high. Further, the residual flexibility is interpreted from the RMSF studies. The RMSF plot for both the studied sites (Site I and Site III) of the BSA-ERL system suggests that the complex formed between BSA and ERL was stable with a fluctuation of less than 3 Å [54].

4. Conclusions

This study examined the influence of flavonoid QUR on the BSA-ERL interaction by multispectroscopic and computational methods. Quercetin presence in the BSA-ERL system reduced the binding constant of the BSA-ERL system to almost half of what was observed in its absence. Thus, it can be concluded that there will be a higher free drug fraction of ERL in the system in the presence of QUR. However, the use of ERL in a therapeutic regimen leads to several adverse events, which in turn are associated with ERL plasma concentrations. Thus, co-administration of QUR and ERL might influence the pharmacokinetics of ERL and needs to be investigated by in vivo studies. Further, QUR is highly metabolized in the human body, necessitating studying the effect of the QUR on co-administered drugs in future studies. Hence, the information gained from such studies can benefit from dose optimization where the two drugs are intended to be co-administered.

Author Contributions: Conceptualization: T.A.W., S.Z.; Methodology: T.A.W., S.Z.; Software: A.H.B.; Formal analysis: N.A.A., M.M.A.; Investigation: A.H.B., T.A.W., S.Z., O.M.A.; Resources: T.A.W.; Writing: S.Z.; T.A.W.; Review & Editing: A.A.K.; Project administration: T.A.W. All authors have read and agreed to the published version of the manuscript.

Funding: Researchers Supporting Project number (RSP-2021/357), King Saud University, Riyadh, Saudi Arabia.

Institutional Review Board Statement: Not applicable.

Informed Consent Statement: Not applicable.

Data Availability Statement: Data will be available on request to corresponding author.

Acknowledgments: The authors extend their appreciation to Researchers Supporting Project number (RSP-2021/357), King Saud University, Riyadh, Saudi Arabia, for funding this work.

Conflicts of Interest: The authors declare no conflict of interest.

Sample Availability: Samples are to be prepared freshly for analysis.

References

1. Peters, T., Jr. *All about Albumin: Biochemistry, Genetics, and Medical Applications*; Academic Press: Cambridge, MA, USA, 1995.
2. Wang, B.-L.; Kou, S.-B.; Lin, Z.-Y.; Shi, J.-H. Investigation on the binding behavior between BSA and lenvatinib with the help of various spectroscopic and in silico methods. *J. Mol. Struct.* **2020**, *1204*, 127521. [CrossRef]
3. Fan, J.; Sun, W.; Wang, Z.; Peng, X.; Li, Y.; Cao, J. A fluorescent probe for site I binding and sensitive discrimination of HSA from BSA. *Chem. Commun.* **2014**, *50*, 9573–9576. [CrossRef] [PubMed]
4. Zhang, Y.-F.; Zhou, K.-L.; Lou, Y.-Y.; Pan, D.-Q.; Shi, J.-H. Investigation of the binding interaction between estazolam and bovine serum albumin: Multi-spectroscopic methods and molecular docking technique. *J. Biomol. Struct. Dyn.* **2017**, *35*, 3605–3614. [CrossRef] [PubMed]
5. Sudlow, G.; Birkett, D.J.; Wade, D.N. Spectroscopic techniques in the study of protein binding. A fluorescence technique for the evaluation of the albumin binding and displacement of warfarin and warfarin-alcohol. *Clin. Exp. Pharmacol. Physiol.* **1975**, *2*, 129–140. [CrossRef] [PubMed]
6. He, X.M.; Carter, D.C. Atomic structure and chemistry of human serum albumin. *Nature* **1992**, *358*, 209–215. [CrossRef] [PubMed]
7. Wani, T.A.; Bakheit, A.H.; Zargar, S.; Alamery, S. Mechanistic competitive binding interaction study between olmutinib and colchicine with model transport protein using spectroscopic and computer simulation approaches. *J. Photochem. Photobiol. A Chem.* **2022**, *426*, 113794. [CrossRef]
8. Jing, J.J.; Liu, B.; Wang, X.; Wang, X.; He, L.L.; Guo, X.Y.; Xu, M.L.; Li, Q.Y.; Gao, B.; Dong, B.Y. Binding of fluphenazine with human serum albumin in the presence of rutin and quercetin: An evaluation of food-drug interaction by spectroscopic techniques. *Luminescence* **2017**, *32*, 1056–1065. [CrossRef]
9. Alanazi, M.M.; Almehizia, A.A.; Bakheit, A.H.; Alsaif, N.A.; Alkahtani, H.M.; Wani, T.A. Mechanistic interaction study of 5,6-Dichloro-2-[2-(pyridin-2-yl)ethyl] isoindoline-1,3-dione with bovine serum albumin by spectroscopic and molecular docking approaches. *Saudi Pharm. J.* **2018**, *27*, 341–347. [CrossRef]
10. Ni, Y.; Su, S.; Kokot, S. Spectrofluorimetric studies on the binding of salicylic acid to bovine serum albumin using warfarin and ibuprofen as site markers with the aid of parallel factor analysis. *Anal. Chim. Acta* **2006**, *580*, 206–215. [CrossRef]
11. Zargar, S.; Alamery, S.; Bakheit, A.H.; Wani, T.A. Poziotinib and bovine serum albumin binding characterization and influence of quercetin, rutin, naringenin and sinapic acid on their binding interaction. *Spectrochim. Acta Part A Mol. Biomol. Spectrosc.* **2020**, *235*, 118335. [CrossRef]
12. Park, J.H.; Liu, Y.; Lemmon, M.A.; Radhakrishnan, R. Erlotinib binds both inactive and active conformations of the EGFR tyrosine kinase domain. *Biochem. J.* **2012**, *448*, 417. [CrossRef] [PubMed]
13. Li, Z.; Xu, M.; Xing, S.; Ho, W.T.; Ishii, T.; Li, Q.; Fu, X.; Zhao, Z.J. Erlotinib effectively inhibits JAK2V617F activity and polycythemia vera cell growth. *J. Biol. Chem.* **2007**, *282*, 3428–3432. [CrossRef] [PubMed]

14. Rosell, R.; Carcereny, E.; Gervais, R.; Vergnenegre, A.; Massuti, B.; Felip, E.; Palmero, R.; Garcia-Gomez, R.; Pallares, C.; Sanchez, J.M. Erlotinib versus standard chemotherapy as first-line treatment for European patients with advanced EGFR mutation-positive non-small-cell lung cancer (EURTAC): A multicentre, open-label, randomised phase 3 trial. *Lancet Oncol.* **2012**, *13*, 239–246. [CrossRef]

15. Rasoulzadeh, F.; Asgari, D.; Naseri, A.; Rashidi, M.R. Spectroscopic studies on the interaction between erlotinib hydrochloride and bovine serum albumin. *DARU J. Pharm. Sci.* **2010**, *18*, 179.

16. Wach, A.; Pyrzyńska, K.; Biesaga, M. Quercetin content in some food and herbal samples. *Food Chem.* **2007**, *100*, 699–704. [CrossRef]

17. Zhang, M.; Swarts, S.G.; Yin, L.; Liu, C.; Tian, Y.; Cao, Y.; Swarts, M.; Yang, S.; Zhang, S.B.; Zhang, K. Antioxidant properties of quercetin. In *Oxygen Transport to Tissue XXXII*; Springer: Berlin/Heidelberg, Germany, 2011; pp. 283–289.

18. Filipa Brito, A.; Ribeiro, M.; Margarida Abrantes, A.; Salome Pires, A.; Jorge Teixo, R.; Guilherme Tralhao, J.; Filomena Botelho, M. Quercetin in cancer treatment, alone or in combination with conventional therapeutics? *Curr. Med. Chem.* **2015**, *22*, 3025–3039. [CrossRef]

19. Poór, M.; Boda, G.; Needs, P.W.; Kroon, P.A.; Lemli, B.; Bencsik, T. Interaction of quercetin and its metabolites with warfarin: Displacement of warfarin from serum albumin and inhibition of CYP2C9 enzyme. *Biomed. Pharmacother.* **2017**, *88*, 574–581. [CrossRef]

20. Almeida, A.F.; Borge, G.I.A.; Piskula, M.; Tudose, A.; Tudoreanu, L.; Valentová, K.; Williamson, G.; Santos, C.N. Bioavailability of quercetin in humans with a focus on interindividual variation. *Compr. Rev. Food Sci. Food Saf.* **2018**, *17*, 714–731. [CrossRef]

21. Ezzati, M.; Yousefi, B.; Velaei, K.; Safa, A. A review on anti-cancer properties of Quercetin in breast cancer. *Life Sci.* **2020**, *248*, 117463. [CrossRef]

22. Ulusoy, H.G.; Sanlier, N. A minireview of quercetin: From its metabolism to possible mechanisms of its biological activities. *Crit. Rev. Food Sci. Nutr.* **2020**, *60*, 3290–3303. [CrossRef]

23. Yu, C.-P.; Wu, P.-P.; Hou, Y.-C.; Lin, S.-P.; Tsai, S.-Y.; Chen, C.-T.; Chao, P.-D.L. Quercetin and rutin reduced the bioavailability of cyclosporine from Neoral, an immunosuppressant, through activating P-glycoprotein and CYP 3A4. *J. Agric. Food Chem.* **2011**, *59*, 4644–4648. [CrossRef] [PubMed]

24. Kumar, K.K.; Priyanka, L.; Gnananath, K.; Babu, P.R.; Sujatha, S. Pharmacokinetic drug interactions between apigenin, rutin and paclitaxel mediated by P-glycoprotein in rats. *Eur. J. Drug Metab. Pharmacokinet.* **2015**, *40*, 267–276. [CrossRef]

25. Nguyen, M.; Staubach, P.; Wolffram, S.; Langguth, P. Effect of single-dose and short-term administration of quercetin on the pharmacokinetics of talinolol in humans–implications for the evaluation of transporter-mediated flavonoid–drug interactions. *Eur. J. Pharm. Sci.* **2014**, *61*, 54–60. [CrossRef] [PubMed]

26. Wang, Y.-H.; Chao, P.-D.L.; Hsiu, S.-L.; Wen, K.-C.; Hou, Y.-C. Lethal quercetin-digoxin interaction in pigs. *Life Sci.* **2004**, *74*, 1191–1197. [CrossRef] [PubMed]

27. Zsila, F.; Bikadi, Z.; Simonyi, M. Probing the binding of the flavonoid, quercetin to human serum albumin by circular dichroism, electronic absorption spectroscopy and molecular modelling methods. *Biochem. Pharmacol.* **2003**, *65*, 447–456. [CrossRef]

28. Vanekova, Z.; Hubcik, L.; Toca-Herrera, J.L.; Furtmuller, P.G.; Valentova, J.; Mucaji, P.; Nagy, M. Study of Interactions between Amlodipine and Quercetin on Human Serum Albumin: Spectroscopic and Modeling Approaches. *Molecules* **2019**, *24*, 487. [CrossRef]

29. Wani, T.A.; Bakheit, A.H.; Zargar, S.; Alanazi, Z.S.; Al-Majed, A.A. Influence of antioxidant flavonoids quercetin and rutin on the in-vitro binding of neratinib to human serum albumin. *Spectrochim. Acta Part A Mol. Biomol. Spectrosc.* **2021**, *246*, 118977. [CrossRef]

30. Alsaif, N.; Wani, T.A.; Bakheit, A.H.; Zargar, S. Multi-spectroscopic investigation, molecular docking and molecular dynamic simulation of competitive interactions between flavonoids (quercetin and rutin) and sorafenib for binding to human serum albumin. *Int. J. Biol. Macromol.* **2020**, *165*, 2451–2461. [CrossRef]

31. Wani, T.A.; Bakheit, A.H.; Al-Majed, A.R.A.; Bhat, M.A.; Zargar, S. Study of the interactions of bovine serum albumin with the new anti-inflammatory agent 4-(1,3-Dioxo-1,3-dihydro-2H-isoindol-2-yl)-N'-[(4-ethoxy-phenyl) methylidene] benzohydrazide using a multi-spectroscopic approach and molecular docking. *Molecules* **2017**, *22*, 1258. [CrossRef]

32. Mishra, B.; Barik, A.; Priyadarsini, K.I.; Mohan, H. Fluorescence spectroscopic studies on binding of a flavonoid antioxidant quercetin to serum albumins. *J. Chem. Sci.* **2005**, *117*, 641–647. [CrossRef]

33. Bolli, A.; Marino, M.; Rimbach, G.; Fanali, G.; Fasano, M.; Ascenzi, P. Flavonoid binding to human serum albumin. *Biochem. Biophys. Res. Commun.* **2010**, *398*, 444–449. [CrossRef] [PubMed]

34. Ehteshami, M.; Rasoulzadeh, F.; Mahboob, S.; Rashidi, M.R. Characterization of 6-mercaptopurine binding to bovine serum albumin and its displacement from the binding sites by quercetin and rutin. *J. Lumin.* **2013**, *135*, 164–169. [CrossRef]

35. Sengupta, B.; Sengupta, P.K. The interaction of quercetin with human serum albumin: A fluorescence spectroscopic study. *Biochem. Biophys. Res. Commun.* **2002**, *299*, 400–403. [CrossRef]

36. Jacob, R.B.; Andersen, T.; McDougal, O.M. Accessible high-throughput virtual screening molecular docking software for students and educators. *PLoS Comp. Biol.* **2012**, *8*, e1002499. [CrossRef] [PubMed]

37. Vina, A. Improving the speed and accuracy of docking with a new scoring function, efficient optimization, and multithreading Trott, Oleg; Olson, Arthur J. *J. Comput. Chem* **2010**, *31*, 455–461.

38. El Yazbi, F.A.; Mahrous, M.E.; Hammud, H.H.; Sonji, G.M.; Sonji, N.M. Kinetic spectrophotometric determination of betaxolol, clopidogrel and imidapril in pharmaceutical preparations. *Curr. Anal. Chem.* **2010**, *6*, 228–236. [CrossRef]
39. Phillips, J.C.; Hardy, D.J.; Maia, J.D.; Stone, J.E.; Ribeiro, J.V.; Bernardi, R.C.; Buch, R.; Fiorin, G.; Hénin, J.; Jiang, W. Scalable molecular dynamics on CPU and GPU architectures with NAMD. *J. Chem. Phys.* **2020**, *153*, 044130. [CrossRef]
40. Xie, L.; Wehling, R.L.; Ciftci, O.; Zhang, Y. Formation of complexes between tannic acid with bovine serum albumin, egg ovalbumin and bovine beta-lactoglobulin. *Food Res. Int.* **2017**, *102*, 195–202. [CrossRef]
41. Al-Mehizia, A.A.; Bakheit, A.H.; Zargar, S.; Bhat, M.A.; Asmari, M.M.; Wani, T.A. Evaluation of Biophysical Interaction between Newly Synthesized Pyrazoline Pyridazine Derivative and Bovine Serum Albumin by Spectroscopic and Molecular Docking Studies. *J. Spectrosc.* **2019**, *2019*, 3848670. [CrossRef]
42. Zargar, S.; Wani, T.A. Exploring the binding mechanism and adverse toxic effects of persistent organic pollutant (dicofol) to human serum albumin: A biophysical, biochemical and computational approach. *Chem.-Biol. Interact.* **2021**, *350*, 109707. [CrossRef]
43. Kameníková, M.; Furtmüller, P.G.; Klacsová, M.; Lopez-Guzman, A.; Toca-Herrera, J.L.; Vitkovská, A.; Devínsky, F.; Mučaji, P.; Nagy, M. Influence of quercetin on the interaction of gliclazide with human serum albumin–spectroscopic and docking approaches. *Luminescence* **2017**, *32*, 1203–1211. [CrossRef] [PubMed]
44. Lakowicz, J.R. *Principles of Fluorescence Spectroscopy*; Springer Science & Business Media: Berlin/Heidelberg, Germany, 2013.
45. Wani, T.A.; Bakheit, A.H.; Zargar, S.; Bhat, M.A.; Al-Majed, A.A. Molecular docking and experimental investigation of new indole derivative cyclooxygenase inhibitor to probe its binding mechanism with bovine serum albumin. *Bioorg. Chem.* **2019**, *89*, 103010. [CrossRef] [PubMed]
46. Poór, M.; Boda, G.; Kunsági-Máté, S.; Needs, P.W.; Kroon, P.A.; Lemli, B. Fluorescence spectroscopic evaluation of the interactions of quercetin, isorhamnetin, and quercetin-3′-sulfate with different albumins. *J. Lumin.* **2018**, *194*, 156–163. [CrossRef]
47. Zargar, S.; Wani, T.A. Protective Role of Quercetin in Carbon Tetrachloride Induced Toxicity in Rat Brain: Biochemical, Spectrophotometric Assays and Computational Approach. *Molecules* **2021**, *26*, 7526. [CrossRef] [PubMed]
48. Timmers, L.; Boons, C.C.; Moes-Ten Hove, J.; Smit, E.F.; van de Ven, P.M.; Aerts, J.G.; Swart, E.L.; Boven, E.; Hugtenburg, J.G. Adherence, exposure and patients' experiences with the use of erlotinib in non-small cell lung cancer. *J. Cancer Res. Clin. Oncol.* **2015**, *141*, 1481–1491. [CrossRef]
49. Kiyohara, Y.; Yamazaki, N.; Kishi, A. Erlotinib-related skin toxicities: Treatment strategies in patients with metastatic non-small cell lung cancer. *J. Am. Acad. Dermatol.* **2013**, *69*, 463–472. [CrossRef]
50. Rabbani, G.; Khan, M.J.; Ahmad, A.; Maskat, M.Y.; Khan, R.H. Effect of copper oxide nanoparticles on the conformation and activity of β-galactosidase. *Colloids Surf. B Biointerfaces* **2014**, *123*, 96–105. [CrossRef]
51. Rabbani, G.; Lee, E.J.; Ahmad, K.; Baig, M.H.; Choi, I. Binding of tolperisone hydrochloride with human serum albumin: Effects on the conformation, thermodynamics, and activity of HSA. *Mol. Pharm.* **2018**, *15*, 1445–1456. [CrossRef]
52. Ahmad, A.; Ahmad, M. Understanding the fate of human serum albumin upon interaction with edifenphos: Biophysical and biochemical approaches. *Pestic. Biochem. Physiol.* **2018**, *145*, 46–55. [CrossRef]
53. Taghipour, P.; Zakariazadeh, M.; Sharifi, M.; Dolatabadi, J.E.N.; Barzegar, A. Bovine serum albumin binding study to erlotinib using surface plasmon resonance and molecular docking methods. *J. Photochem. Photobiol. B Biol.* **2018**, *183*, 11–15. [CrossRef]
54. Wani, T.A.; Bakheit, A.H.; Al-Majed, A.A.; Altwaijry, N.; Baquaysh, A.; Aljuraisy, A.; Zargar, S. Binding and drug displacement study of colchicine and bovine serum albumin in presence of azithromycin using multispectroscopic techniques and molecular dynamic simulation. *J. Mol. Liq.* **2021**, *333*, 115934. [CrossRef] [PubMed]

Article

Deep Learning and Structure-Based Virtual Screening for Drug Discovery against NEK7: A Novel Target for the Treatment of Cancer

Mubashir Aziz [1], Syeda Abida Ejaz [1,*], Seema Zargar [2], Naveed Akhtar [3], Abdullahi Tunde Aborode [4], Tanveer A. Wani [5,*], Gaber El-Saber Batiha [6], Farhan Siddique [7,8], Mohammed Alqarni [9] and Ashraf Akintayo Akintola [10]

[1] Department of Pharmaceutical Chemistry, Faculty of Pharmacy, The Islamia University of Bahawalpur, Bahawalpur 63100, Pakistan; mubashirali035@gmail.com

[2] Department of Biochemistry, College of Science, King Saud University, P.O. Box 22452, Riyadh 11451, Saudi Arabia; szargar@ksu.edu.pk

[3] Department of Pharmaceutics, Faculty of Pharmacy, The Islamia University of Bahawalpur, Bahawalpur 63100, Pakistan; naveed.akhtar@iub.edu.pk

[4] Department of Chemistry, Mississippi State University, Starkville, MS 39759, USA; abdullahiaborodet@gmail.com

[5] Department of Pharmaceutical Chemistry, College of Pharmacy, King Saud University, P.O. Box 2457, Riyadh 11451, Saudi Arabia

[6] Department of Pharmacology and Therapeutics, Faculty of Veterinary Medicine, Damanhour University, Damanhour 22511, AlBeheira, Egypt; gaberbatiha@gmail.com

[7] Laboratory of Organic Electronics, Department of Science and Technology, Linköping University, SE-60174 Norrköping, Sweden; drfarhansiddique@gmail.com

[8] Department of Pharmacy, Royal Institute of Medical Sciences (RIMS), Multan 60000, Pakistan

[9] Department of Pharmaceutical Chemistry, College of Pharmacy, Taif University, P.O. Box 11099, Taif 21944, Saudi Arabia; m.alqarni@tu.edu.sa

[10] Department of Biomedical Convergence Science and Technology, Kyungpook National University, Daegu 41566, Korea; ashraf.akintola@gmail.com

* Correspondence: abida.ejaz@iub.edu.pk or abidaejaz2010@gmail.com (S.A.E.); twani@ksu.edu.sa (T.A.W.); Tel.: +92-062-9250245 (S.A.E.); Fax: +92-062-9250245 (S.A.E.)

Citation: Aziz, M.; Ejaz, S.A.; Zargar, S.; Akhtar, N.; Aborode, A.T.; A. Wani, T.; Batiha, G.E.-S.; Siddique, F.; Alqarni, M.; Akintola, A.A. Deep Learning and Structure-Based Virtual Screening for Drug Discovery against NEK7: A Novel Target for the Treatment of Cancer. *Molecules* **2022**, *27*, 4098. https://doi.org/10.3390/molecules27134098

Academic Editor: Peng Zhan

Received: 26 May 2022
Accepted: 18 June 2022
Published: 25 June 2022

Publisher's Note: MDPI stays neutral with regard to jurisdictional claims in published maps and institutional affiliations.

Abstract: NIMA-related kinase7 (NEK7) plays a multifunctional role in cell division and NLRP3 inflammasone activation. A typical expression or any mutation in the genetic makeup of NEK7 leads to the development of cancer malignancies and fatal inflammatory disease, i.e., breast cancer, non-small cell lung cancer, gout, rheumatoid arthritis, and liver cirrhosis. Therefore, NEK7 is a promising target for drug development against various cancer malignancies. The combination of drug repurposing and structure-based virtual screening of large libraries of compounds has dramatically improved the development of anticancer drugs. The current study focused on the virtual screening of 1200 benzene sulphonamide derivatives retrieved from the PubChem database by selecting and docking validation of the crystal structure of NEK7 protein (PDB ID: 2WQN). The compounds library was subjected to virtual screening using Auto Dock Vina. The binding energies of screened compounds were compared to standard Dabrafenib. In particular, compound **762** exhibited excellent binding energy of −42.67 kJ/mol, better than Dabrafenib (−33.89 kJ/mol). Selected drug candidates showed a reactive profile that was comparable to standard Dabrafenib. To characterize the stability of protein–ligand complexes, molecular dynamic simulations were performed, providing insight into the molecular interactions. The NEK7–Dabrafenib complex showed stability throughout the simulated trajectory. In addition, binding affinities, pIC50, and ADMET profiles of drug candidates were predicted using deep learning models. Deep learning models predicted the binding affinity of compound **762** best among all derivatives, which supports the findings of virtual screening. These findings suggest that top hits can serve as potential inhibitors of NEK7. Moreover, it is recommended to explore the inhibitory potential of identified hits compounds through in-vitro and in-vivo approaches.

Keywords: NEK7; virtual screening; DFTs; deep learning; molecular dynamics; drug design; drug repurposing; structural-based; cancer

---

1. Introduction

Cancer is the most common cause of death, with a high mortality rate worldwide causing 10 million fatalities per year. Cancer is characterized by unregulated cell growth and rapid proliferation [1]. Uncontrolled cell proliferation, aggregation, and an aberrant cell cycle are hallmarks of human cancer. Typically, cell division is controlled by several regulatory factors, including protein kinases [2]. Among all known protein kinases, NIMA (never in mitosis, gene A) related kinase7 (NEK7) plays a multifunctional role [3], including centrosome duplication, intracellular protein transport, mitotic spindle assembly, DNA repair, and cytokinesis [4–7].

NEK7 is a highly conserved serine/threonine kinase consisting of approximately 302 amino acids [6]. NEK7 is structurally related to NEK6, which shares 85% amino acid sequence identity. However, NEK7 is involved in critical roles that NEK6 cannot take over. NEK7 is centrosome-localized and is known to be highly expressed in a variety of vital organs such as the heart, lung, fat, brain, liver, and spleen [8]. It enhances the centrosome duplication efficiency by promoting the pericentriolar material at the centrosome during the S and G1 phases [3].

In addition, NEK7 also encourages the proliferation of resting cells, which indicates its high-level involvement in various cancer types, including non-small lung cancer, breast cancer, NLRP3-related inflammatory disease, and gastric cancer progression [9]. NEK7 also has a promising role in growth and survival. NEK9 regulates the activation of NEK7 during mitosis, which promotes spindle assembly, centrosome separation, and mitotic division of the cell [7].

Besides promoting the proliferation of various resting cells, NEK7 is also involved in the progression and development of fatal inflammatory diseases, including Alzheimer's disease, auto-immune disorders, inflammatory bowel disease, gout, and tumor formation [10]. Researchers have reported the involvement of NEK7 in the activation of NLRP3 inflammasome via ROS species formation, lysosomal destabilization, and potassium efflux. Stimulation of inflammatory mediators by NEK7 induces fibrosis and diabetic retinopathy and leads to hepatic carcinoma [10]. In brief, any mutation or atypical expression of NEK7 leads to the development of cellular oncogenesis and may provoke a fatal inflammatory response, causing tumorigenesis of multiple organs. These findings lend testimony to the involvement of NEK7 in the progression and development of numerous deadly diseases.

NEK7 is a promising target for multiple diseases, primarily cancer-related therapy research. NEK7 came into consideration two decades ago [2], but it has yet to be explored as a therapeutic target for preventing and treating NEK7-related diseases. A few medications have recently been developed to target the NEK7-mediated inflammasome pathway, but the mechanism and treatment outcomes are not specific and consistent [2].

Moreover, there is no FDA-approved medication that can selectively inhibit the expression of NEK7. Only Dabrafenib has shown activity against BRAF-mutant melanoma, which expresses more NEK9 [1]. These findings indicate that no published work has reported the selective and potent inhibitors of NEK7. As a result, the current study seeks more specific inhibitors that will provide a beneficial treatment option for NEK7-related cancer malignancies.

The current study focused on structure-based virtual screening (SBVS) of 1200 compounds library and drug repurposing of FDA-approved drug Dabrafenib. Dabrafenib demonstrated inhibitory potential against NEK9 with an IC50 value ranging from 1–9 nM [11,12]. Dabrafenib is comprised of benzene sulphonamide scaffolds. The basic sulphonamide group occurs in numerous biological active compounds [12], including anti-microbial [13], anti-tumor [14], anti-thyroid [15], antibiotics [16], and carbonic anhydrase inhibitors [17].

Clinically, sulphonamide-possessing drugs are used to treat lower urinary tract infections, whereas aromatic or hetero-aromatic sulphonamide derivatives possess a wide range of biological activities, including anti-tumor, anti-rheumatic, anti-microbial, and anti-inflammatory [18–21]. These findings have provided a strong rationale to retrieve structural analogues of Dabrafenib containing a basic sulphonamide nucleus.

The library of 1200 structural analogues of Dabrafenib was retrieved from the PubChem database and subjected to the in-silico drug discovery process. The discovery of a new anti-cancer agent is an extensive and laborious process. Thus, computer-aided drug design (CADD) [22] methods could serve as an alternative drug development strategy [23]. Among in-silico approaches, drug repurposing is an advanced tool for revisiting the activities of already approved drugs [24], which can save time and money [25].

The current study was focused to revisit the activity of Dabrafenib against NEK7 protein [26]. In addition, structure-based virtual screening (SBVS) [26] of 1200 structural analogues of Dabrafenib was carried out using molecular docking [27] and deep learning models [28]. SBVS is an advanced technology for the identification of potential hits with significant pharmacological properties against multiple molecular targets. Several robust docking programs are available for docking purposes in commercial and academic settings. In the present study, the Auto Dock Vina was used for virtual screening [29,30]. Moreover, density functional theory studies were conducted to explore the chemical reactivity profile of top-ranked analogues obtained through virtual screening. The structural geometry optimization and frequency calculations were performed. In addition, frontier molecular orbital (FMO) analysis and global reactivity descriptors were also determined. The efficacy of any drug is determined by its interaction with targeted biomolecules. Deep learning algorithms [31] were used for prediction of binding affinity and pIC_{50} values of top hits obtained via virtual screening. Predicted values of top hits were compared to in-vitro activity of Dabrafenib.

Furthermore, in-silico ADMET properties were also determined using a message-passing neural network (MPNN). The MPNN model is widely used for prediction of molecular properties such as blood brain barriers, human intestinal absorption, and solubility profiles [32]. The molecular docking approach only provides a static view of the molecular interactions of the complex. Still, to determine the stability of the protein–ligand complex, molecular dynamic simulations (MD simulations) have been performed to determine the stability, which provide significant insight into the molecular interactions of top-ranked complexes under accelerated conditions. Top hits obtained through structure based virtual screening a shown in Figure 1. All hits shared the same pharmacophore with standard Dabrafenib.

This is the first comprehensive computational study for the identification of selective inhibitors of the NEK7 protein. The current study has utilized the latest computational approaches, suggesting identified hits as a new strategy for treatment of NEK7-associated malignancies. Findings of the current study suggest the exploration of the inhibiting potential of these hits at the molecular level using in-vitro and in-vivo experimental techniques.

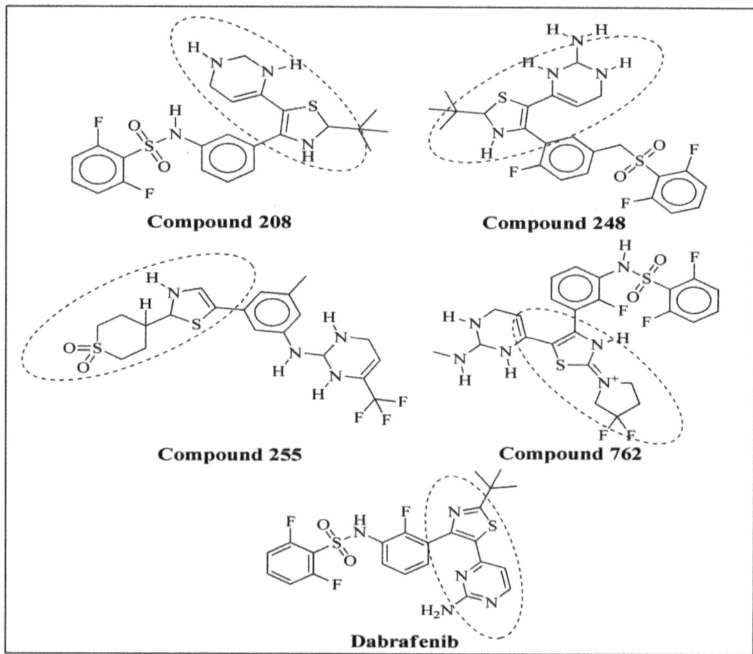

Figure 1. Top Hits obtained through SBVS. All hits were sharing same basic Pharmacophore with standard Dabrafenib.

2. Experimental

2.1. Computational Studies

2.1.1. Density Functional Theory Calculation

The ground state electronic energy is ascertained by electron density of the compound [33]. The electron density defines the number of electrons, nuclear charge and position of the nuclei in a compound [34]. Variation in electron density yields different ground state energy, and both of these properties are related by density functional theory methods [35]. DFT methods are based on suggestions that electron density can be accurately assumed by the set of specific orbitals using an exchange correlation such as B3LYP [36]. Based on their computational accuracy, DFT methods are a reliable and efficient approach for correct estimation of electronic properties of the compound [37]. The structural geometries of selected compounds were optimized through DFT studies. DFT calculations were executed on Guassian09 program [38] using B3LYPfunctional correlation and 6-31G* as a basis set [39]. It is a compelling theory to calculate the electronic structure of atoms and molecules. Gauss View 6 was used for visualization of output files [40]. In addition, DFT/B3LYP method was employed for generation of Frontier molecular orbitals (FMOs), electrostatic surface potential map and global and local reactivity of descriptors. After completion of calculations, the output log file was visualized in Gauss View 6 for determination of optimization energy, dipole moment, frequency and polarizability [41].

2.1.2. Structure Based Virtual Screening (SBVS)

Drug candidates were retrieved from the PubChem database (https://pubchem.ncbi.nlm.nih.gov/) (accessed on 28 April 2022) to create the ligand library. There were 1200 structural analogues of Dabrafenib in the library. PyRx software was used to prepare the compounds library, which was converted to pdbqt format for virtual screening using Auto Dock Vina. The MMFF94 force field was used to minimize the energy of ligands. The crystallographic structure of the targeted protein was retrieved from Protein Data Bank (https://www.rcsb.org/) (accessed on 1 May 2022) (PDB ID: 2WQN). After that, MGL tools were used to prepare macromolecule, which included removing Het atoms and water molecules, and the addition of polar hydrogen. The protein was examined for any missing residues. Furthermore, Kollman's charges were used to neutralize protein, and Gasteiger charges were calculated. Finally, for virtual screening of the compound library using Auto Dock Vina, a 1-angstrom grid box was built centered on the crystalline structure of protein at the point of co-crystal ligand (ADP) binding-site coordinates. The central xyz axis of the grid box was set to $80 \times 80 \times 80$. Virtual screening was carried out after the targeted protein was prepared utilizing Auto Dock Vina's script-based technique. The exhaustiveness was set to 5 and the number of nodes was set to 20. The virtual screening was repeated twice to ensure the accuracy of docking results. In addition, docking protocol was validated by re-docking the co-crystal ligand with targeted protein. A RMSD value of less than 2 angstrom indicates the reliability of the docking pose. After completion of virtual screening, the output findings of the virtual screening module were analyzed and docking scores of drug candidates were compared to standard Dabrafenib. Only four compounds were found to have higher docking scores than standard Dabrafenib. The top hits were subjected to further analysis using deep learning algorithms. Deep learning models were used to predict drug affinity and determine the stability of protein–ligand complexes.

2.1.3. Molecular Dynamics Simulation

The molecular docking experiment provided an initial static protein–ligand complex for molecular dynamic studies. Desmond, a package from Schrödinger LLC [42], was used to run molecular dynamic simulation for 100 ns. Molecular docking studies provide insight into the binding state of ligand with protein. Docking produces the static orientation of a ligand molecule inside active pockets of targeted protein [43], and MD simulations measure the average displacement of atoms with respect to a reference. MD simulations provide information about the stability of the best complex [44,45].

Maestro or Protein Preparation Wizard were employed for processing of the protein–ligand complex. The system was prepared in the system builder tool of the Desmond package. The system was solvated by Monte-Carlo equilibration, TIP3P solvent model extended 10.0 angstrom in each direction. The counter NaCl ions at a concentration of 0.15 M were added to neutralize the system. The optimized potential for liquid simulation (OPLS 2005) [46] was used as a forcefield to generate parameter files [46]. The pressure control was conducted through the Martyna−Tuckerman−Klein chain coupling scheme with a coupling constant of 2 ps [47], whereas the Noose–Hoover chain coupling scheme was used for temperature control [48]. The energy minimization was performed for 20,000 steps in order to remove any intra-molecular steric clashes. Initially, the system was equilibrated (NVT ensemble) for 1 ns, and afterwards the NPT ensemble was performed for an additional 1 ns at 300 K temperature and 1 bar pressure. Finally, production run was performed for 100 ns under periodic boundaries conditions. The Particle Mesh Ewald (PME) method [49] was used to determine electrostatic interactions [50]. The Verlet/Leapfrog algorithm was used for numerical integration. A time step of 1 fs was used for minimization and a time step of 2 fs was used for molecular dynamic simulation [51]. Thermal MM-GBSA.py script [52,53] was used to calculate the ligand strain and ligand-binding free energy for docked conformations over a 100 ns period [54].

2.1.4. Prediction of Binding Affinities, pIC$_{50}$ and ADMET Properties Using Deep Learning Models

Dabrafenib, which has been approved by the FDA, has been found to be effective against BRAF-mutant melanoma with a high level of NEK9 protein expression. Dabrafenib's inhibitory concentration was in the nanomolar range, 1–9 nM [1]. The drug's effectiveness is largely determined by its binding affinity (IC$_{50}$) and ADMET profile. Therefore, we have employed deep learning models to predict IC$_{50}$, pIC$_{50}$, and ADMET properties of top hits acquired through virtual screening in order to provide a direct comparison of binding affinities of top hits with standard Dabrafenib. Predicting the binding affinity and ADMET characteristics in silico, rather than using an experimental method, is a promising alternative. Deep learning (DP) models were used to predict drug target interactions (DTI) in the current work, which were formulated on encoder and decoder architectures. A DL model takes the SMILES string and amino acid sequence of the targeted protein as input and uses over 17 state-of-the-art DP learning techniques to predict drug efficacy indicators (Figure 2). The MPNN-CNN deep learning algorithms were used for affinity prediction in this work, while the MPNN model was used for ADMET predictions [32].

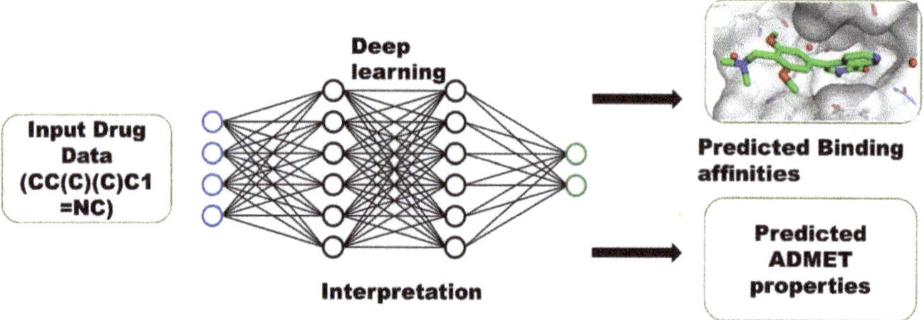

Figure 2. Implementation of Deep learning Model.

3. Results and Discussion

The 1200-compound library was retrieved from the PubChem database and subjected to SBVS and the FDA drug Dabrafenib. Dabrafenib was maintained as the standard drug to which docking scores of 1200 compounds were compared. It was observed that only four combinations have better docking scores and binding affinity than standard Dabrafenib. These four compounds were considered top hits and subjected to further analysis, including geometry optimization and FMO analysis via density functional theory studies. Moreover, IC50, pIC50, and ADMET properties of the top four compounds were also predicted using deep learning models.

3.1. Density Functional Theory Studies (DFTs)

Quickly calculating physicochemical properties of atoms, bonds and molecules is necessary to process thousands or millions of structures in data mining investigations. Calculations in quantum chemistry based on ab initio and density functional theory (DFT) yield increasingly reliable assessments of many characteristics [55]. The B3LYP hybrid functional is likely the most popular DFT functional, and its cost-effectiveness has been widely acknowledged. Nonetheless, DFT computations are still too computationally expensive to be conducted on single workstations or tiny clusters in less than a few hours [56].

3.1.1. Optimized Geometries

In the present study, geometries of FDA-approved drug Dabrafenib and top hits were completely optimized using the DFT/B3LYP method and 6-31G* as a basis set. No negative frequencies were obtained after the geometry optimization, which demonstrates that current geometries are true local minima. Optimized structures of drug candidates are presented in Figure 3.

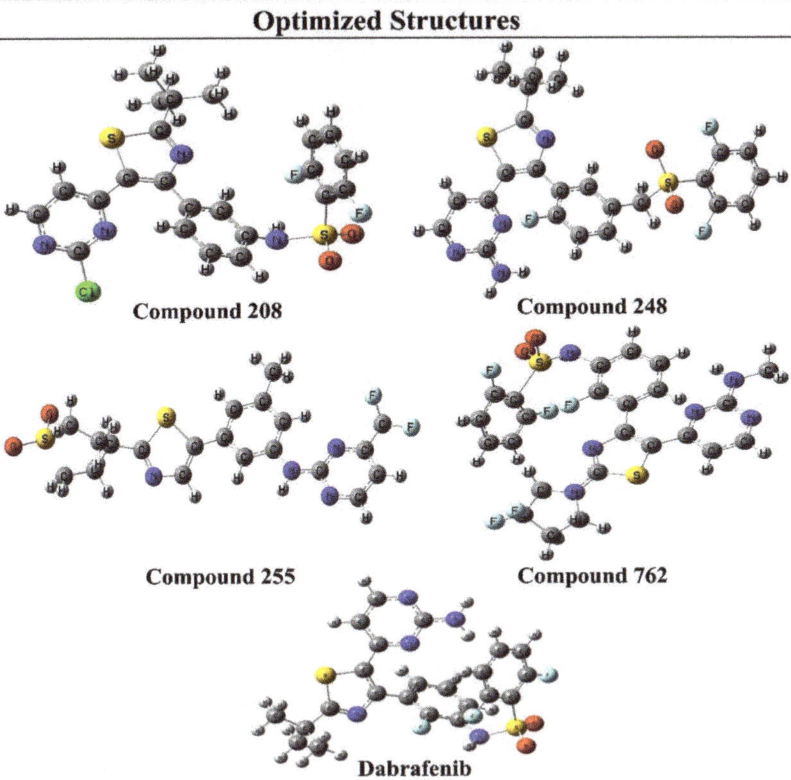

Figure 3. Optimized structures of selected compounds.

The compound **762** showed high value for polarizability and dipole moment, which indicates its high polarity and chemical reactivity. Optimization and polarizability values of top hits and Dabrafenib are given in Table 1.

Table 1. Energetic parameters of top hits and standard Dabrafenib.

Compound	Optimization Energy (Hartree)	Polarizability (α) (a.u.)	Dipole Moment (Debye)
Compound **208**	−2712.903	340.588	13.330
Compound **248**	−2391.895	345.671	7.283
Compound **255**	−2238.320	304.820	8.827
Compound **762**	−2699.752	360.213	10.042
Dabrafenib	−2407.203	319.254	6.682

3.1.2. Frontier Molecular Orbital (FMOs)

The way a molecule interacts with other species is determined by its frontier molecular orbitals. The highest occupied molecular orbital, or HOMO, is the outermost orbital-bearing electrons, and it primarily works as an electron donor. The lowest unoccupied molecular orbital, or LUMO, is the innermost orbital with free electron acceptor sites. The ionization potential ought to be proportional to the HOMO energy, while the LUMO energy should be proportional to the electron affinity (Table 2).

Table 2. ΔE_{gap} of HOMO/LUMO orbitals of selected compounds.

Compound	E_{HOMO}(eV)	E_{LUMO}(eV)	ΔE_{gap}(eV)	Potential Ionization I (eV)	Affinity A (eV)
Compound **208**	−0.234	−0.099	0.135	0.234	0.099
Compound **248**	−0.222	−0.070	0.152	0.222	0.070
Compound **255**	−0.228	−0.083	0.145	0.228	0.083
Compound **762**	−0.216	−0.089	0.127	0.216	0.089
Dabrafenib	−0.233	−0.074	0.159	0.233	0.074

The energy gap is the difference in energy between the HOMO and LUMO orbitals, and it is the most key variable in predicting the stability of a molecule. The HOMO–LUMO energy gap is used to evaluate the chemical reactivity and kinetic stability of the molecule. A soft molecule is a structure with a narrow gap that has a higher degree of polarization. As a measure of electron conductivity, the energy difference between HOMO and LUMO was recently employed to illustrate the bioactivity of intramolecular charge transfer (ICT). The stronger the chemical reactivity and the less stable the kinetics, the smaller the gap. The compound **762** has the narrowest energy gap at 0.127 eV among all the compounds. Compound **248** has a greater energy gap, measuring 0.152 eV. Thus, it shows compound **762** is chemically more reactive than all other compounds, which are comparatively the stable ones. In addition, the electron density of HOMO orbitals for Dabrafenib was localized over morpholine and piperdinyl rings, whereas electron density of LUMO is localized to the carbonitrile and benzocarbazole moiety of the drug. FMOs orbitals are shown in Figure 4.

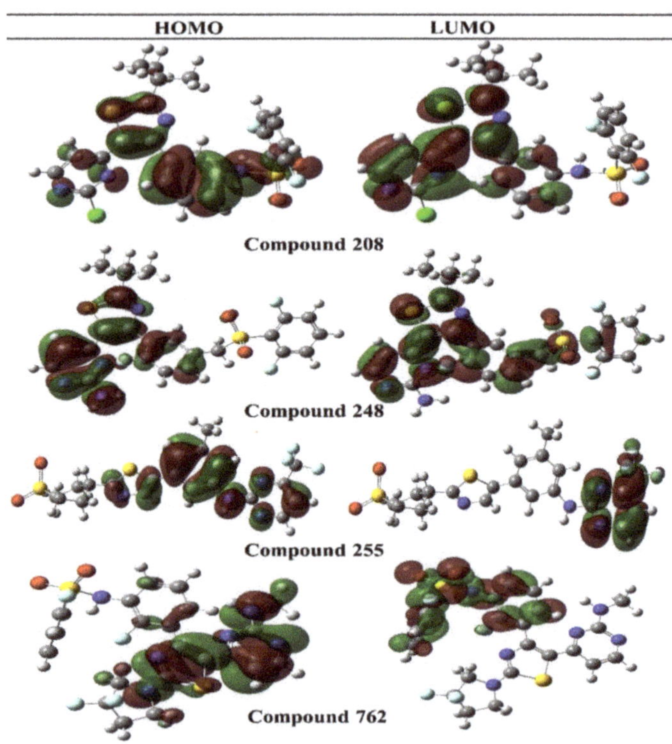

Figure 4. HOMO–LUMO structures of the selected compounds.

3.1.3. Global and Local Reactivity Descriptors

The HOMO and LUMO frontier orbitals are used to predict chemical reactivity. The HOMO orbital energy of a compound is significantly correlated with its vulnerability to electrophilic attack and ionization potential. A compound's LUMO orbital energy is a reliable predictor of electron affinity and nucleophilic attack. The energy of LUMO is proportional to its electron affinity, indicating that it is susceptible to nucleophile attack. The frontier molecular orbital energies are also related to the hard and soft characteristics of a molecule. Hard nucleophiles have a low HOMO, whereas soft nucleophiles have a high HOMO. Similarly, hard electrophiles have a high LUMO energy, whereas soft electrophiles have a low LUMO energy. According to the frontier theory of electron reactivity, the chemical reaction occurs at the point where the HOMO and LUMO have the most overlap. All reactions require the HOMO density of the donor molecule, while all reactions require the LUMO density of the acceptor molecule. The frontier orbital densities of individual atoms can be used to quantify their reactivity inside a molecule. Chemical behavior is frequently predicted using electronegativity and hardness. Compound **248** presented greater energy gaps, indicating it to be the tougher among all compounds. Compound **208** demonstrated the greatest electrophilicity index value of 0.207 eV. This indicates that compound **208** is an excellent electrophile among all the other compounds. The HOMO–LUMO energy gap for Dabrafenib was found to be 0.159 eV. Dabrafenib showed a softness value of 6 (Table 3). The Koopman's theorem was used to express ionization energy and electron affinity of drug candidates.

$$I = -EHOMO: A = -ELUMO$$

Table 3. Global reactivity descriptors.

Compound	Hardness (η)	Softness (S)	Electronegativity (X)	Chemical Potential (μ)	Electrophilicity Index (ω)
Compound **208**	0.067	7.433	0.167	−0.167	0.207
Compound **248**	0.076	6.566	0.147	−0.147	0.141
Compound **255**	0.072	6.901	0.156	−0.156	0.167
Compound **762**	0.063	7.880	0.153	−0.153	0.184
Dabrafenib	0.080	6.280 [38]	0.154	−0.154	0.149

Compound	Electrodonating power (ω^-)	Electroaccepting power (ω^+)	Net Electrophilicity($\Delta\omega^\pm$)
Compound **208**	0.299	0.132	0.432
Compound **248**	0.224	0.077	0.301
Compound **255**	0.254	0.098	0.352
Compound **762**	0.268	0.116	0.384
Dabrafenib	0.236	0.082	0.318 [38]

We evaluated the following parameters by using their respective formulas: Hardness: $\eta = 1/2(ELUMO - EHOMO)$; Softness: $S = 1/2\eta$; Electronegativity: $\chi = -1/2(ELUMO + EHOMO)$; Chemical potential: $\mu = -\chi$; Electrophilicity index: $\omega = \mu/2\eta$.

3.2. Structure Based Virtual Screening and Predicted Binding Affinities

Initially, the Molecular docking methodology was validated by redocking a co-crystal ligand with targeted protein using the same coordinates. RMSD values of less than 2 angstrom were obtained, which demonstrate the successful validation of the docking protocol and can be used to describe ligand poses with specificity and accuracy. Afterward, virtual screening was conducted with 1200 ligands library and Dabrafenib. The coordinates of co-crystal ligand were used to dock the ligand library and Dabrafenib. Out of

1200 drug candidates, only four drug candidates showed excellent binding energies that were even better than Dabrafenib. The binding energies of top hits and Dabrafenib are tabulated in Table 4. In particular, compound **762** showed a maximum binding energy of -42.67 kJ/mol and exhibited strong binding affinity with NEK7 protein when subjected to the DL prediction model.

Table 4. Binding energies and Predicted binding affinities via Deep learning model.

Compound	Binding Energies (kJ/mol)	Predicted Binding Affinity (IC_{50}) nM	pIC_{50} (Predicted via Deep Learning Model)
Compound **208**	-33.47	206.26	6.69
Compound **248**	-35.56	268.80	6.57
Compound **255**	-35.98	283	6.55
Compound **762**	-42.67	61.74	7.21
Dabrafenib	-33.89	1-9 (Experimental) [1]	—

Among the four top hits, compound **208** exhibited promising hydrophobic and hydrophilic interactions. The amino acid residues involved in important molecular interactions were as follows: ASP115, ARG121, GLY117, ASP179, PHE168, ILE195, ALA114, ALA61, ILE40, and ASP118. It was observed that two hydrogen bonds were involved in stabilizing the protein–ligand complex. One hydrogen bond was observed with ASP118 with a bond length of 2.97 angstrom, while the second hydrogen bond was observed with GLY117 having a bond length of 3.14 angstrom. Important residues of the active site were engaged in hydrophobic interactions, including van der Waals interactions, pi-alkyl and alkyl–alkyl interactions. The docking score of compound **208** was found to be -33.47 kJ/mol. Similarly, compound **248** exhibited stronger molecular interactions with the following amino acid residues: ARG121, ASP118, ALA165, LYS63, ASN166, ASP179, PHE168, LEU111, VAL48, ALA116, ASP115, and GLY117. It was discovered that important amino acid residues of the NEK7 protein's DLG/DFG motifs were involved in interactions. Furthermore, the amino acids LEU111 and LYS163 interacted via hydrophobic bonds. Two important hydrogen bonds were contributing toward the stability of conformations. One hydrogen bond engaged GLY117 residues with a bond length of 2.2 angstroms. Another hydrogen bond was engaging ASN166 amino acid with a bond length of 3.34 angstroms. Among hydrophobic interactions, van der Waals interactions played a pivotal role in stabilizing the complex. The docking score of the compound **248** was -35.56 kJ/mol. The putative 2D and 3D binding modes of compounds **208** and **248** are shown in Figure 5.

Another important top hit was compound **255**, which exhibited potent molecular interactions with amino acid residues of the active site. It was the second-best drug candidate obtained via virtual screening. Amino acid residues involved in bonding and nonbonding interactions were as follows: PHE45, SER46, LYS63, ALA114, ASP115, GLU112, PHE168, ASP179, VAL48, GLY43, and GLN44. It was observed that two important hydrogen bonds with short bond lengths were contributing toward stabilizing the complex. One hydrogen bond occurs between the electronegative oxygen atom of the compound **255** and the SER46 residue of the targeted protein. Moreover, the second hydrogen bond was engaged in GLY117 with a bond length of 3.16 angstroms. As shown in Figure 5, amino acid residues from the active site were involved in hydrophobic interactions with compound **255**.

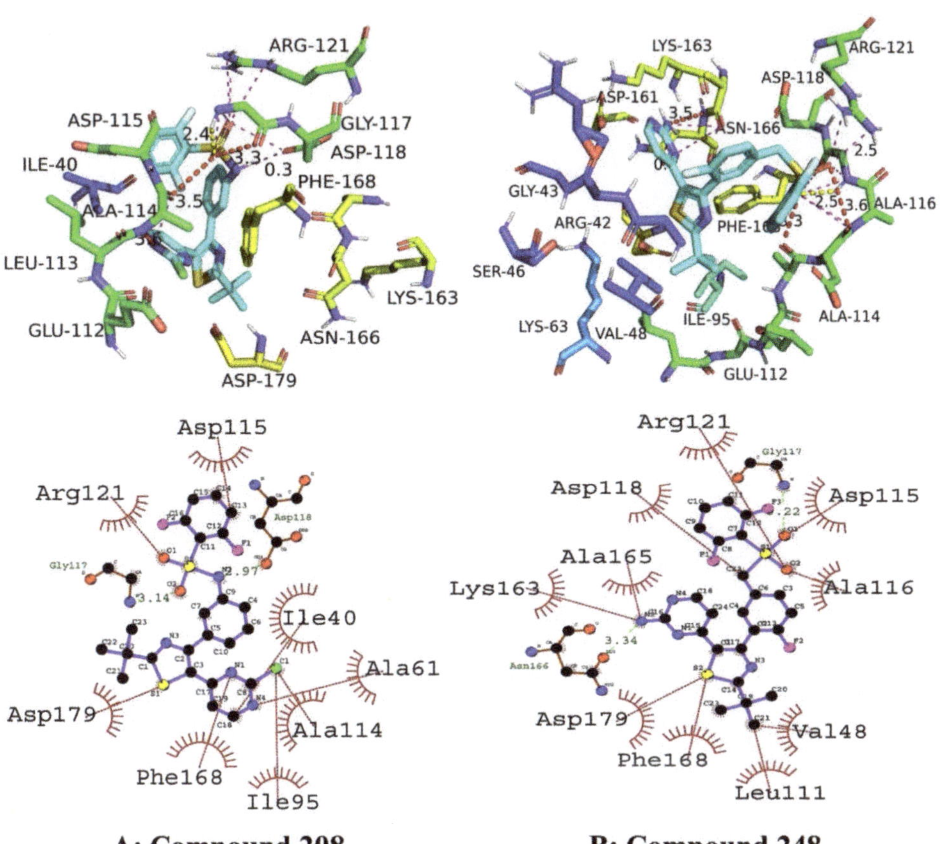

Figure 5. The putative 2D and 3D binding mode of compound **208** (**A**) and **248** (**B**). Green dashes are indicating hydrogen bonding whereas red dashes are indicating hydrophobic interactions.

Now referring to the top hit obtained through SBVS, namely compound **762**, It has shown excellent docking scores and demonstrated significant binding affinity obtained through deep learning models. It was observed that compound **762** was engaged in three hydrogen bonds of moderate-to-strong strength. One hydrogen bond occurred between the pentazole ring of the compound **762** and the electronegative oxygen atom of TYR201. The bond length of interaction was 3.08 angstroms. Similarly, the second hydrogen bond engaged SER234 residues with a surprisingly smaller bond length of 2.92 angstroms. These interactions lend enough testimony to stronger molecular interactions and more stabilized protein–ligand complexes. Furthermore, the third and last hydrogen bond occurred between TYR237 and compound **762** with a bond length of 3.02 angstroms. All three amino acid residues involved in hydrogen bonding belong to the activation loop of the NEK7 protein. The remaining active site residues, ILE123, GLU228, PHE236, MET203, PRO200, LEU246 and LEU232, engaged in hydrophobic interactions with compound **762**. The docking score and binding affinity (IC_{50}) were predicted to be best among all top hits, i.e., −42.67 kJ/mol and 61.74 nM, respectively. Compound **762** could be a promising drug candidate for the treatment of NEK7-associated malignancies. The binding interactions of compounds **255** and **762** are shown in Figure 6.

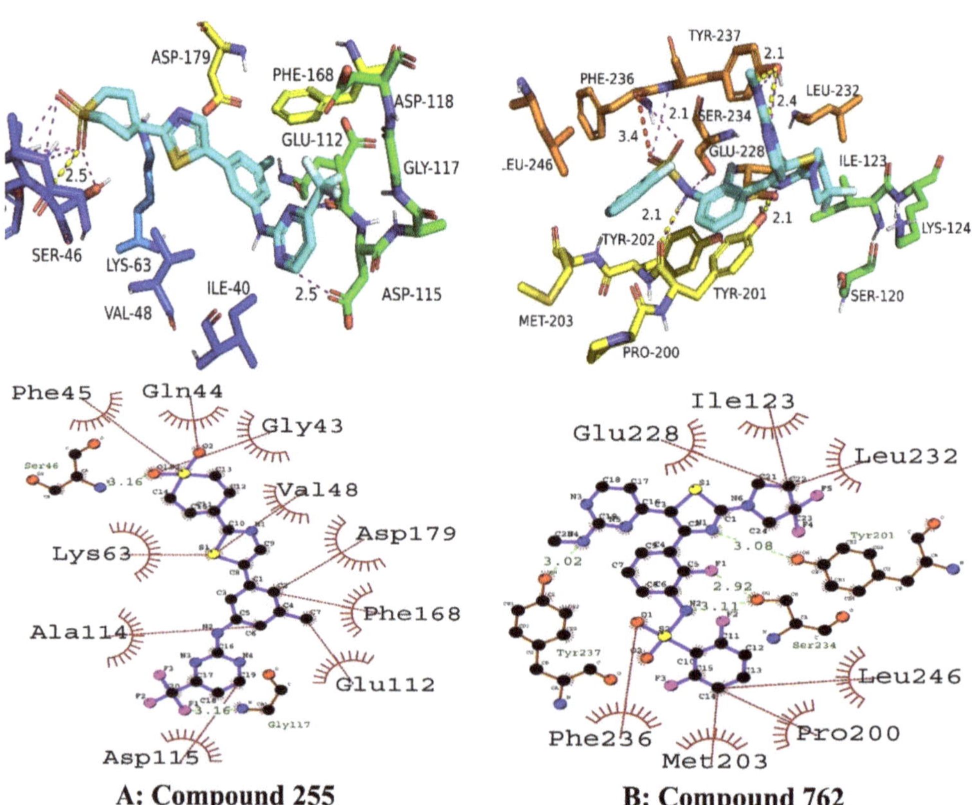

Figure 6. The putative 2D and 3D binding mode of compound **255** (**A**) and **762** (**B**). Green dashes are indicating hydrogen bonding whereas red dashes are indicating hydrophobic interactions.

The bonding and non-bonding interactions of standard Dabrafenib was involving important amino acid residues of NEK7 activation loop. ARG50, LYS38, ALA165, ILE40, GLY117, ASP115, PHE168, LEU111, LEU112, ALA114, LEU113, ALA161, ASP179, and ILE95 were the amino acid residues implicated in molecular interactions with Dabrafenib. Dabrafenib exhibited significant molecular interactions, which contributed towards complex binding affinity. The strong interactions were observed with targeted protein and sulphonamide rings. The sulphonamide ring was implicated in several important stabilizing contacts, including conventional hydrogen bonding with ASP115 of the activation loop, Pi-cation interaction with ARG50, and interactions with ILE40 and ASP115 by two fluorine atoms connected to the ring. PHE168 formed pi-cation and pi-pi T-shaped contacts with the butylthiazole ring, whereas the pyrimidine ring produced conventional hydrogen bonds with GLU112 and ASP179, a carbon–hydrogen connection with ALA114, and a pi-alkyl interaction with ALA161. Due to important chemical interactions, Dabrafenib has a good binding energy of −33.89 kJ/mol. van der Walls interactions are essential hydrophobic interactions that have been observed with the amino acids LYS38, ALA165, GLY117, LEU111, LEU113, and ILE95. Figure 7 depicts the probable binding mode of Dabrafenib with NEK7.

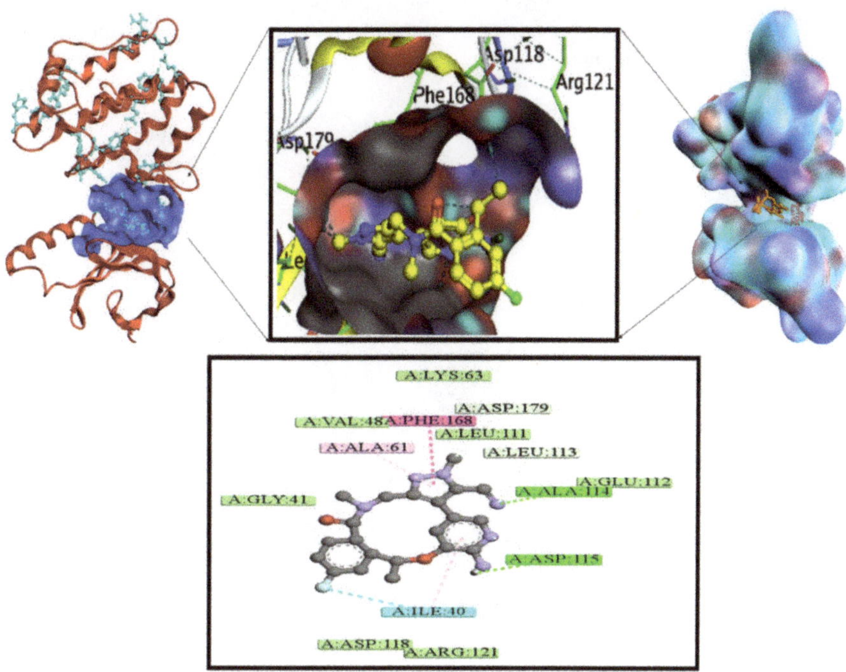

Figure 7. 2D and 3D interactions of NEK7–Dabrafenib complex.

3.3. Electrostatic Surface Potential Map

Investigating the electrostatic surface potential (ESP) map is a key activity in drug design as it determines the chemical reactivity of the compound and its ability to produce important molecular interactions. It is an effective way to visualize the molecular reactivity and evaluate the nature of ligand-binding with a targeted protein. The ESP map is depicted by different colored regions depending upon the electronegativity of the compound. The highly electronegative part is represented by the color red, whereas the electropositive part is represented by the color blue. The QM calculations were performed using DFTs at the B3LYP/6-31G* level of theory. Figure 8 depicts the ESP potential and the nature of ligand-binding with the targeted protein. In this study, the contribution of the electronegative oxygen atom in all interactions is indicated by the color red, whereas the contribution of the nitrogen atom is provided in the color blue. Considering the electrostatic surface potential map, the contribution of oxygen atoms toward interaction potential is higher than that of nitrogen atoms. It was observed that in the case of compound 208, the electronegative oxygen atom was acting as a hydrogen acceptor and was producing strong hydrogen bonding with GLY117. Similarly, in compound **248**, **255** and **762**, electronegative oxygen atoms were involved in stronger intermolecular interactions. In contrast, nitrogen atom was involved in hydrogen bonding by donating the hydrogen bond for example, in case of compound **208**, nitrogen was donating hydrogen bond to ASP118 residue. In addition, the docked conformation of ligands on the protein surface is also represented by different colored regions (Figure 8). The red surface indicates the hydrogen bond acceptor region, while the blue surface indicates the hydrogen bond donor locations. Whereas the grey color areas indicate the hydrophobic interactions, including van der Waals interactions. The red-colored surface area of protein is buried by nitrogen atoms as they act as proton donors, whereas the blue-colored protein surface is buried by electronegative atoms such as oxygen, fluorine and chlorine, which acts as a hydrogen bond acceptor. It can be observed that the grey surface area of protein is mostly involved in hydrophobic interactions, and these regions are buried by alkyl, phenyl rings and other hydrophobic groups present in all compounds.

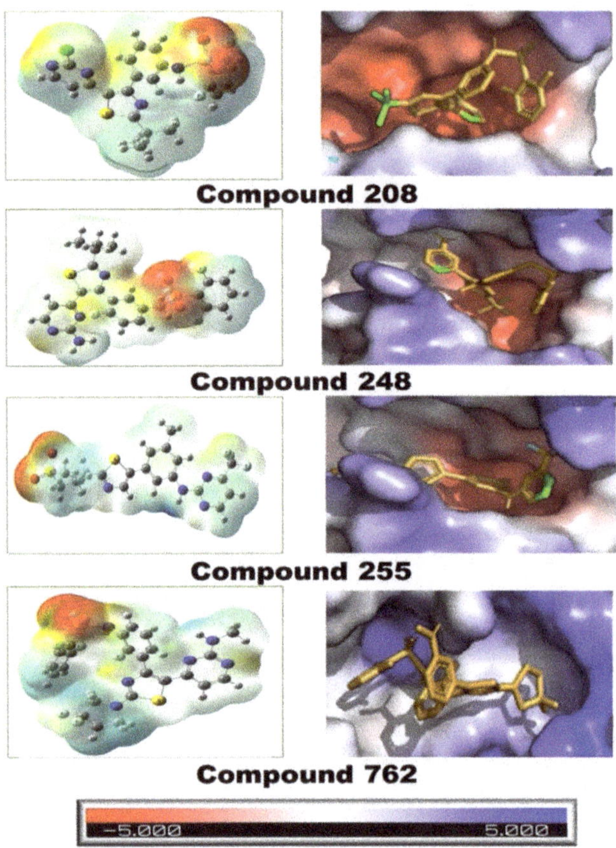

Figure 8. Electrostatic surface potential map of all ligand complexes.

3.4. Buried Surface Area (BSA)

Molecular interactions are the critical factors in determining the stability of protein–ligand complexes. Molecular interactions existing between protein–ligand complexes can be modelled by taking into account the physicochemical properties and complementarity of the shape of the binding interface. A useful method for determining the complementarity of the shape and extent of molecular interactions is the estimation of the buried surface area (BSA) of a protein–ligand complex. In the current study, the BSA of best complexes was calculated using a new Shrake–Ruply algorithm-based tool (*dr_sasa*) [57] used for calculating the solvent accessible surface area (SASA), buried surface area (BSA), and contact surface area (CSA). All four top compounds (**208**, **248**, **255**, and **762**) were subjected to the calculation of BSA. It was observed that the targeted NEK7 protein was buried up to 80% and 70% by compounds **208** and **248**, respectively. In particular, amino acid residues ILE40 and PHE168 were strongly buried by compound **208** (49 Å^2). Compound **248**, on the other hand, was strongly engaging the ARG121 and PHE168 with BSA of 39 Å^2 and 41.3 Å^2 respectively. The detailed buried surface area of both compounds is shown in Figure 9.

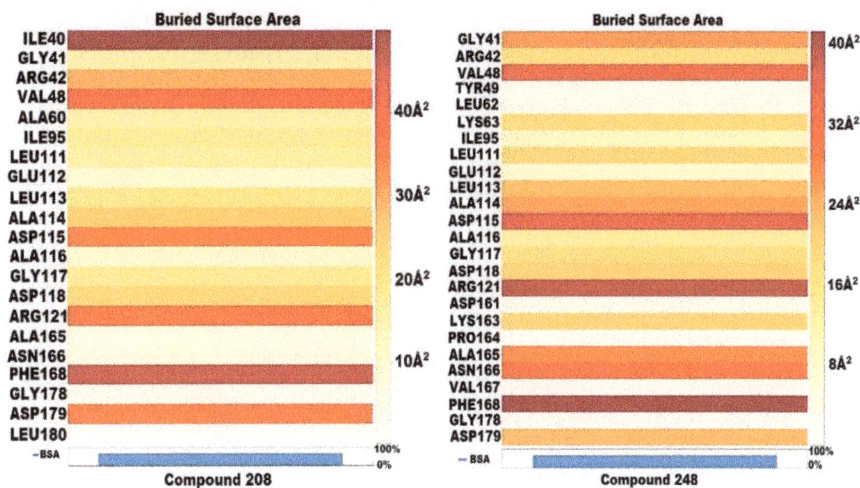

Figure 9. Buried surface area (BSA) of compound **208** and **248**.

In terms of compounds **255** and **762**, it was observed that both compounds significantly engaged the amino acid residues of the target protein. Compound **255**, in particular, was burying the surface area of the NEK7 protein by up to 360 Å^2. The BSA of compound **255** with VAL48, LYS63, ALA114, and PHE168 was 168, 212, 187, and 351 Å^2, respectively, which was the best among all top hits. These values demonstrate the strong nature of molecular interactions existing between the target protein and compound **255**. In the case of compound **762**, important amino acid residues were buried by compound **762**. In particular, TYR201, TYR237, and MET241 were significantly buried by compound **762** with BSA of 64, 72, and 25.6 Å^2. Moreover, it was worth noticing that the major contributing atoms were oxygen, nitrogen, fluorine, sulphur, and chlorine, which were involved in increasing the contact surface area of compounds with a targeted protein. The BSA of compounds **255** and **762** is shown in Figure 10.

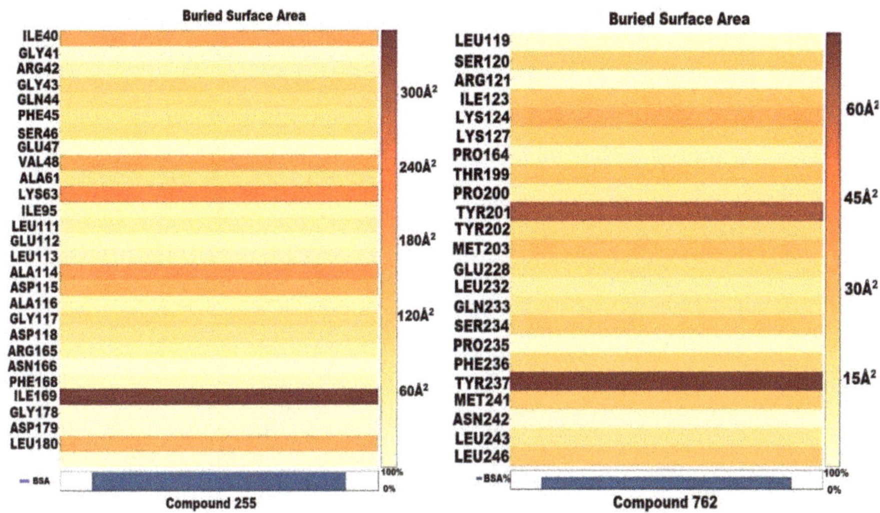

Figure 10. Buried surface area (BSA) of compound **255** and **762**.

3.5. Molecular Dynamic Simulation

The molecular docking technique is comparatively rapid and imprecise. The docking deficiencies and flexibility of protein may interfere with protein–ligand complex. However, molecular dynamic simulations are computationally expensive and time-consuming but provide reliable and accurate illustration of protein displacement. Considering these facts, molecular dynamic simulations were undertaken using Desmond software package [58,59]. The root-mean-square deviation (RMSD) patterns provide significant insight into average change in displacement of atoms with respect to a frame. The RMSD trajectory provides information about the structural configuration of protein. It is computed for each frame of the trajectory. In order to gain insight into the structure of a protein, it is important to monitor the protein's RMSD. Plotting the RMSD of the ligand is possible once the protein-ligand complex is aligned on a reference protein backbone and the RMSD of the ligand heavy atoms is measured. It is likely that the ligand has diffused from its initial binding site if measured values exceed the protein's RMSD by a substantial margin. Molecular dynamic trajectory analysis is also used to determine the root-mean-square fluctuation of the targeted protein.

3.5.1. RMSD Analysis of Protein and Protein–Ligand Complexes

The RMSD patterns for C-alpha atoms of NEK7 protein were estimated in order to determine the effect of the bounded drug on the conformational stability of NEK7 protein. Figure 11 is displaying the progression of RMSD values for the C-alpha atoms of NEK7 as a function of time. The 2wqn–Dabrafenib complex reaches the equilibrium after around 5 nanoseconds of simulation, and although side chain residues displayed fluctuations, they remained in the permissible range of 1–4 angstroms, which can be considered insignificant [60]. The NEK7–Dabrafenib complex showed slight fluctuations after 50 ns, which again became stable after 60 ns of simulation and remained equilibrated throughout the simulated trajectory. RMSD fluctuation was observed from 70 to 90 ns, which is due to the decrease in the number of contacts during this time, but after 90 ns, the number of contacts with amino acid residues increased and RMSD pattern became stable. It demonstrate the existence of stable molecular interactions. After being equilibrated, NEK7 RMSD values fluctuated within 2 angstrom. After 80 ns, protein RMSD showed slight fluctuation up to 2.5 angstrom and dropped again after 95. The average RMSD value for the protein–ligand complex and NEK7 protein is tabulated in Table 5.

Table 5. Average values obtained from MD simulations.

Protein-Ligand Complex	Average Protein RMSD (Å)	Average Protein RMSF (Å)	Average Protein-Ligand Complex RMSD (Å)	Average Radius of Gyration (Å)	Average SASA (Residue Wise) (Å^2)
NEK7–Dabrafenib complex	1.97	0.87	3.89	19.76	282.72

These findings suggest that the ligands stayed firmly bound to the receptor throughout the simulation period. Moreover, small RMSD patterns indicate the fewer structural rearrangements and lesser conformational changes within binding site residues [61].

It is beneficial to identify local differences in the protein chain by using the root-mean-square fluctuation (RMSF). Figure 12 peaks on the RMSF graph represent the regions of the protein that change the most during the simulation. The average RMSF for the NEK7 backbone was 0.87 angstrom, indicating the fewer structural rearrangements (Table 5). The N- and C-terminal ends of proteins are more likely to undergo alteration than any other portion of the protein. In the range of amino acids from 180 to 220, the RMSF value fluctuated, as can be seen in the RMSF graph. These residues are found in the C-terminal lobe. The protein's structure, such as its alpha and beta helices and strands, tends to be stiffer and less variable than its unstructured component. According to MD trajectories, the residues with the highest peaks are found in loop areas or the N- and C-terminal regions. Binding site residues with low RMSF values imply a stable ligand–protein interaction.

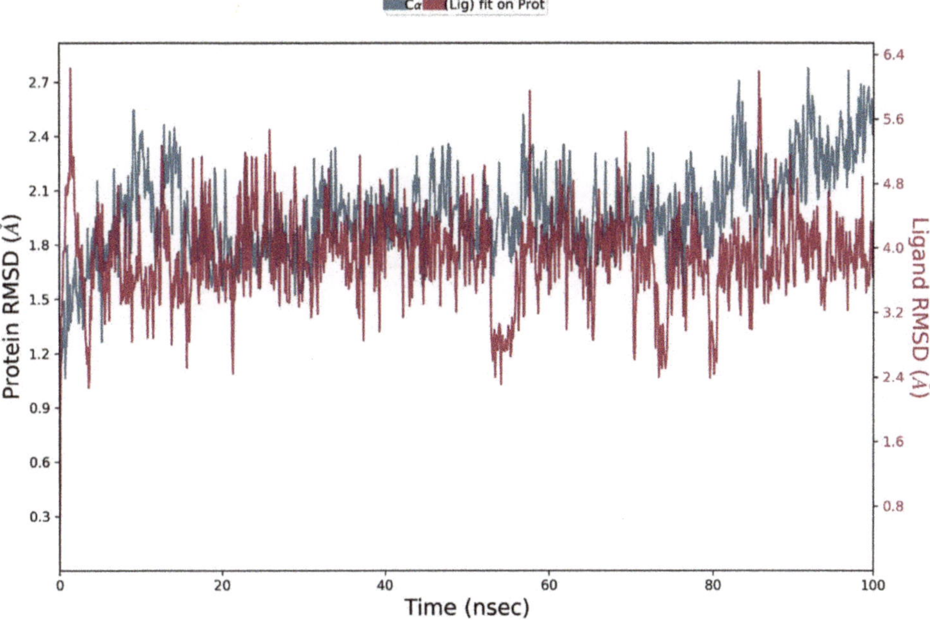

Figure 11. Residue wise root-mean-square deviation (RMSD) of the C-alpha atoms of NEK7 (2wqn) and Dabrafenib Complex.

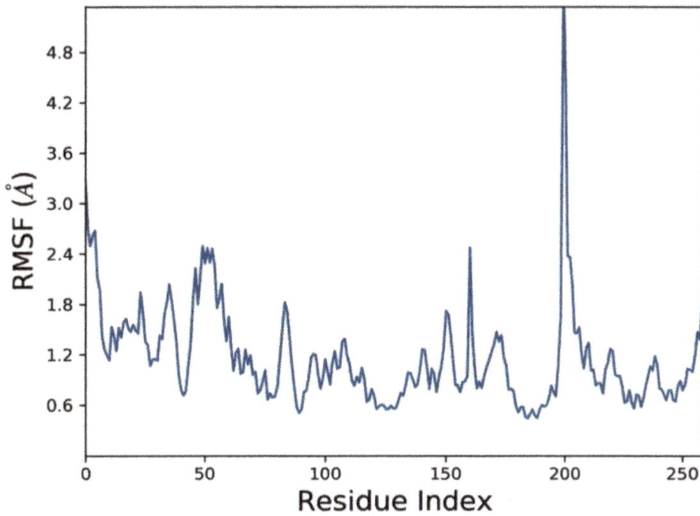

Figure 12. Root-mean-square fluctuations (RMSF) of the C-alpha atoms of NEK7 (2wqn).

The contact profiles of NEK7–Dabrafenib were computed from simulated trajectories, as shown in Figure 13. FDA-approved drug Dabrafenib interacted with ILE40, LYS163 and

ARG121 through Water Bridge and hydrogen bonding. The amino acid residues, ILE40, LYS163, and ARG42, were involved in H-bonding. During MD simulations, 12 hydrogen bonds were found to be dominant with significant occupancy. Details of hydrogen bonding is given in Table 6.

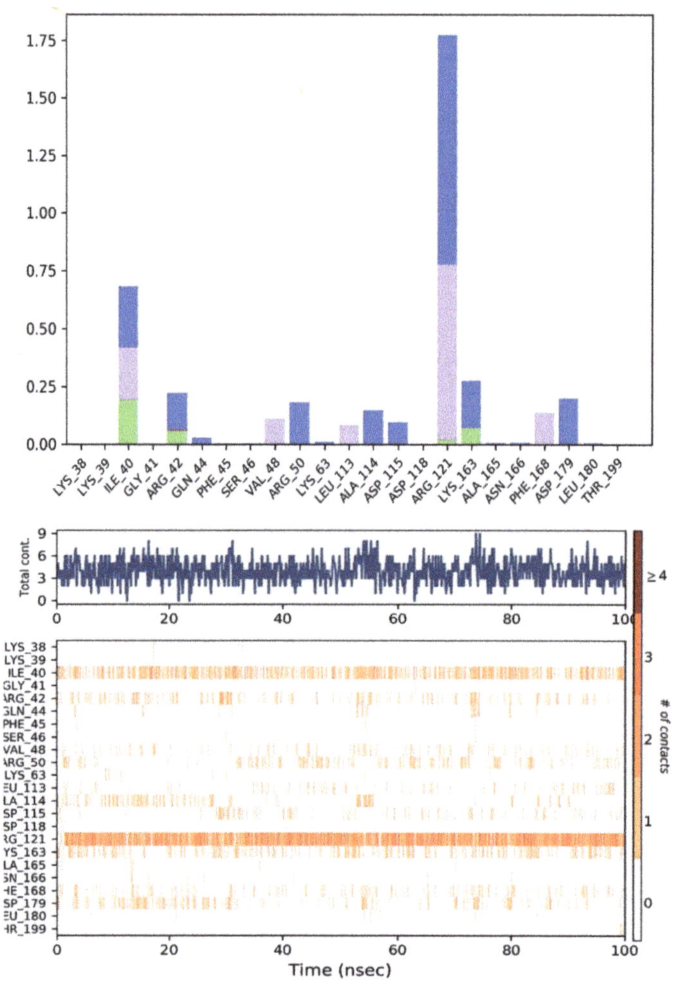

Figure 13. NEK7-Dabrafenib Contact histogram.

Table 6. Important Hydrogen bonding observed during MD simulations.

Sr No.	Hydrogen Donor	Hydrogen Acceptor
1	ALA114-Main	LIGAND-Side
2	LIGAND-Side	LEU113-Side
3	LIGAND-Side	ASP115-Side
4	ARG121-Side	LIGAND-Side
5	LIGAND-Side	ASP179-Side
6	LIGAND-Side	GLU112-Main

Table 6. *Cont.*

Sr No.	Hydrogen Donor	Hydrogen Acceptor
7	GLY117-Main	LIGAND-Side
8	ASP115-Main	LIGAND-Side
9	LIGAND-Side	ALA114-Main
10	LIGAND-Side	ILE40-Main
11	LIGAND-Side	ARG42-Main
12	GLY41-Main	LIGAND-Side

The hydrogen bond with ALA114 existed for more than 25% of simulation time. Hydrophobic interactions existed between VAL48, LEU113, VAL48, and PHE168. These molecular interactions contributed towards stabilizing the protein–ligand complex.

3.5.2. Radius of Gyration (Rg) and Solvent-Accessible Surface Area of Protein (SASA)

Radius of gyration (Rg) is measure of protein compactness, stability, integrity and foldness of protein backbone. The Rg trajectory for NEK7 is depicted in Figure 14. Trajectory analysis for the radius of gyration revealed that protein retained compactness throughout the simulated trajectory, and only slight fluctuations were observed around 30 ns, which stabilized after a short period of time.

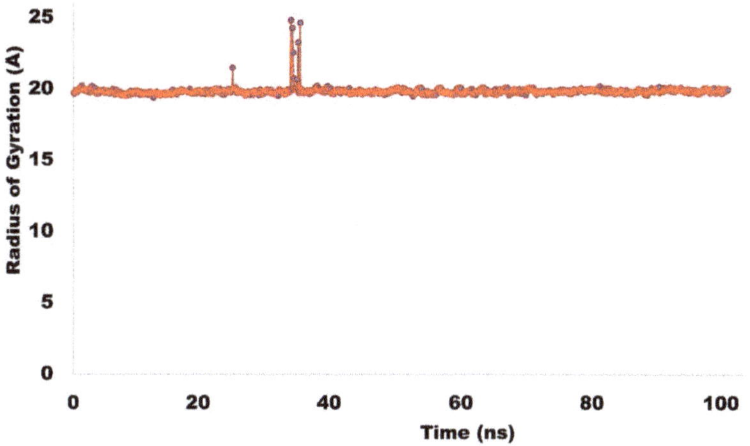

Figure 14. Radius of gyration (NEK7).

Solvent-accessible surface area (SASA) is the area of protein that is accessible by the solvent. The higher the value for SASA, the lower the stability of the protein. In the current study, residue wise SASA was calculated and ranged between 180 to 350 Å^2, which is quite acceptable. The average SASA value was computed to be 282.72 Å^2 (Table 5). The residue wise SASA of targeted protein is shown in Figure 15.

3.5.3. Principle Component Analysis (PCA)

It is an essential multivariate statistical technique used to describe the protein dynamics in a spatial scales. It is a linear relationship that extracts essential features of protein using covariance and/or correlation matrices. These matrices are derived from the atomic coordinate that represents the accessible degree of freedom (DOF) of the protein in a simulated trajectory. In the current study, Pearson's cross-correlation matrix was employed as it can normalize the large protein variables and prevent high atomic variations that can skew the results. In addition, eigenvectors with a specific variance value also play an important role in characterizing the motion of protein in spatial scales. In the current study, essential dynamics of protein were calculated by applying PCA analysis to the protein trajectory. It was observed that different variables were forming tight clusters with narrow angles, which indicates that they were correlated with vectors (PCs) [62]. PCs are the vectors that are used to describe protein motion with respect to variables. Two PCs are used in the current study to characterize the protein motion. In Figure 16, it can be observed that PC1 and PC2 are clearly indicating the behavior of various variables. Distribution on the scatter plot indicates the protein components are tightly clustered with small angles.

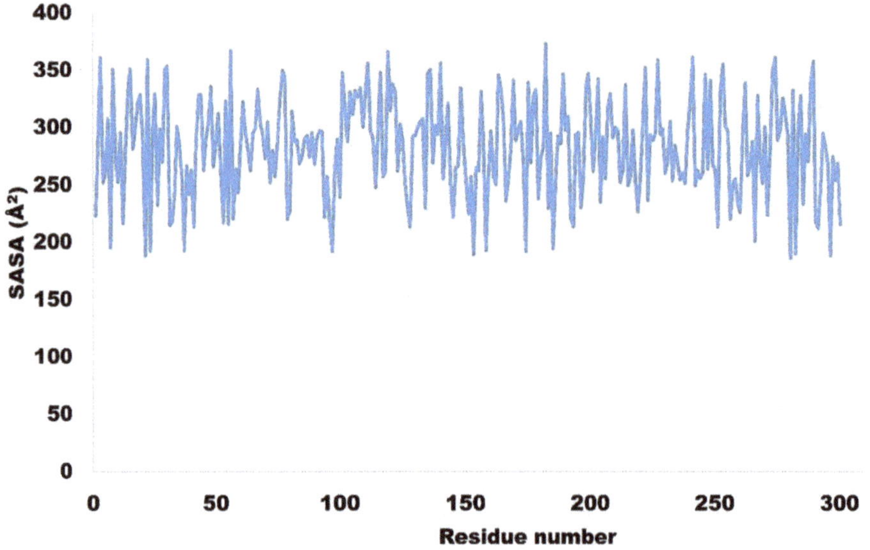

Figure 15. Residue wise solvent accessible surface area (SASA) for NEK7.

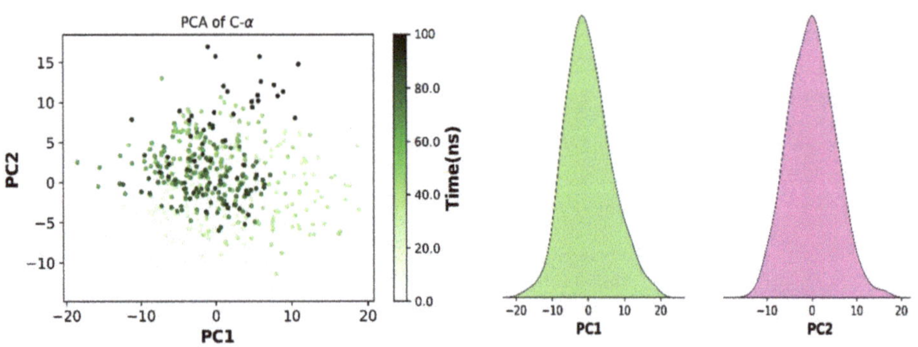

Figure 16. The correlation between protein variables and two top PCs.

Correlation matrices are also the correlation coefficients between variables and PCs. In Pearson's cross-correlation, the percent of variance in a protein variable is explained by PCs. Figure 17 is depicting the Pearson correlation graph for NEK7 variables.

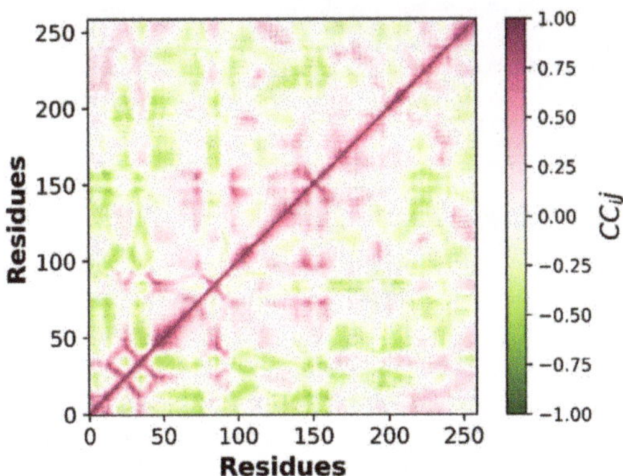

Figure 17. Pearson correlation graph for NEK7 variables.

3.5.4. MM-GBSA Energy Calculations

Molecular docking is a robust technique for determining the binding orientation of a protein–ligand complex. However, it is still lacking in its ability to correctly identify the binding affinities of docked ligands. In order to determine correct binding energies of docked conformations, MM-GBSA energy calculations were performed, which are an efficient and reliable method for the determination of binding free energies. The MM-GBSA method provides free energy calculations by taking into account all hydrophobic, hydrophilic and electrostatic interactions [63]. After energy calculations, values obtained were more negative and showed stronger binding affinities, as compared to the docking scores obtained from molecular docking. The following equation was used to calculate binding free energy [64];

$$\Delta G_{bind} = \Delta E_{mm} + \Delta G_{sol} + \Delta G_{SA}$$

The MM-GBSA energies for the protein–ligand complex was determined through the Thermal_mmgbsa script of Schrodinger. MM-GBSA energies are tabulated in Table 7.

Table 7. MM-GBSA binding energies of Dabrafenib docked at active site of NEK7.

	Binding Free Energy ΔG_{bind} (kcal/mol)	$\Delta E_{coulomb}$ (kcal/mol)	$\Delta E_{covalent}$ (kcal/mol)	$\Delta E_{H\text{-bond}}$ (kcal/mol)	ΔE_{vdW} (kcal/mol)	Lipophilic Energy (kcal/mol)	Sol_GB (kcal/mol)
Dabrafenib	−50.44	31.05	11.71	−0.18	−38.88	−33.98	−17.74

3.5.5. MM-PBSA Energy Calculations

In MMPBSA energy analysis, the free binding energies of protein, ligand and protein–ligand complex are estimated by following equation;

$$G = E_{bnd} + E_{el} + E_{vdW} + G_{pol} + G_{np} - TS$$

where, E_{bnd} I refers to bond energy, E_{el} refers to electrostatic energy and E_{vdW} represents van der Waals interactions. In the current study, Poisson–Boltzmann calculations were performed using the internal PBSA solver in mmpbsa_py_energy. All units are represented in kcal/mol. MM-PBSA energy analysis is given in Table 8.

Table 8. MM-PBSA binding energies of Dabrafenib docked at active site of NEK7.

	Binding Free Energy ΔG_{bind} (kcal/mol)	ΔE_{vdW} (kcal/mol)	E_{el} (kcal/mol)	E_{NPOLAR} (kcal/mol)	E_{PB} (kcal/mol)	E_{DISPER} (kcal/mol)
Dabrafenib	−56.12	−38.27	−17.84	−26.71	32.37	47.53

3.6. ADMET Profile

In-silico ADMET properties of top-ranked hits were determined by deep learning models; more than 17 models were employed at the backend, which provided predictions on the ADME profile of each hit. It is an important part in drug development that can identify the desired pharmacological properties of compounds. In the current study, the message passing neural network (MPNN) is employed for the determination of ADMET properties. It was observed that compound **762** showed the lowest clinical toxicity value of 0.28%. The ADMET profile of top hits is tabulated in Table 9.

Table 9. ADMET properties of top hits predicted via MPNN model.

Property Predicted	Compound 208	Compound 248	Compound 255	Compound 762
Solubility	−4.50 log mol/L	−4.05 log mol/L	−3.19 log mol/L	−3.10 log mol/L
Lipophilicity	1.82 (log-ratio)	1.89 (log-ratio)	1.40 (log-ratio)	1.88 (log-ratio)
(Absorption) Caco-2	−5.14 cm/s	−5.20 cm/s	−5.14 cm/s	−5.21 cm/s
(Absorption) HIA	91.18%	89.62%	89.69%	92.89%
(Absorption) Pgp	10.18%	13.10%	5.84%	11.42%
(Absorption) Bioavailability F20	76.34%	75.94%	75.46%	76.42%
(Distribution) BBB	76.85%	76.66%	93.85%	86.17%
(Distribution) PPBR	79.65%	77.11%	63.16%	80.66%
(Metabolism) CYP2C19	81.26%	72.00%	37.21%	24.15%
(Metabolism) CYP2D6	62.01%	51.90%	25.60%	13.64%
(Metabolism) CYP3A4	74.21%	56.97%	60.78%	22.83%
(Metabolism) CYP1A2	36.70%	10.29%	9.08%	21.05%
(Metabolism) CYP2C9	17.86%	6.99%	4.13%	8.60%
(Excretion) Half life	8.06 h	8.01 h	7.86 h	7.88 h
(Excretion) Clearance	8.23 mL/min/kg	8.26 mL/min/kg	8.10 mL/min/kg	8.54 mL/min/kg
Clinical Toxicity	14.92%	15.59%	24.33%	0.28%

4. Conclusions

In the current study, structure-based virtual screening of a 1200-compound library and Dabrafenib was carried out using Auto Dock Vina. These compounds are in the early stages of drug development, and the in-silico approach used in this study was contributing toward investigating the inhibiting potential of these compounds through molecular docking, DFTs, and MD simulation, as well as determining the drug-like properties of these compounds through deep learning models. The FDA-approved drug, Dabrafenib, was considered as a standard drug to which in-silico findings could be compared. SBVS findings discovered four important hits having better binding energies as compared to standard Dabrafenib. In addition, the chemical reactivity profiles of top hits were determined via DFT studies. Findings from DFT studies revealed the reactive nature of the compounds. Moreover, the current study has utilized deep learning models for prediction of binding affinity, pIC_{50}, and ADMET properties. It was observed that compound **762** showed good binding affinity and demonstrated a promising ADMET profile. Moreover, molecular dynamics simulations were performed to determine the stability of the protein–ligand complex under accelerated conditions. It was observed that the ligand remained significantly attached to the protein-activation loop, suggesting potential inhibiting activity of the compound. In short, the findings of the current study identify top hits that could prove an effective treatment strategy for NEK7-associated cancer malignancies. These findings will assist researchers to develop newer leads without consuming much time and money. Further experimental studies are also recommended for future prospects.

Author Contributions: Conceptualization, M.A. (Mubashir Aziz), S.A.E. and T.A.W.; methodology, M.A. (Mubashir Aziz), F.S., M.A. (Mohammed Alqarni) and A.A.A.; software, G.E.-S.B.; validation, M.A. (Mubashir Aziz) and S.A.E.; formal analysis, S.Z. and G.E.-S.B.; investigation, M.A. (Mubashir Aziz), A.T.A. and T.A.W.; resources, M.A. (Mubashir Aziz), G.E.-S.B., M.A. (Mohammed Alqarni) and A.A.A.; data curation, F.S.; writing—original draft preparation, M.A. (Mubashir Aziz) and S.A.E.; writing—review and editing, M.A. (Mubashir Aziz), S.A.E., N.A. and T.A.W.; visualization, S.A.E.; supervision, S.A.E.; project administration, S.A.E.; funding acquisition, T.A.W. and S.Z. All authors have read and agreed to the published version of the manuscript.

Funding: This research was funded by King Saud University, Riyadh Saudi Arabia project number (RSP-2021/357).

Institutional Review Board Statement: Not Applicable.

Informed Consent Statement: Not Applicable.

Data Availability Statement: Not Applicable.

Acknowledgments: The authors extend their appreciation to researchers supporting project number (RSP-2021/357), King Saud University, Riyadh Saudi Arabia for funding this research.

Conflicts of Interest: The authors declare no conflict of interest.

Sample Availability: Not Applicable.

References

1. Phadke, M.; Remsing Rix, L.L.; Smalley, I.; Bryant, A.T.; Luo, Y.; Lawrence, H.R.; Schaible, B.J.; Chen, Y.A.; Rix, U.; Smalley, K.S. Dabrafenib inhibits the growth of BRAF-WT cancers through CDK16 and NEK9 inhibition. *Mol. Oncol.* **2018**, *12*, 74–88. [CrossRef] [PubMed]
2. Sun, Z.; Gong, W.; Zhang, Y.; Jia, Z. Physiological and pathological roles of mammalian NEK7. *Front. Physiol.* **2020**, *11*, 1608. [CrossRef] [PubMed]
3. Loncarek, J.; Hergert, P.; Magidson, V.; Khodjakov, A. Control of daughter centriole formation by the pericentriolar material. *Nat. Cell Biol.* **2008**, *10*, 322–328. [CrossRef] [PubMed]
4. Tan, R.; Nakajima, S.; Wang, Q.; Sun, H.; Xue, J.; Wu, J.; Hellwig, S.; Zeng, X.; Yates, N.A.; Smithgall, T.E.; et al. Nek7 protects telomeres from oxidative DNA damage by phosphorylation and stabilization of TRF1. *Mol. Cell* **2017**, *65*, 818–831.e5. [CrossRef]
5. Haq, T.; Richards, M.W.; Burgess, S.G.; Gallego, P.; Yeoh, S.; O'Regan, L.; Reverter, D.; Roig, J.; Fry, A.M.; Bayliss, R. Mechanistic basis of Nek7 activation through Nek9 binding and induced dimerization. *Nat. Commun.* **2015**, *6*, 8771. [CrossRef]
6. Fry, M.A.; Bayliss, R.; Roig, J. Mitotic regulation by NEK kinase networks. *Front. Cell Dev. Biol.* **2017**, *5*, 102. [CrossRef]
7. Hauwermeiren, V.F.; Lamkanfi, M. The NEK-sus of the NLRP3 inflammasome. *Nat. Immunol.* **2016**, *17*, 223–224. [CrossRef]
8. Xu, J.; Lu, L.; Li, L. NEK7: A novel promising therapy target for NLRP3-related inflammatory diseases. *Acta Biochim. Biophys. Sin.* **2016**, *48*, 966–968. [CrossRef]
9. Gupta, A.; Tsuchiya, Y.; Ohta, M.; Shiratsuchi, G.; Kitagawa, D. NEK7 is required for G1 progression and procentriole formation. *Mol. Biol. Cell* **2017**, *28*, 2123–2134. [CrossRef]
10. Liu, G.; Chen, X.; Wang, Q.; Yuan, L. NEK7: A potential therapy target for NLRP3-related diseases. *BioScience Trends* **2020**, *14*, 74–82. [CrossRef]
11. García-Galán, M.J.; Díaz-Cruz, M.S.; Barceló, D. Identification and determination of metabolites and degradation products of sulfonamide antibiotics. *Trends Anal. Chem.* **2008**, *27*, 1008–1022. [CrossRef]
12. Supuran, C.T.; Casini, A.; Scozzafava, A.J.M. Protease inhibitors of the sulfonamide type: Anticancer, antiinflammatory, and antiviral agents. *Med. Res. Rev.* **2003**, *23*, 535–558. [CrossRef]
13. Scozzafava, A.; Owa, T.; Mastrolorenzo, A.; Supuran, C.T. Anticancer and antiviral sulfonamides. *Curr. Med. Chem.* **2003**, *10*, 925–953. [CrossRef]
14. Reddy, N.S.; Mallireddigari, M.R.; Cosenza, S.; Gumireddy, K.; Bell, S.C.; Reddy, E.P.; Reddy, M.R. Synthesis of new coumarin 3-(N-aryl) sulfonamides and their anticancer activity. *Bioorganic Med. Chem. Lett.* **2004**, *14*, 4093–4097. [CrossRef]

15. Bilbao-Ramos, P.; Galiana-Roselló, C.; Dea-Ayuela, M.A.; González-Alvarez, M.; Vega, C.; Rolón, M.; Pérez-Serrano, J.; Bolás-Fernández, F.; González-Rosende, M.E. Nuclease activity and ultrastructural effects of new sulfonamides with antileishmanial and trypanocidal activities. *Parasitol. Int.* **2012**, *61*, 604–613. [CrossRef]

16. Dauvergne, J.; Wellington, K.; Chibale, K. Unprecedented observation of sulfonamides in the transesterification of N-unsubstituted carbamates with sulfonyl chlorides. *Tetrahedron Lett.* **2004**, *45*, 43–47. [CrossRef]

17. Yasuhara, A.; Kameda, M.; Sakamoto, T. Selective monodesulfonylation of N, N-disulfonylarylamines with tetrabutylammonium fluoride. *Chem. Pharm. Bull.* **1999**, *47*, 809–812. [CrossRef]

18. O'Connell, J.F.; Rapoport, H. 1-Benzenesulfonyl-and 1-p-toluenesulfonyl-3-methylimidazolium triflates: Efficient reagents for the preparation of arylsulfonamides and arylsulfonates. *J. Org. Chem.* **1992**, *57*, 4775–4777. [CrossRef]

19. Chandrasekhar, S.; Mohapatra, S. Neighbouring group assisted sulfonamide cleavage of Sharpless aminols under acetonation conditions. *Tetrahedron Lett.* **1998**, *39*, 695–698. [CrossRef]

20. Gleckman, R.; Alvarez, S.; Joubert, D.W. Drug therapy reviews: Trimethoprim-sulfamethoxazole. *Am. J. Hosp. Pharm.* **1979**, *36*, 893–906. [CrossRef]

21. Bushby, S.R.M.; Hitchings, G.H. Trimethoprim, a sulphonamide potentiator. *Br. J. Pharmacol. Chemother.* **1968**, *33*, 72–90. [CrossRef]

22. Song, C.M.; Lim, S.J.; Tong, J.C. Recent advances in computer-aided drug design. *Briefings Bioinform.* **2009**, *10*, 579–591. [CrossRef]

23. Macalino, S.J.Y.; Gosu, V.; Hong, S.; Choi, S. Role of computer-aided drug design in modern drug discovery. *Arch. Pharmacal Res.* **2015**, *38*, 1686–1701. [CrossRef]

24. Rifaioglu, A.S.; Atas, H.; Martin, M.J.; Cetin-Atalay, R.; Atalay, V.; Doğan, T. Recent applications of deep learning and machine intelligence on in silico drug discovery: Methods, tools and databases. *Brief. Bioinform.* **2019**, *20*, 1878–1912. [CrossRef]

25. Shoichet, B.K. Virtual screening of chemical libraries. *Nature* **2004**, *432*, 862–865. [CrossRef]

26. Trott, O.; Olson, A.J. AutoDock Vina: Improving the speed and accuracy of docking with a new scoring function, efficient optimization, and multithreading. *J. Comput. Chem.* **2010**, *31*, 455–461. [CrossRef]

27. Friesner, R.A.; Banks, J.L.; Murphy, R.B.; Halgren, T.A.; Klicic, J.J.; Mainz, D.T.; Repasky, M.P.; Knoll, E.H.; Shelley, M.; Perry, J.K.; et al. Glide: A new approach for rapid, accurate docking and scoring. 1. Method and assessment of docking accuracy. *J. Med. Chem.* **2004**, *47*, 1739–1749. [CrossRef]

28. Sánchez-Linares, I.; Pérez-Sánchez, H.; Cecilia, J.M.; García, J.M. High-throughput parallel blind virtual screening using BINDSURF. *BMC Bioinform.* **2012**, *13*, 1–14. [CrossRef]

29. Imbernón, B.; Cecilia, J.M.; Pérez-Sánchez, H.; Giménez, D. METADOCK: A parallel metaheuristic schema for virtual screening methods. *Int. J. High Perform. Comput. Appl.* **2018**, *32*, 789–803. [CrossRef]

30. Ban, T.A. The role of serendipity in drug discovery. *Dialogues Clin. Neurosci.* **2022**, *8*, 335–344. [CrossRef]

31. Huang, K.; Fu, T.; Khan, D.; Abid, A.; Abdalla, A.; Abid, A.; Glass, L.M.; Zitnik, M.; Xiao, C.; Sun, J. Moldesigner: Interactive design of efficacious drugs with deep learning. *arXiv* **2020**, arXiv:03951.

32. Huang, K.; Fu, T.; Glass, L.M.; Zitnik, M.; Xiao, C.; Sun, J. DeepPurpose: A deep learning library for drug–target interaction prediction. *Bioinformatics* **2020**, *36*, 5545–5547. [CrossRef] [PubMed]

33. Hohenberg, P.; Kohn, W. Inhomogeneous electron gas. *Phys. Rev.* **1964**, *136*, B864. [CrossRef]

34. Calais, J.L. Orthonormalization and symmetry adaptation of crystal orbitals. *Int. J. Quantum Chem.* **1985**, *28*, 655–667. [CrossRef]

35. Rodríguez, J.I.; Ayers, P.W.; Götz, A.W.; Castillo-Alvarado, F.D.L. Virial theorem in the Kohn–Sham density-functional theory formalism: Accurate calculation of the atomic quantum theory of atoms in molecules energies. *J. Chem. Phys.* **2009**, *131*, 021101. [CrossRef]

36. Ziegler, T.J.C.R. Approximate density functional theory as a practical tool in molecular energetics and dynamics. *Chem. Rev.* **1991**, *91*, 651–667. [CrossRef]

37. Bartolotti, L.J.; Flurchick, K. An introduction to density functional theory. *Rev. Comput. Chem.* **1996**, *7*, 187–260.

38. Aziz, M.; Ejaz, S.A.; Tamam, N.; Siddique, F.; Riaz, N.; Qais, F.A.; Chtita, S.; Iqbal, J. Identification of potent inhibitors of NEK7 protein using a comprehensive computational approach. *Sci. Rep.* **2022**, *12*, 6404. [CrossRef]

39. Del Bene, J.E.; Person, W.B.; Szczepaniak, K. Properties of Hydrogen-Bonded Complexes Obtained from the B3LYP Functional with 6-31G (d, p) and 6-31+ G (d, p) Basis Sets: Comparison with MP2/6-31+ G (d, p) Results and Experimental Data. *J. Phys. Chem.* **1995**, *99*, 10705–10707. [CrossRef]

40. Hanwell, M.D.; Curtis, D.E.; Lonie, D.C.; Vandermeersch, T.; Zurek, E.; Hutchison, G.R. Avogadro: An advanced semantic chemical editor, visualization, and analysis platform. *J. Chemin.* **2012**, *4*, 1–17. [CrossRef]

41. Azarakhshi, F.; Khaleghian, M.; Farhadyar, N. DFT study and NBO analysis of conformational properties of 2-Substituted 2-Oxo-1, 3, 2-dioxaphosphorinanes and their dithia and diselena analogs. *Lett. Org. Chem.* **2015**, *12*, 516–522. [CrossRef]

42. Aziz, M.; Ejaz, S.A.; Rehman, H.M.; Al-Buriahi, M.S.; Siddique, F.; Somaily, H.H.; Alrowaili, Z.A. Identification of NEK7 Inhibitors: Structure based Virtual Screening, Molecular Docking, Density functional theory calculations and Molecular Dynamics Simulations. *Res. Sq.* **2022**. *preprint*. [CrossRef]

43. Ferreira, L.G.; Ricardo, N. Molecular Docking and Structure-Based Drug Design Strategies. *Molecules* **2015**, *20*, 13384–13421. [CrossRef]

44. Hildebrand, P.W.; Rose, A.; Tiemann, J. Bringing Molecular Dynamics Simulation Data into View. *Trends Biochem. Sci.* **2019**, *44*, 902–913. [CrossRef]

45. Rasheed, M.A.; Iqbal, M.N.; Saddick, S.; Ali, I.; Khan, F.S.; Kanwal, S.; Ahmed, D.; Ibrahim, M.; Afzal, U.; Awais, M. Identification of lead compounds against Scm (fms10) in *Enterococcus faecium* using computer aided drug designing. *Life* **2021**, *11*, 77. [CrossRef]

46. Shivakumar, D.; Williams, J.; Wu, Y.; Damm, W.; Shelley, J.; Sherman, W. Prediction of absolute solvation free energies using molecular dynamics free energy perturbation and the OPLS force field. *J. Chem. Theory Comput.* **2010**, *6*, 1509–1519. [CrossRef]

47. Martyna, G.J.; Tobias, D.J.; Klein, M.L. Constant pressure molecular dynamics algorithms. *J. Chem. Phys.* **1994**, *101*, 4177–4189. [CrossRef]

48. Hoover, W.G. Canonical dynamics: Equilibrium phase-space distributions. *Phys. Rev. A* **1985**, *31*, 1695. [CrossRef]

49. Luty, B.A.; Davis, M.E.; Tironi, I.G.; Van Gunsteren, W.F. A comparison of particle-particle, particle-mesh and Ewald methods for calculating electrostatic interactions in periodic molecular systems. *Mol. Simul.* **1994**, *14*, 11–20. [CrossRef]

50. Zhang, Y.; Zhang, T.J.; Tu, S.; Zhang, Z.H.; Meng, F.H. Identification of Novel Src Inhibitors: Pharmacophore-Based Virtual Screening, Molecular Docking and Molecular Dynamics Simulations. *Molecules* **2020**, *25*, 4094. [CrossRef]

51. Humphreys, D.D.; Friesner, R.A.; Berne, B.J. A Multiple-Time-Step Molecular Dynamics Algorithm for Macromolecules. *J. Phys. Chem.* **1994**, *98*, 6885–6892. [CrossRef]

52. Jacobson, M.P.; Pincus, D.L.; Rapp, C.S.; Day, T.J.; Honig, B.; Shaw, D.E.; Friesner, R.A. A hierarchical approach to all-atom protein loop prediction. *Proteins Struct. Funct. Bioinform.* **2004**, *55*, 351–367. [CrossRef]

53. Jacobson, M.P.; Friesner, R.A.; Xiang, Z.; Honig, B. On the role of the crystal environment in determining protein side-chain conformations. *J. Mol. Biol.* **2002**, *320*, 597–608. [CrossRef]

54. Bowers, K.J.; Chow, D.E.; Xu, H.; Dror, R.O.; Eastwood, M.P.; Gregersen, B.A.; Klepeis, J.L.; Kolossvary, I.; Moraes, M.A.; Sacerdoti, F.D.; et al. Scalable algorithms for molecular dynamics simulations on commodity clusters. In Proceedings of the 2006 ACM/IEEE Conference on Supercomputing, Tampa, FL, USA, 11–17 November 2006.

55. Qu, X.; Latino, D.A.; Aires-De-Sousa, J. A big data approach to the ultra-fast prediction of DFT-calculated bond energies. *J. Chemin.* **2013**, *5*, 34. [CrossRef]

56. Cohen, A.J.; Mori-Sánchez, P.; Yang, W. Challenges for Density Functional Theory. *Chem. Rev.* **2011**, *112*, 289–320. [CrossRef]

57. Ribeiro, J.; Ríos-Vera, C.; Melo, F.; Schüller, A. Calculation of accurate interatomic contact surface areas for the quantitative analysis of non-bonded molecular interactions. *Bioinformatics* **2012**, *35*, 3499–3501. [CrossRef]

58. Azam, F.; Alabdullah, N.H.; Ehmedat, H.M.; Abulifa, A.R.; Taban, I.; Upadhyayula, S. NSAIDs as potential treatment option for preventing amyloid β toxicity in Alzheimer's disease: An investigation by docking, molecular dynamics, and DFT studies. *J. Biomol. Struct. Dyn.* **2018**, *36*, 2099–2117. [CrossRef]

59. Hospital, A.; Goñi, J.R.; Orozco, M.; Gelpí, J. Molecular dynamics simulations: Advances and applications. *Adv. Appl. Bioinform. Chem. AABC* **2015**, *8*, 37. [PubMed]

60. Choudhary, M.I.; Shaikh, M.; tul-Wahab, A.; ur-Rahman, A. In silico identification of potential inhibitors of key SARS-CoV-2 3CL hydrolase (Mpro) via molecular docking, MMGBSA predictive binding energy calculations, and molecular dynamics simulation. *PLoS ONE* **2020**, *15*, e0235030. [CrossRef] [PubMed]

61. Katari, S.K.; Natarajan, P.; Swargam, S.; Kanipakam, H.; Pasala, C.; Umamaheswari, A. Inhibitor design against JNK1 through e-pharmacophore modeling docking and molecular dynamics simulations. *J. Recept. Signal Transduct.* **2016**, *36*, 558–571. [CrossRef] [PubMed]

62. David, C.C.; Jacobs, D.J. Principal component analysis: A method for determining the essential dynamics of proteins. In *Protein Dynamics*; Humana Press: Totowa, NJ, USA, 2014; pp. 193–226.

63. Vijayakumar, B.; Parasuraman, S.; Raveendran, R.; Velmurugan, D. Identification of natural inhibitors against angiotensin I converting enzyme for cardiac safety using induced fit docking and MM-GBSA studies. *Pharmacogn. Mag.* **2014**, *10* (Suppl. S3), S639.

64. Lyne, P.D.; Lamb, A.M.L.; Saeh, J.C. Accurate Prediction of the Relative Potencies of Members of a Series of Kinase Inhibitors Using Molecular Docking and MM-GBSA Scoring. *J. Med. Chem.* **2006**, *49*, 4805–4808. [CrossRef]

Article

Mechanistic Insight into Binding of Huperzine A with Human Serum Albumin: Computational and Spectroscopic Approaches

Anas Shamsi [1,2,*], Moyad Shahwan [2,3], Mohd Shahnawaz Khan [4], Fahad A. Alhumaydhi [5], Suliman A. Alsagaby [6], Waleed Al Abdulmonem [7], Bekhzod Abdullaev [8] and Dharmendra Kumar Yadav [9,*]

1   Centre for Interdisciplinary Research in Basic Sciences, Jamia Millia Islamia, Jamia Nagar, New Delhi 110025, India
2   Centre of Medical and Bio-Allied Health Sciences Research, Ajman University, Ajman P.O. Box 346, United Arab Emirates; moyad76@hotmail.com
3   College of Pharmacy & Health Sciences, Ajman University, Ajman P.O. Box 346, United Arab Emirates
4   Department of Biochemistry, College of Sciences, King Saud University, Riyadh 11451, Saudi Arabia; moskhan@ksu.edu.sa
5   Department of Medical Laboratories, College of Applied Medical Sciences, Qassim University, Buraidah 52571, Saudi Arabia; f.alhumaydhi@qu.edu.sa
6   Department of Medical Laboratory Sciences, College of Applied Medical Sciences, Majmaah University, Majmaah 11932, Saudi Arabia; s.alsaqaby@mu.edu.sa
7   Department of Pathology, College of Medicine, Qassim University, Buraidah 51452, Saudi Arabia; waleedmonem@qumed.edu.sa
8   Scientific Department, Akfa University, Tashkent 100022, Uzbekistan; b.abdullaev@akfauniversity.org
9   College of Pharmacy, Gachon University of Medicine and Science, Hambakmoeiro, Yeonsu-gu, Incheon 21924, Korea
*   Correspondence: anas.shamsi18@gmail.com (A.S.); dharmendra30oct@gmail.com (D.K.Y.)

Citation: Shamsi, A.; Shahwan, M.; Khan, M.S.; Alhumaydhi, F.A.; Alsagaby, S.A.; Al Abdulmonem, W.; Abdullaev, B.; Yadav, D.K. Mechanistic Insight into Binding of Huperzine A with Human Serum Albumin: Computational and Spectroscopic Approaches. *Molecules* **2022**, *27*, 797. https://doi.org/10.3390/molecules27030797

Academic Editor: Anna Maria Almerico

Received: 1 December 2021
Accepted: 14 January 2022
Published: 25 January 2022

Publisher's Note: MDPI stays neutral with regard to jurisdictional claims in published maps and institutional affiliations.

Abstract: Human serum albumin (HSA) is the most abundant protein in plasma synthesized by the liver and the main modulator of fluid distribution between body compartments. It has an amazing capacity to bind with multiple ligands, offering a store and transporter for various endogenous and exogenous compounds. Huperzine A (HpzA) is a natural sesquiterpene alkaloid found in *Huperzia serrata* and used in various neurological conditions, including Alzheimer's disease (AD). This study elucidated the binding of HpzA with HSA using advanced computational approaches such as molecular docking and molecular dynamic (MD) simulation followed by fluorescence-based binding assays. The molecular docking result showed plausible interaction between HpzA and HSA. The MD simulation and principal component analysis (PCA) results supported the stable interactions of the protein–ligand complex. The fluorescence assay further validated the in silico study, revealing significant binding affinity between HpzA and HSA. This study advocated that HpzA acts as a latent HSA binding partner, which may be investigated further in AD therapy in experimental settings.

Keywords: Huperzine A; molecular dynamics simulation; fluorescence spectroscopy; human serum albumin; neurodegenerative disorders; drug–protein interactions

1. Introduction

HSA is a major transporter and the most abundant protein in the plasma. It is responsible for balancing the osmotic pressure and is a major regulator of fluid distribution in the body. HSA exhibits astonishing ligand-binding capabilities, acting as a warehouse and transporter of many endo- and exogenous compounds [1]. This promiscuous, nonspecific affinity can lead to sudden changes in concentrations caused by displacement when two or more compounds compete for binding to the same molecular site [2]. HSA, a major carrier in the human circulatory system performs a key function of circulating various compounds such as drugs [3]. HSA interactions with drugs alter proteins' pharmacokinetics and pharmacodynamics as well. Free drugs diffuse inactively across the membranes via specific transporters to interact with their respective protein targets [4]. A pivotal step in the domain of drug discovery is the investigation of the pharmacokinetics and pharmacodynamics of drugs. Pharmacological profiling of drugs offers understanding of the interactions of vital therapeutic drugs or derivatives with either plasma or target tissue proteins [5]. The binding of a drug to HSA is a critical factor determining its pharmacological profiling and distribution [6]. In medicinal chemistry, studies pertaining to plasma proteins and drugs binding are attracting researchers across the globe, because these studies provide a platform to study drugs' behavior and action, thereby delineating their transport and distribution characteristics in the circulatory system. Furthermore, it is also vital to study the protein–protein or protein–drug interactions to make progressive inroads in the advancements made in pharmaceutical industry [7]. HSA is the main plasma protein, and it is imperative to study binding of drugs with HSA. Thus, this study aimed at investigating the binding mechanism of HpzA to HSA. HSA is a one-chained polypeptide weighing 66.5 kDa. The 585 amino acid polypeptide assembles in a heart-shaped structure [1]. The protein displays two specific sites for drug binding as specified using Sudlow's fluorescent probe displacement method. According to the Sudlow's classification, there are two main binding sites for drug ligands of HSA, namely, subdomain IIA (site IIA) and IIIA (site IIIA) [8]. Site IIA, positioned in subdomain IIA, concerns with the selective binding of heterocyclic anions, while aromatic carboxylates bind to the site IIIA that is located in subdomain IIIA,. For example, ibuprofen binds to site IIIA and warfarin to site IIA [9]. The binding of all the ligands cannot be considered under Sudlow's model. Site IB, a third binding pocket, located in subdomain IB, is a hydrophobic D-shaped cavity, is also used for interaction by some compounds such as bilirubin [10]. A study [11] suggested site IB as the third major site having the potential to bind drug ligands of HSA.

AD, the sixth leading cause of claiming lives worldwide, is a progressive, neurodegenerative disorder. The condition is characterized by impairments in mental functions, amnesia, and dementia [12]. The major hallmarks of the condition are accumulated amyloid β-peptide (Aβ) in plaques, the formation of neurofibrillary tangles (NFTs), and degeneration of neurons. The mechanism by which neurons degenerate is still unclear and needs extensive research. Aβ accumulation and deposition can cause neurodegeneration by many mechanisms, such as inflammation, oxidative stress, and apoptosis. In addition to the genetic factors, environmental factors such as neurotoxins, stress, etc. play a major role in AD progression [13]. Reactive oxygen species (ROS) creates a condition of oxidative stress that leads to cell death, oxidative bursts, accumulation of spare free metals, etc. These have been hypothesized as major mediators in AD progression and neurodegeneration. Postmortem reports have indicated apoptosis as a major event in neurodegeneration and AD [14].

HpzA, a Lycopodium alkaloid, is a Chinese medicine isolated from *Qiang Ceng Ta*, the whole plant of *Huperzia serrata*. The plant belongs to the Huperziaceae family, and it has gained wide popularity due to its anticholinesterase (AChE) and anti-AD properties. The alkaloid also acts as a remedy for various ailments such as strains and swellings and neurological disorders such as schizophrenia and neurodegenerative disorders [15]. HpzA, a well-known AChE inhibitor, has properties superior to those of FDA-approved drugs such as donepezil, galantamine, etc. [16]. Figure S1 describes the 2D and 3D structural features of HpzA. Extensive clinical trials for the HpzA conducted in China showed enhanced memory in elderly individuals, subjects with issues related to amnesia, and AD patients. The naturally extracted ligand has also been through various extensive clinical trials on vascular dementia, showing improved cognitive function in subjects with AD [17]. In this introductory study, we report the plausible binding of HpzA with HSA. A molecular docking study was performed primarily to explore the possible interactions and binding affinity of HpzA towards HSA and was followed by MD simulation studies for 100 ns. Computational studies suggested plausible binding of HpzA with HSA, which was further validated by fluorescence-based binding assays. Thus, this study delineates the mechanism of binding of HpzA with HSA by employing spectroscopic and in silico approaches.

2. Results

2.1. Molecular Docking

The functional activity of a protein changes broadly upon conformational alterations induced by ligand binding. The molecular docking approach can systematically consider potential binding modes for a protein–ligand system [18]. The blind docking approach was used to find all possible interactions between HpzA and HSA. HpzA has multiple binding sites on has, with varying affinities. All the docking sites and their docking energies are depicted in Figures S2 and S3, respectively. Out of nine docked conformations of HpzA in the splitting process, the preferable docked conformation was taken and explored in detail. The selected conformation of the docked HpzA showed a considerable binding score of −7.2 kcal/mol. It had LE and pKi values 0.4 (kcal/mol/non-H atom) and 5.28, respectively. The docking score affirmed HpzA as a plausible binding partner of HSA. Figure 1 shows the binding mode of HpzA with HSA. HpzA was shown to have a decent structural complementarity with good binding affinity. It showed various interactions with the key residues of the HSA binding pocket, such as one hydrogen bond with His266 and a few other noncovalent interactions with important HSA residues (Figure 1). The binding site of HpzA is located in domain II of HSA, but Tyr172, Tyr174, and Glu177 of domain I also participate in the interaction. HpzA binds at the Sudlow site I of HSA. The surface representation of HSA showed HpzA fitted inside the deep cavity of the binding site, which might cause significant changes to the conformational activity of HSA (Figure 1).

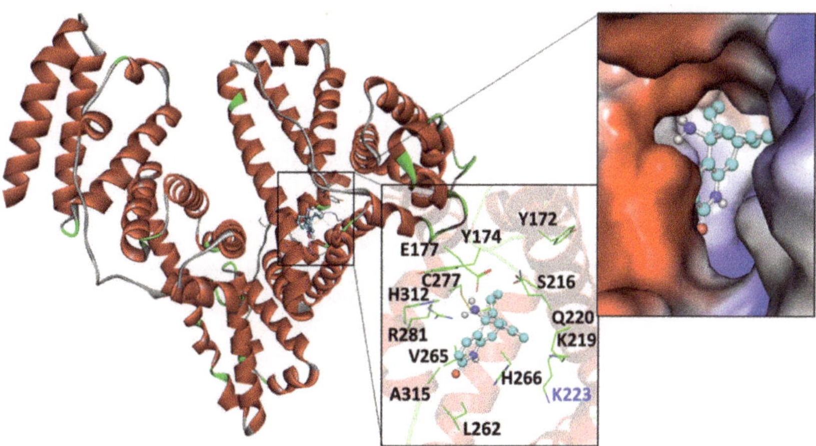

Figure 1. The binding of HpzA with HSA. Ribbon and surface representation of HSA protein showing various interactions and docking fit of HpzA (UniProt ID: P02768).

The selected docking pose of HpzA with HSA was further subjected to a detailed analysis of possible interactions. The generated 2D plot showed that HpzA occupied the hydrophobic pocket of HSA bordered by residues Tyr174, Glu177, Lys219, Gln220, Lys223, Leu262, His266, Cys269, Arg281, His312, and Ala315 (Figure 2A). The plot suggested that HpzA was tightly bound with HSA with the help of one hydrogen bond with His266 and several hydrophobic contacts. HpzA showed multiple van der Waals interactions with Glu177, Gln220, Leu262, and His312 of HSA. The phenolic moiety at the end of HpzA occupied a crucial binding cleft of HSA formed by different residues (Figure 2B). As a whole, the HpzA binding with HSA suggested that it could disrupt the structural conformation of the protein, which may further result in altering HSA's binding capability.

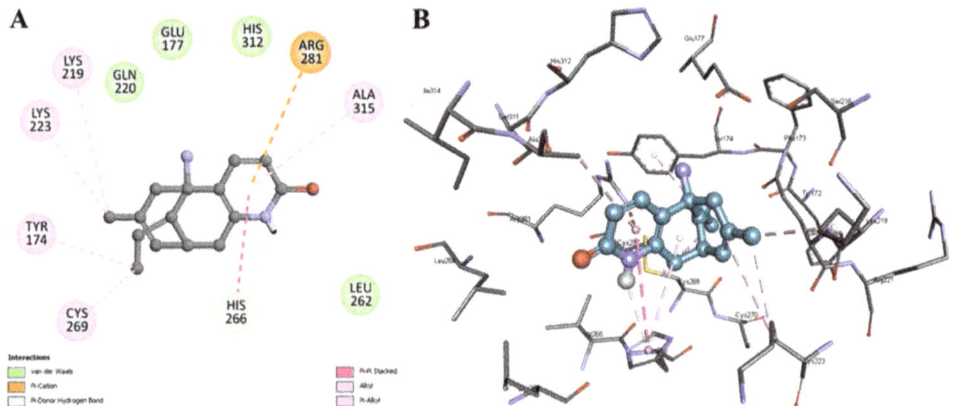

Figure 2. Detailed interactions of HSA with HpzA as (**A**) 2D and (**B**) 3D plots.

2.2. MD Simulations

Many studies have used the MD simulation approach to investigate ligand binding with proteins [7,19–21]. All-atom MD simulations were carried out for the docked complex of HSA–HpzA and compared with those for the free state of HSA. Plots for root mean square deviation (RMSD), root mean square fluctuation (RMSF), radius of gyration (R_g), and solvent accessible surface area (SASA) fluctuations, along with hydrogen bond analysis, were generated and analyzed in detail. The analysis of RMSD assists in understanding the stability of a protein and protein–ligand complex. To explore structural deviations in HSA, we used RMSD examination of the simulated protein and protein–ligand complex. As examined, both HSA and the HSA–HpzA complex reached the equilibrium phase without any major shift in the RMSD pattern (Figure 3A, left panel). The protein–ligand complex was stable throughout 100 ns simulation. The analysis of the generated plots suggested a small amount of fluctuation at the beginning in the RMSD values of the HSA–HpzA that was possibly due to the initial adjustment of the system. As a whole, HSA showed a small number of fluctuations in its backbone in the simulation, but no significant shift was observed (Figure 3A, right panel).

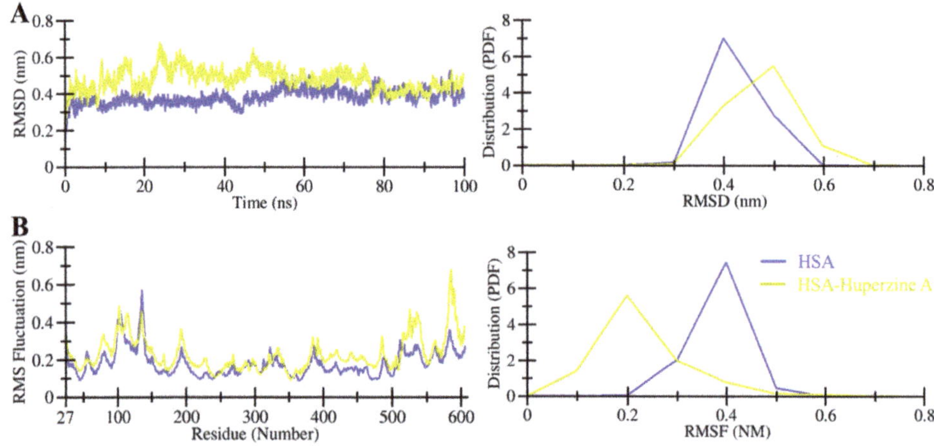

Figure 3. Dynamics performance of HSA. (**A**) RMSD plot of HSA and its docked complex with HpzA. (**B**) Residual fluctuation plot of HSA and its docked complex with HpzA.

Exploring RMSFs is helpful in understanding the fluctuations in the protein residues during the course of the simulation [22,23]. The RMSFs were recorded for each residue in HSA and plotted for analysis purposes. The RMSF analysis emphasized possible residual movements in HSA in free and ligand-bound states (Figure 3B). The compared RMSF values between the free and ligand-bound HSA suggested that the residual fluctuations were decreased and compacted the ligand-bound state of HSA during the simulation (Figure 3B, left panel). The results suggested that the fluctuations in the residues were minimized upon binding of HpzA. However, a few increased fluctuations were observed in the HSA residues corresponding to the loop region far from the ligand site. RMSF values distribution plot revealed that the residual fluctuations were minimized after HpzA binding, which further confirmed a robust constancy of the protein–ligand complex (Figure 3B, right panel).

Calculating and examining the R_g of a protein is useful to measure the shape of the protein based on its hydrodynamic radius [18,23–25]. The R_gs of HSA and the HSA–HpzA docked complex were calculated to assess their structural compactness during the simulation (Figure 4A). The average R_g values of free HSA and the HSA–HpzA docked complex were 2.76 nm and 2.80 nm, respectively. The pattern of R_g values of HSA was like that of the R_g values of HSA and the HSA–HpzA docked complex, but with a minor increase (Figure 4A, left panel). This minor increase could have resulted from the occupancy of intramolecular space of HSA by HpzA. The distribution plot of R_g values also indicated a slight increase in the R_g of the HSA–HpzA docked complex compared to that of the free state of HSA (Figure 4A, right panel).

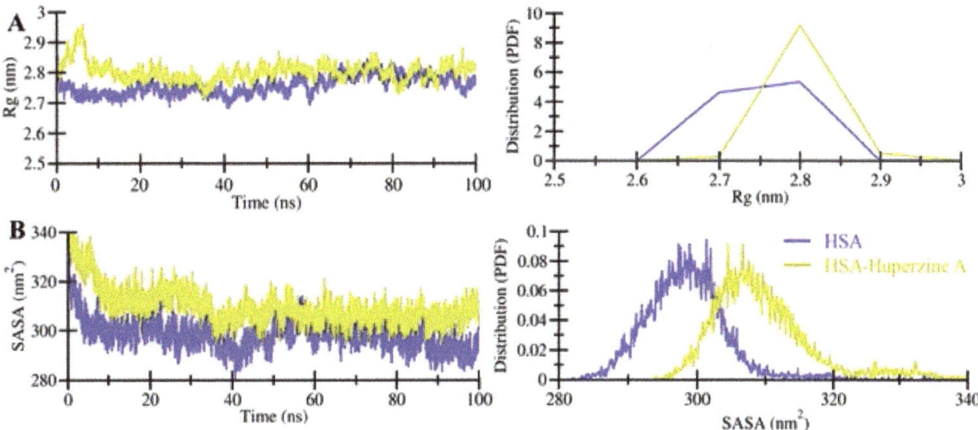

Figure 4. Structural compactness of HSA as a function of time. (**A**) Time evolution of the radius of gyration. (**B**) SASA plot of HSA as a function of time.

SASA delineates the folding mechanism of a protein structure by analyzing the protein's solvent accessibility [26]. SASA values were calculated and plotted to investigate the surface area of HSA and the HSA–HpzA docked complex during the simulation, the (Figure 4B). The results indicated that the SASA values were minorly increased for HSA in ligand-bound state. This slight increase in the values of SASA indicated that a number of internal residues might be exposed to the surface post the binding of HpzA (Figure 4B, left panel). The distribution plot of SASA values also indicated a slight increase in SASA of the HSA–HpzA docked complex (Figure 4B, right panel).

Dynamics of Intra-/Intermolecular Hydrogen Bonding

Hydrogen bonds are necessary for maintaining the structural conformation of a protein [27]. Analysis of hydrogen bonding is useful to assess the stability of a protein and protein–ligand complex [27]. The stability of HSA and the HSA–HpzA docked complex was examined by exploring the formation of intramolecular hydrogen bonds (Figure 5). The analysis of the intramolecular hydrogen bonds formed within HSA indicated the formation of several hydrogen bonds, which maintained the integrity of HSA's three-dimensional structure. The average hydrogen bonds estimated within HSA before and after the HpzA binding were 480 and 467, respectively. The plot indicated a significant decline in hydrogen bonds in HSA when it formed a complex with HpzA, which was possibly due to the ligand's occupancy of intramolecular space in the binding pocket of HSA. The binding of HpzA might cause a hindrance in the formation of intramolecular hydrogen bonding within HSA. However, the plot suggested that this decrement in the formation did not lead to any major shift, which resulted in maintaining the geometry of the HSA structure during the simulation. The PDF distribution also indicated a decrease in intramolecular hydrogen bonding within HSA when bound to HpzA (Figure 5B).

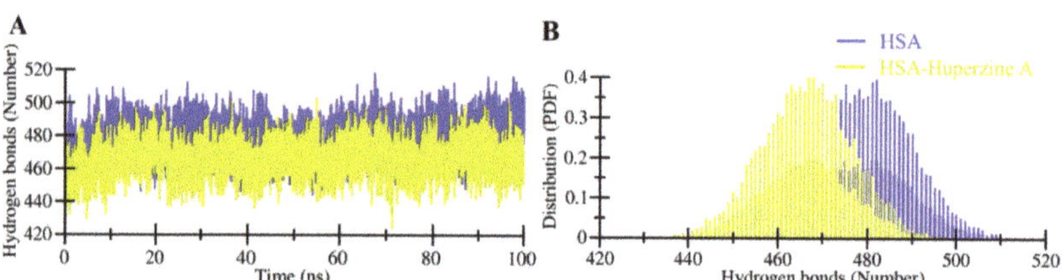

Figure 5. Intramolecular hydrogen bonds during the simulation. (**A**) Time evolution and stability of hydrogen bonds formed intramolecular within HSA. (**B**) The probability of distribution of hydrogen bonding as PDF.

The formation of intermolecular hydrogen bonds plays a crucial role in the stability and directionality of a protein–ligand complex [28]. The intermolecular hydrogen bonds can be explored to measure the strength of a ligand towards the binding pocket of a protein. The hydrogen bonds formed between HpzA and HSA were recorded and plotted to explore their formation and breakdown during the simulation (Figure 6A). Analysis of the generated plot indicated the formation of up to three hydrogen bonds during the simulation time. The analysis clearly indicated that at least one hydrogen bond was formed during the simulation time. This strong hydrogen bond could be correlated with the molecular docking observation that His266 was involved in the conventional hydrogen bond formation. The analysis suggested the importance of intermolecular hydrogen bonding in the ligand binding to form a stable complex. The PDF plot indicated that at least one hydrogen bond was formed with higher probability and stability in the HSA–HpzA docked complex (Figure 6B).

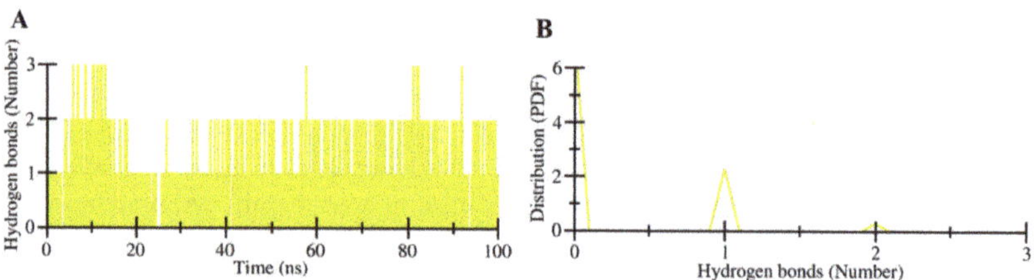

Figure 6. Intermolecular hydrogen bonds during the simulation. (**A**) Time evolution of hydrogen bonds formed between HpzA and HSA. (**B**) The probability of distribution of hydrogen bonds.

2.3. Principal Component and Free Energy Landscape Analyses

Principal component analysis (PCA) is an arithmetic technique to reduce the complexity of MD trajectories by obtaining the collective motion of Cα atoms. It is vital to evaluate the stability of proteins and protein–ligand complexes [29,30]. PCA was performed to evaluate the conformational projection and transition dynamics of HSA with HpzA. The 2D projections of the first two principal components (PCs), PC1 and PC2, for HSA and HSA–HpzA is presented in Figure 7A. The analysis revealed that the 2D projection of HSA showed smaller phase space than that of HSA–HpzA. The projection of the first two eigenvectors (EVs), EV1 and EV2, showed the time evolution of the HSA projection, which indicated a similar pattern of trajectories (Figure 7B). The PCA showed that HSA and HSA–HpzA had similar correlated motions without significant change. However, the ligand-bound HSA explored a wider phase space on both EVs, which indicated its higher dynamics. The PCA confirmed that the HSA–HpzA complex was quite stable during the simulation and behaved like the free state of HSA.

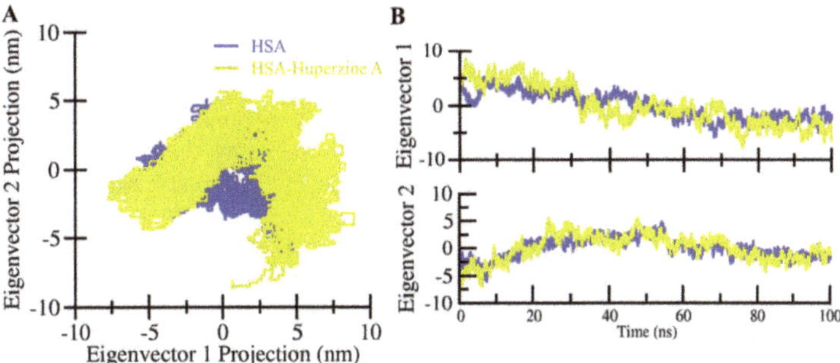

Figure 7. Principal component analysis. (**A**) Two-dimensional projections of HSA trajectories on eigenvector 1 and eigenvector 2. (**B**) Time evaluation of conformational projections of trajectories on eigenvector 1 and eigenvector 2.

Free energy landscape (FEL) analysis based on PCA has been utilized in representing protein conformations in a time evolution manner [31]. The FEL distinguishes the kinetic and thermodynamic estates of proteins and protein–ligand complexes. The FELs depend on the prospect of an arrangement of specific data points and converting them to free-energy values. FELs analysis was performed to investigate the conformational stability underlying HSA–HpzA binding. Figure 8 shows the FEL plots projected onto PC1 and PC2 of HSA and the HSA–HpzA complex for Cα atoms. The centralized blue zones suggested that the subsequent system was stable at that moment. The dimension and appearance of the minimum energy space, called the global minimum (in blue) in the FELs, indicated that the HSA–HpzA system was more stable than HSA. HSA and HSA–HpzA displayed a single global minimum confined within a large basin. The FEL analysis further validated our previous observations that HSA, when bound with HpzA, was stable and reached a more thermodynamically stable conformational state (Figure 8B).

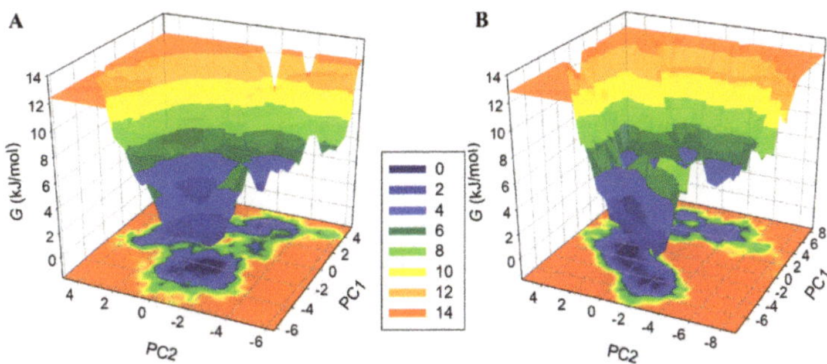

Figure 8. The Gibbs free energy landscapes of (**A**) free HSA and (**B**) HSA–HpzA obtained during 100 ns MD simulation.

2.4. Fluorescence-Based Binding

Fluorescence spectroscopy serves as an important technique to provide an insight into the protein–ligand interactions, revealing different binding parameters that give an idea about the strength of interaction between a protein and ligand [32]. Intrinsic fluorescence depicts changes in the local microenvironment of aromatic amino acid residues and this serves as an important technique in determining the protein-ligand complex formation [33,34]. When a decrease in the fluorescence of native protein is observed with increasing concentration of the ligands, it is referred to as fluorescence quenching [35]. Fluorescence quenching was performed at three different temperatures (15, 20, and 25 °C). Fluorescence emission spectra of free HSA and HSA with different HpzA concentrations (0–11 μM) at various temperatures are shown in Figure 9A–C. Native HSA showed fluorescence emission maxima at 346 nm. There was a noticeable decrease HSA's fluorescence intensity with increasing HpzA concentration, with no peak shift. Fluorescence quenching can be either static, dynamic, or a combination of both [36]. The deviation of the binding parameters with temperature delineates the operative quenching for a particular interaction. The quenching data obtained were fitted into the Stern–Volmer, modified Stern–Volmer, and van 't Hoff equations to obtain various binding and thermodynamic parameters for HSA–HpzA interaction as per previously published reports [37,38]. Table 1 shows the obtained values of the Stern–Volmer constant (K_{sv}) at various temperatures, which were found to increase with increasing temperature, revealing the existence of the dynamic mode. Moreover, to confirm the quenching mode, a biomolecular quenching rate constant (K_q) was calculated using $K_{sv} = K_q/\tau_0$ ($\tau_0 = 2.7 \times 10^{-9}$ s), and the value was found to be higher than that of the maximum dynamic quenching constant (nearly 10^{10} M^{-1} s^{-1}) [39], confirming the existence of a combination of static and dynamic quenching. Figure 10A shows the experimental fitting obtained in accordance with the modified Stern–Volmer equation. The slope of the plot gives the number of binding sites (n) while the intercept gives the binding constant (K). HpzA binds to HSA with a high binding affinity, ($K = 9.3 \times 10^5$ M^{-1} at 25 °C). The value of K was found to increase at a higher temperature suggesting that the HSA–HpzA complex is more stable at high temperatures. These observations affirm in silico results advocating the significant binding affinity between HpzA and HSA. Thermodynamic parameters were also found for the HSA–HpzA complex fitting the obtained data in the van 't Hoff equation. Figure 10B shows the van 't Hoff plot obtained for the HSA–HpzA complex. Table 1 shows the obtained thermodynamic parameters for the complex. Negative ΔH and ΔS implied the van der Waals and hydrogen bonding as the dominant forces driving the complex formation. Moreover, a negative ΔG implied the spontaneous nature of the reaction.

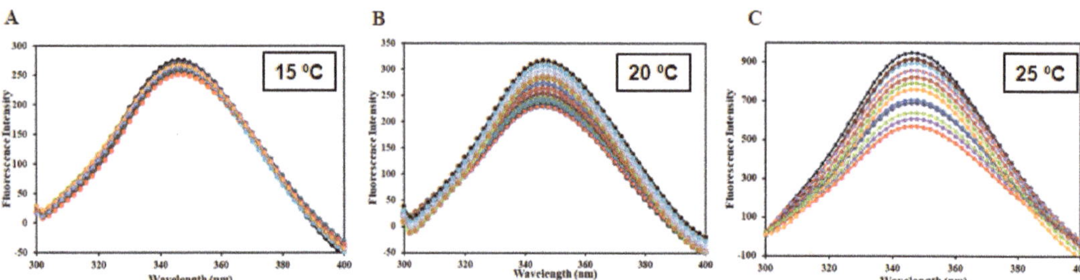

Figure 9. Fluorescence emission spectra of HSA in the absence and presence of varying HpzA concentrations (0–11 μM) at (**A**) 15 °C, (**B**) 20 °C, and (**C**) 25 °C.

Table 1. Binding and thermodynamic parameters obtained for the HSA–HpzA complex from fluorescence quenching studies.

Temperature (°C)	K_{sv} (10^4 M^{-1})	K 10^5 M^{-1}	n	ΔG kcal mol^{-1}	ΔS cal mol^{-1} K^{-1}	ΔH kcal mol^{-1}	$T\Delta S$ kcal mol^{-1}
15	1.8	0.36	1.07	−7.78			−47.59
20	3.33	1.17	1.10	−6.96	−165.26	−55.38	−48.42
25	5.19	9.35	1.25	−6.13			−49.24

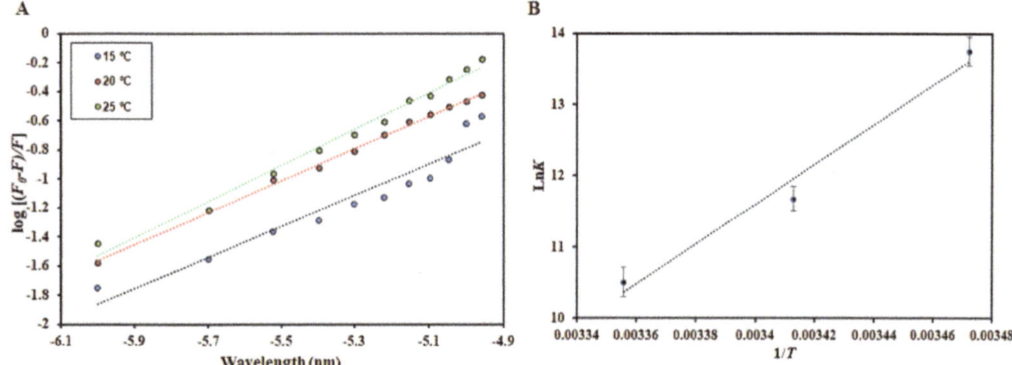

Figure 10. (**A**) Modified Stern–Volmer plot of the HSA–HpzA complex at different temperatures. (**B**) van 't Hoff plot of the HSA–HpzA complex.

3. Materials and Methods

3.1. Materials

HSA (fatty acid-free) and HpzA (42643) were purchased from Sigma-Aldrich Co. (St. Louis, MO, USA). We used double distilled for the preparation of all buffers. HSA stock solution (75 μM) was made in 20 mM sodium phosphate buffer, pH 7.4. Appropriate blanks were used as control, and the reported spectra are the subtracted spectra. We used analytical grade chemicals for buffer preparations.

3.2. Receptor and Ligand Preparation

HSA structure was obtained from the Protein Data Bank with PDB-ID: 6HSC in a three-dimensional state. The structure of HpzA was taken in SDF format from the PubChem database with PubChem CID: 854026 in a three-dimensional state. The water molecules and other cocrystallized ligands (i.e., aristolochic acid, myristic acid, and 1,2-ethanediol) present in the protein structure were removed PDB file and optimized through the Swiss-PDB-Viewer tool (2021) [40]. The PDB files of receptor HSA and SDF file of the ligand HpzA were converted in the required format, i.e., PDBQT by InstaDock [41]. Appropriate atom types were assigned to both protein and ligand structures before performing docking.

3.3. Molecular Docking

The prepared files of the receptor and ligand structures were utilized for docking study using the InstaDock software [41]. InstaDock was used to calculate binding free energy between the protein–ligand complex by employing the AutoDock Vina [42] scoring function. The grid box for docking search was set blindly to facilitate free moving and searching on HSA by HpzA. The grid coordinates were summarized as 86 Å × 62 Å × 76 Å, centralized at 25.53, 9.51, and 20.27 for X, Y, and Z, respectively. The spacing of 1 Å was utilized in the docking box with default parameters. The docking result was saved in a separate directory for further analysis of the protein–ligand complex. The negative decimal logarithms of the inhibition constant as pKi and ligand efficiency were also estimated from the docking result through the inbuilt program of InstaDock.

3.4. Interaction Analysis

To explore the possible interactions between HpzA and HSA, all the docked conformations for the ligand output file were separated. The interaction analysis of HpzA with HSA was carried out through the Discovery Studio Visualizer (2021) [43] and PyMOL (2021) [44] tools. First, the binding pose of HpzA with HSA was selected based on the specific interaction, and then the HSA–HpzA protein–ligand complex was generated for further studies in MD simulation.

3.5. MD Simulations

GROMACS [45] v.2020 beta was utilized to execute the all-atom MD simulations for examining the reliability of the HSA–HpzA docking complex in solvent conditions. The Gromacs parameters for HpzA were generated from the PRODRG server [46] to prepare a protein–ligand complex system. Free HSA and the docked complex were implanted in the simple point charge (SPC) solvent model. HSA and the docked complex, HSA–HpzA, were neutralized with an appropriate number of counter ions (Na^+ and Cl^-). The salt concentration was kept at physiological condition, i.e., 0.15 M. Both systems were stipulated on periodic boundary conditions, and the SHAKE algorithm was applied for limiting the movement of all bonds. The energy minimization of both the systems was performed using the steepest descent algorithm with and without solute restraints [18]. Then, 1000 ps simulations were performed in NVT and NPT ensembles at temperature 300 K. To control the temperature and pressure during the simulation, the Berendsen thermostats and barostats were applied. Both the relaxed systems were subjected to a final MD run for 100 ns with a time step of 2 fs [47]. For analysis purposes, RMSD, RMSF, R_g, SASA, and hydrogen bonding were recorded. The recorded trajectories were checked for the stability of the HSA–HpzA docked complex in comparison with that of the free state of HSA.

3.6. Principal Component Analysis and Free Energy Landscape

Principal component analysis (PCA) is a comprehensive and useful tool to investigate conformational changes of proteins [29]. The theoretical notes on PCA have been explained in various preceding reports [29,31,48]. To realistically discover the variations in the structural configuration of HSA and the HSA–HpzA docked complex, PCA was employed on the equilibrized trajectories from the MD study. The PCA was performed using the GROMACS utilities by calculating the eigenvalues (EVs) and their projection along the two principal components (PC1 and PC2). FEL analysis has been widely used in examining the folding mechanisms and overall stability of protein and protein–ligand complexes [49]. FELs for HSA and the HSA–HpzA docked complex were generated through the *gmx sham* tool of the GROMACS package.

3.7. Fluorescence-Based Assay

Fluorescence-based binding was carried out on a Jasco spectrofluorometer FP 6200 (Jasco, Tokyo, Japan). HSA was excited at 280 nm with the emission range set at 300–400 nm. We analyzed the quenching data as per earlier published studies [32,50].

All the experiments were performed in triplicates.

4. Conclusions

The possible binding of HpzA with HSA was explored by utilizing molecular docking, MD simulation, PCA, and FEL analyses. Docking results indicated that the binding of HpzA with HSA showed an appreciable binding affinity and many intermolecular interactions. MD trajectory analyses (i.e., RMSD, RMSF, R_g, SASA, and hydrogen bonding) suggested that the HSA–HpzA docked complex was quite stable with minimal conformational alterations. Moreover, PCA and FEL analyses of HSA and HSA–HpzA confirmed that both the systems were stable without any significant change. However, a minor change in the shape and position of the global minima of HSA when bound to HpzA indicated that the HpzA might alter the conformational shape of the binding site of HSA. Fluorescence-based binding ascertained the actual binding affinity between HpzA and HSA, suggesting that HpzA binds to HSA with a significant affinity, validating the in silico observations. Overall, this study reinforces the idea that HSA–HpzA interactions can be explored in AD. The results contribute in many ways to our knowledge and provide a base for setting up an experimental platform of HpzA interactions with HSA to explore them in AD management.

Supplementary Materials: The following supporting information can be downloaded online. Figure S1: Molecular structure of Huperzine A: (A) 2D, (B) 3D stick, and (C) 3D ball and stick model of Huperzine A. Figure S2: Different docked conformations of HpzA with HSA. Figure S3: Energy table of different docked conformations of HpzA with HSA.

Author Contributions: Conceptualization, A.S. and D.K.Y.; methodology, A.S., M.S.K., M.S. and F.A.A.; software, S.A.A., W.A.A. and A.S.; validation; A.S., M.S.K., M.S. and D.K.Y.; formal analysis, A.S., F.A.A., B.A. and S.A.A.; writing—original draft and preparation; A.S., M.S.K., M.S. and F.A.A.; writing—review and editing; D.K.Y. and A.S.; visualization, A.S. and M.S.; project administration, A.S., M.S.K. and M.S.; Funding acquisition, A.S. and D.K.Y. All authors have read and agreed to the published version of the manuscript.

Funding: M.S.K. acknowledges the generous support from the Research Supporting Project (RSP-2021/352) by King Saud University, Riyadh, Saudi Arabia.

Institutional Review Board Statement: Not applicable.

Informed Consent Statement: Not applicable.

Data Availability Statement: The data presented in this study are contained within the article and Supplementary Materials.

Conflicts of Interest: The authors declare no conflict of interest.

Sample Availability: Not applicable.

References

1. Fanali, G.; di Masi, A.; Trezza, V.; Marino, M.; Fasano, M.; Ascenzi, P. Human serum albumin: From bench to bedside. *Mol. Asp. Med.* **2012**, *33*, 209–290. [CrossRef] [PubMed]
2. Rimac, H.; Debeljak, Z.; Bojić, M.; Miller, L. Displacement of Drugs from Human Serum Albumin: From Molecular Interactions to Clinical Significance. *Curr. Med. Chem.* **2017**, *24*, 1930–1947. [CrossRef]
3. He, X.M.; Carter, D.C. Atomic structure and chemistry of human serum albumin. *Nature* **1992**, *358*, 209–215. [CrossRef]
4. Kragh-Hansen, U. Molecular aspects of ligand binding to serum albumin. *Pharmacol. Rev.* **1981**, *33*, 17–53. [PubMed]
5. Yadav, D.K. Recent Advances on Small Molecule Medicinal Chemistry to Treat Human Diseases-Part III. *Curr. Top. Med. Chem.* **2021**, *21*, 1517–1518. [CrossRef]
6. Shamsi, A.; Ahmed, A.; Khan, M.S.; Al Shahwan, M.; Husain, F.M.; Bano, B. Understanding the binding between Rosmarinic acid and serum albumin: In vitro and in silico insight. *J. Mol. Liq.* **2020**, *311*, 113348. [CrossRef]
7. Yadav, D.K.; Kumar, S.; Teli, M.K.; Kim, M. Ligand-based pharmacophore modeling and docking studies on vitamin D receptor inhibitors. *J. Cell. Biochem.* **2020**, *121*, 3570–3583. [CrossRef]
8. Sudlow, G.; Birkett, D.J.; Wade, D.N. Further characterization of specific drug binding sites on human serum albumin. *Mol. Pharmacol.* **1976**, *12*, 1052–1061.
9. Dockal, M.; Carter, D.C.; Rüker, F. Conformational Transitions of the Three Recombinant Domains of Human Serum Albumin Depending on pH. *J. Biol. Chem.* **2000**, *275*, 3042–3050. [CrossRef]
10. Fani, N.; Bordbar, A.; Ghayeb, Y. Spectroscopic, docking and molecular dynamics simulation studies on the interaction of two Schiff base complexes with human serum albumin. *J. Lumin.* **2013**, *141*, 166–172. [CrossRef]
11. Zsila, F. Subdomain IB Is the Third Major Drug Binding Region of Human Serum Albumin: Toward the Three-Sites Model. *Mol. Pharm.* **2013**, *10*, 1668–1682. [CrossRef] [PubMed]
12. Bartus, R.T.; Dean, R.L., III; Beer, B.; Lippa, A.S. The cholinergic hypothesis of geriatric memory dysfunction. *Science* **1982**, *217*, 408–414. [CrossRef]
13. Behl, C.; Moosmann, B. Antioxidant neuroprotection in Alzheimer's disease as preventive and therapeutic approach. *Free Radic. Biol. Med.* **2002**, *33*, 182–191. [CrossRef]

14. Eckert, A.; Marques, C.A.; Keil, U.; Schüssel, K.; Müller, W.E. Increased apoptotic cell death in sporadic and genetic Alzheimer's disease. *Ann. N. Y. Acad. Sci.* **2003**, *1010*, 604–609. [CrossRef]

15. Ma, X.; Tan, C.-H.; Zhu, D.; Gang, D.R.; Xiao, P. Huperzine A from Huperzia species—An ethnopharmacolgical review. *J. Ethnopharmacol.* **2007**, *113*, 15–34. [CrossRef]

16. Wang, H.; Tang, X. Anticholinesterase effects of huperzine A, E2020, and tacrine in rats. *Zhongguo yao li xue bao = Acta Pharmacol. Sin.* **1998**, *19*, 27–30.

17. Li, H.; Min, Q. Huperzine A improved the cognition of vascular dementia: A report of 30 patients in therapeutics. *Chin. J. Clin. Rehabil.* **2001**, *5*, 59.

18. Yadav, D.K.; Kumar, S.; Saloni; Misra, S.; Yadav, L.; Teli, M.; Sharma, P.; Chaudhary, S.; Kumar, N.; Choi, E.H.; et al. Molecular insights into the interaction of RONS and Thieno [3, 2-c] pyran analogs with SIRT6/COX-2: A molecular dynamics study. *Sci. Rep.* **2018**, *8*, 4777. [CrossRef]

19. Tiwari, M.K.; Coghi, P.; Agrawal, P.; Shyamlal, B.R.K.; Jun Yang, L.; Yadav, L.; Peng, Y.; Sharma, R.; Yadav, D.K.; Sahal, D.; et al. Design, Synthesis, Structure-Activity Relationship and Docking Studies of Novel Functionalized Arylvinyl-1, 2, 4-Trioxanes as Potent Antiplasmodial as well as Anticancer Agents. *ChemMedChem* **2020**, *15*, 1216–1228. [CrossRef]

20. Shyamlal, B.R.K.; Yadav, L.; Tiwari, M.K.; Mathur, M.; Prikhodko, J.I.; Mashevskaya, I.V.; Yadav, D.K.; Chaudhary, S. Synthesis, Bioevaluation, Structure-Activity Relationship and Docking Studies of Natural Product Inspired (Z)-3-benzylideneisobenzofuran-1(3H)-ones as Highly Potent antioxidants and Antiplatelet agents. *Sci. Rep.* **2020**, *10*, 2307. [CrossRef]

21. Teli, M.K.; Kumar, S.; Yadav, D.K.; Kim, M. In silico identification of prolyl hydroxylase inhibitor by per-residue energy decomposition-based pharmacophore approach. *J. Cell. Biochem.* **2021**, *122*, 1098–1112. [CrossRef] [PubMed]

22. Mohammad, T.; Siddiqui, S.; Shamsi, A.; Alajmi, M.F.; Hussain, A.; Islam, A.; Ahmad, F.; Hassan, M.I. Virtual Screening Approach to Identify High-Affinity Inhibitors of Serum and Glucocorticoid-Regulated Kinase 1 among Bioactive Natural Products: Combined Molecular Docking and Simulation Studies. *Molecules* **2020**, *25*, 823. [CrossRef] [PubMed]

23. Gadhe, C.G.; Kim, M.-H. Insights into the binding modes of CC chemokine receptor 4 (CCR4) inhibitors: A combined approach involving homology modelling, docking, and molecular dynamics simulation studies. *Mol. Biosyst.* **2015**, *11*, 618–634. [CrossRef]

24. Lobanov, M.Y.; Bogatyreva, N.S.; Galzitskaya, O.V. Radius of gyration as an indicator of protein structure compactness. *Mol. Biol.* **2008**, *42*, 623–628. [CrossRef]

25. Gadhe, C.G.; Balupuri, A.; Cho, S.J. In silico characterization of binding mode of CCR8 inhibitor: Homology modeling, docking and membrane based MD simulation study. *J. Biomol. Struct. Dyn.* **2015**, *33*, 2491–2510. [CrossRef]

26. Ito, A.; Mukaiyama, A.; Itoh, Y.; Nagase, H.; Thøgersen, I.B.; Enghild, J.J.; Sasaguri, Y.; Mori, Y. Degradation of Interleukin 1β by Matrix Metalloproteinases. *J. Biol. Chem.* **1996**, *271*, 14657–14660. [CrossRef] [PubMed]

27. Hubbard, R.E. Hydrogen Bonds in Proteins: Role and Strength. In *eLS*; John Wiley & Sons, Ltd.: Hoboken, NJ, USA, 2001. [CrossRef]

28. Hubbard, R.E.; Kamran Haider, M. Hydrogen Bonds in Proteins: Role and Strength. In *eLS*; John Wiley & Sons, Ltd.: Hoboken, NJ, USA, 2010. [CrossRef]

29. Maisuradze, G.G.; Liwo, A.; Scheraga, H.A. Principal Component Analysis for Protein Folding Dynamics. *J. Mol. Biol.* **2009**, *385*, 312–329. [CrossRef]

30. Fatima, S.; Mohammad, T.; Jairajpuri, D.S.; Rehman, T.; Hussain, A.; Samim, M.; Ahmad, F.; Alajmi, M.F.; Hassan, I. Identification and evaluation of glutathione conjugate gamma-l-glutamyl-l-cysteine for improved drug delivery to the brain. *J. Biomol. Struct. Dyn.* **2020**, *38*, 3610–3620. [CrossRef]

31. Altis, A.; Otten, M.; Nguyen, P.H.; Hegger, R.; Stock, G. Construction of the free energy landscape of biomolecules via dihedral angle principal component analysis. *J. Chem. Phys.* **2008**, *128*, 245102. [CrossRef] [PubMed]

32. Anwar, S.; Shamsi, A.; Shahbaaz, M.; Queen, A.; Khan, P.; Hasan, G.M.; Islam, A.; Alajmi, M.F.; Hussain, A.; Ahmad, F.; et al. Rosmarinic Acid Exhibits Anticancer Effects via MARK4 Inhibition. *Sci. Rep.* **2020**, *10*, 10300. [CrossRef]

33. Soares, S.; Mateus, N.; de Freitas, V. Interaction of Different Polyphenols with Bovine Serum Albumin (BSA) and Human Salivary α-Amylase (HSA) by Fluorescence Quenching. *J. Agric. Food Chem.* **2007**, *55*, 6726–6735. [CrossRef]

34. Klajnert, B.; Stanisławska, L.; Bryszewska, M.; Pałecz, B. Interactions between PAMAM dendrimers and bovine serum albumin. *Biochim. Biophys. Acta (BBA)-Proteins Proteom.* **2003**, *1648*, 115–126. [CrossRef]

35. Shamsi, A.; Al Shahwan, M.; Ahamad, S.; Hassan, M.I.; Ahmad, F.; Islam, A. Spectroscopic, calorimetric and molecular docking insight into the interaction of Alzheimer's drug donepezil with human transferrin: Implications of Alzheimer's drug. *J. Biomol. Struct. Dyn.* **2020**, *38*, 1094–1102. [CrossRef]

36. Waseem, R.; Anwar, S.; Khan, S.; Shamsi, A.; Hassan, I.; Anjum, F.; Shafie, A.; Islam, A.; Yadav, D.K. MAP/Microtubule Affinity Regulating Kinase 4 Inhibitory Potential of Irisin: A New Therapeutic Strategy to Combat Cancer and Alzheimer's Disease. *Int. J. Mol. Sci.* **2021**, *22*, 10986. [CrossRef]

37. Anwar, S.; Mohammad, T.; Shamsi, A.; Queen, A.; Parveen, S.; Luqman, S.; Hasan, G.M.; Alamry, K.A.; Azum, N.; Asiri, A.M.; et al. Discovery of Hordenine as a Potential Inhibitor of Pyruvate Dehydrogenase Kinase 3: Implication in Lung Cancer Therapy. *Biomedicines* **2020**, *8*, 119. [CrossRef] [PubMed]

38. Anwar, S.; Shamsi, A.; Kar, R.K.; Queen, A.; Islam, A.; Ahmad, F.; Hassan, I. Structural and biochemical investigation of MARK4 inhibitory potential of cholic acid: Towards therapeutic implications in neurodegenerative diseases. *Int. J. Biol. Macromol.* **2020**, *161*, 596–604. [CrossRef]

39. Shamsi, A.; Mohammad, T.; Anwar, S.; Nasreen, K.; Hassan, I.; Ahmad, F.; Islam, A. Insight into the binding of PEG-400 with eye protein alpha-crystallin: Multi spectroscopic and computational approach: Possible therapeutics targeting eye diseases. *J. Biomol. Struct. Dyn.* **2020**, 1–11. [CrossRef] [PubMed]

40. Guex, N.; Peitsch, M. Swiss-Model and the Swiss-Pdb Viewer: An environment for comparative protein modeling. *Electrophoresis* **1997**, *18*, 2714–2723. [CrossRef]

41. Mohammad, T.; Mathur, Y.; Hassan, I. InstaDock: A single-click graphical user interface for molecular docking-based virtual high-throughput screening. *Briefings Bioinform.* **2021**, *22*, bbaa279. [CrossRef] [PubMed]

42. Trott, O.; Olson, A.J. AutoDock Vina: Improving the speed and accuracy of docking with a new scoring function, efficient optimization, and multithreading. *J. Comput. Chem.* **2010**, *31*, 455–461. [CrossRef]

43. Biovia, D.S. *Discovery Studio Modeling Environment*; Dassault Systèmes: San Diego, CA, USA, 2015.

44. DeLano, W.L. Pymol: An open-source molecular graphics tool. *CCP4 Newsl. Protein Crystallogr.* **2002**, *40*, 82–92.

45. Van Der Spoel, D.; Lindahl, E.; Hess, B.; Groenhof, G.; Mark, A.E.; Berendsen, H.J. GROMACS: Fast, flexible, and free. *J. Comput. Chem.* **2005**, *26*, 1701–1718. [CrossRef]

46. Van Aalten, D.M.F.; Bywater, R.; Findlay, J.B.C.; Hendlich, M.; Hooft, R.W.W.; Vriend, G. PRODRG, a program for generating molecular topologies and unique molecular descriptors from coordinates of small molecules. *J. Comput. Mol. Des.* **1996**, *10*, 255–262. [CrossRef]

47. Yadav, D.K.; Kumar, S.; Choi, E.H.; Chaudhary, S.; Kim, M.H. Computational Modeling on Aquaporin-3 as Skin Cancer Target: A Virtual Screening and Molecular Dynamic Simulation Study. *Front Chem.* **2019**, *8*, 250. [CrossRef]

48. Naqvi, A.A.T.; Mohammad, T.; Hasan, G.M.; Hassan, I. Advancements in Docking and Molecular Dynamics Simulations Towards Ligand-receptor Interactions and Structure-function Relationships. *Curr. Top. Med. Chem.* **2018**, *18*, 1755–1768. [CrossRef]

49. Papaleo, E.; Mereghetti, P.; Fantucci, P.; Grandori, R.; De Gioia, L. Free-energy landscape, principal component analysis, and structural clustering to identify representative conformations from molecular dynamics simulations: The myoglobin case. *J. Mol. Graph. Model.* **2009**, *27*, 889–899. [CrossRef]

50. Shamsi, A.; Anwar, S.; Mohammad, T.; Alajmi, M.F.; Hussain, A.; Rehman, M.T.; Hasan, G.M.; Islam, A.; Hassan, M.I. MARK4 inhibited by AChE inhibitors, donepezil and Rivastigmine tartrate: Insights into Alzheimer's disease therapy. *Biomolecules* **2020**, *10*, 789. [CrossRef]

Article

Structural Characterization of Ectodomain G Protein of Respiratory Syncytial Virus and Its Interaction with Heparan Sulfate: Multi-Spectroscopic and In Silico Studies Elucidating Host-Pathogen Interactions

Abu Hamza [1], Abdus Samad [1], Md. Ali Imam [1], Md. Imam Faizan [2], Anwar Ahmed [3], Fahad N. Almajhdi [3,4], Tajamul Hussain [3], Asimul Islam [1,*] and Shama Parveen [1,*]

[1] Centre for Interdisciplinary Research in Basic Sciences, Jamia Millia Islamia, New Delhi 110025, India; abuhamza830@gmail.com (A.H.); a.samad0195@gmail.com (A.S.); ali.imamuit@gmail.com (M.A.I.)
[2] Multidisciplinary Centre for Advance Research and Studies, Jamia Millia Islamia, New Delhi 110025, India; faizan086@gmail.com
[3] Center of Excellence in Biotechnology Research, College of Science, King Saud University, Riyadh 11451, Saudi Arabia; anahmed@ksu.edu.sa (A.A.); majhdi@ksu.edu.sa (F.N.A.); thussain@ksu.edu.sa (T.H.)
[4] Department of Botany & Microbiology, College of Science, King Saud University, Riyadh 11451, Saudi Arabia
*   Correspondence: aislam@jmi.ac.in (A.I.); sparveen2@jmi.ac.in or shamp25@yahoo.com (S.P.)

Citation: Hamza, A.; Samad, A.; Imam, M.A.; Faizan, M.I.; Ahmed, A.; Almajhdi, F.N.; Hussain, T.; Islam, A.; Parveen, S. Structural Characterization of Ectodomain G Protein of Respiratory Syncytial Virus and Its Interaction with Heparan Sulfate: Multi-Spectroscopic and In Silico Studies Elucidating Host-Pathogen Interactions. *Molecules* **2021**, *26*, 7398. https://doi.org/10.3390/molecules26237398

Academic Editors: Tanveer A. Wani, Seema Zargar and Afzal Hussain

Received: 11 November 2021
Accepted: 2 December 2021
Published: 6 December 2021

Publisher's Note: MDPI stays neutral with regard to jurisdictional claims in published maps and institutional affiliations.

Abstract: The global burden of disease caused by a respiratory syncytial virus (RSV) is becoming more widely recognized in young children and adults. Heparan sulfate helps in attaching the virion through G protein with the host cell membrane. In this study, we examined the structural changes of ectodomain G protein (edG) in a wide pH range. The absorbance results revealed that protein maintains its tertiary structure at physiological and highly acidic and alkaline pH. However, visible aggregation of protein was observed in mild acidic pH. The intrinsic fluorescence study shows no significant change in the λ_{max} except at pH 12.0. The ANS fluorescence of edG at pH 2.0 and 3.0 forms an acid-induced molten globule-like state. The denaturation transition curve monitored by fluorescence spectroscopy revealed that urea and GdmCl induced denaturation native (N) $\leftrightarrow$ denatured (D) state follows a two-state process. The fluorescence quenching, molecular docking, and 50 ns simulation measurements suggested that heparan sulfate showed excellent binding affinity to edG. Our binding study provides a preliminary insight into the interaction of edG to the host cell membrane via heparan sulfate. This binding can be inhibited using experimental approaches at the molecular level leading to the prevention of effective host–pathogen interaction.

Keywords: RSV; ectodomain G protein; heparan sulfate; protein–ligand interaction; fluorescence quenching; molecular docking; molecular dynamic simulation

1. Introduction

Respiratory syncytial virus (RSV) is a lower respiratory tract pathogen that causes pneumonia and bronchiolitis in children, especially those younger than five years of age. It can also affect the elderly and immunocompromised individuals [1–3]. It infects about 70% of children less than one-year-old, 2–3% of whom are hospitalized. By the age of 2, almost all the children get infected due to RSV [4]. At present, there is no licensed vaccine or any therapeutics available against RSV. The only available preventive measure is the injection of monoclonal antibodies cocktail (palivizumab) specific to the fusion glycoprotein, which may decrease the severity of disease and hospitalization of infants [5]. In the developmental race for RSV vaccines, one of the major concerns is whether to incorporate the G glycoprotein in the subunit of live vaccines or not. Some scientists claim that the presence of G protein causes a proinflammatory response, as seen in the formalin-inactivated vaccine. On the other hand, some claim that it induces neutralization antibodies, which may be helpful in protection against RSV [6,7]. However, some monoclonal antibodies (mAbs) which target the G protein neutralize the infection of RSV on human airway epithelial cells and reduce the viral load and disease titer in an animal model [8–10]. Additionally, anti-G protein mAbs re-establish the Th1/Th2 cytokine level and suppress the pulmonary inflammation, mucus production and pro-inflammatory cytokines [11–14].

RSV is a negative-sense, single-stranded RNA virus enclosed in an envelope. The 15.2 kb genome of the RSV encodes the 11 proteins, out of which three are structural proteins SH (small hydrophobic), G (attachment glycoprotein), and F (fusion glycoprotein) that are embedded in the lipid envelope and play a significant role in viral entry, attachment, and fusion respectively. The attachment G glycoprotein is a type II membrane protein with an N-terminal cytoplasmic domain (1–36 aa), transmembrane domain (37–66 aa), and extracellular ectodomain (67–298 aa). The extracellular domain of the G protein comprises a central conserved domain (CCD), composed of 13 amino acids (164–176 aa) which are conserved among all the RSV isolates [15]. The cluster of four cysteine residues is present at 173, 176, 182, and 186 positions joined by two disulfide bonds formed between the Cys173–186 and Cys176–182 residues. The third and fourth cysteine residues form the CX3C motif (182–186 aa), which helps in the attachment of RSV to the susceptible human airway epithelium cell with the interaction of CX3CR1, a chemokine receptor [16,17]. A positively charged heparin-binding domain (184–198 aa) of the G protein also facilitates the attachment of the RSV via cell surface glycosaminoglycans (GAGs) [18,19]. Heparan sulfate (HS) is a type of GAGs molecule which exhibit various biological activities, most of which facilitate attachment with protein [20]. The HS molecules are present in almost all types of mammalian cells and act as a coreceptor for a number of viruses [21]. Recently, studies have suggested that HS act as a coreceptor for the Spike protein of SARS CoV-2, which makes this interaction an attractive target for SARS CoV-2 infection [22].

Protein is an essential biological macromolecule with enormous importance in every physiochemical process. The native or folded form of proteins is necessary to perform their biological function [23]. Conditions responsible for a protein to remain in the native conformation in the solution are pH, temperature, ionic strength, cofactors, and chaperon proteins [24,25]. In some cases, the stability of the protein is also controlled by the addition of osmolytes viz., trehalose and proline that stabilize the native conformation of the protein [26,27]. However, urea and guanidium chloride (GdmCl) are the major chemical denaturant used for proteins [28]. The stability of the protein depends on the noncovalent interaction, which can be disrupted using these denaturants [29]. Several studies are available which determine protein stability by studying denaturation curves [30,31]. Due to the unavailability of the crystal structure of G protein, limited information is available about its biophysical properties. Some studies reported that the G protein behaves differently in different environmental conditions such as pH and temperature. Currently, there is no information available that reports the folding and unfolding behaviour of G protein and determine its structural and functional stability. In this report, we, for the first time, determine the structural conformation change of ectodomain edG in a wide range of pH and determine the protein stability in the presence of denaturant (urea and GdmCl). To understand how different environmental conditions affect the structure and function of edG, it is important to understand the basic mechanism of host–pathogen interactions.

Here, we expressed and purified the ectodomain G protein (edG). The purified protein was used to determine the structural and conformational stability in different environmental conditions (pH, urea, GdmCl). Further, we reported the binding interaction of heparan sulphate with edG by fluorescence quenching, molecular docking, and molecular dynamics (MD) simulation studies.

2. Results and Discussion

The G protein of RSV plays a vital role in the host–pathogen interaction, which is mediated by pH-dependent or independent pathways [30,32]. Hence, the role of G protein is very crucial for attachment to the host cell receptors. Since all the biological processes are controlled by the pH of the environments, it eventually determines its stability. In this study, we performed the pH-based and chemical-induced denaturation study to examine its effect on the structural and conformational stability of edG in in vitro conditions. Further, we also determined the binding interaction studies of heparan sulfate with edG by fluorescence quenching and in silico approaches.

2.1. Structural and Conformational Stability Measurements of edG

2.1.1. Absorbance Measurements

To monitor the effect of pH on the tertiary structure of edG, we measured the absorbance spectra at a wide range of pH. The side chain of aromatic amino acids of protein acts as chromophore as it contains the conjugated double bond system [33]. The maximum absorption (λ_{max}) in the UV region of the protein is extremely sensitive due to the change in the local milieu of the aromatic amino acid as it increases the solvent exposure that leads to a blue shift in the λ_{max} [34]. The edG has five tyrosine and one tryptophan residues that give rise to maximum absorption (λ_{max}) at 278 nm wavelength. The change in the molar absorbance coefficient of native and denatured protein at 278 nm ($\Delta\varepsilon_{278}$) plotted as a function of pH, which probes the environmental changes of the aromatic group of the proteins and hence the native biological structure of the protein [35].

We did not find any significant change in the spectra of edG on increasing the pH from pH 2.0–3.0 and 8.0–12.0, which indicates intact native structure at these pH ranges (Figure 1). The protein does not show any scattering signals during the pH-based studies except at pH 4.0–6.0, indicating good solubility of the protein. However, the acid-induced loss of tertiary interactions occurs due to aggregation at pH values 4.0–6.0 that shows the complete distortion of spectra with a noticeable scattering signal in the range of 320–340 nm. Our finding from the spectral study confers that the edG is stable at physiological pH and highly acidic and basic pH values, but the loss in structure was observed at mildly acidic pH (4.0–6.0). From our study, we also conclude that at highly acidic and basic pH values, the edG does not show any type of aggregation; however, at mild acidic pH, a visible aggregation of protein was observed. The plot of $\Delta\varepsilon_{278}$ versus pH shows no remarkable change in molar absorption coefficient in highly acidic and alkaline pH values (see inset of Figure 1). Hence, the loss of edG tertiary interaction at mildly acidic pH may be correlated with its biological activity, as suggested in our earlier publication [30].

2.1.2. Fluorescence Measurements

Fluorescence studies provide information about the protein's tertiary structure. The aromatic amino acid residue is extremely sensitive to its local environment, and it is responsible for a protein's intrinsic fluorescence measurements. A characteristic redshift in λ_{max} is indicative of the increased interaction of aromatic amino acid residue upon the unfolding of the protein in the solvent [36]. The edG contains five tyrosine and one tryptophan residue, which enables us to perform the intrinsic fluorescence to observe the effect of pH on the tertiary structure of the protein. The changes in λ_{max} were plotted as a function of pH to determine the alteration in the microenvironment of the buried aromatic amino acid of the protein. The changes in intrinsic fluorescence of edG at the wide range of pH are shown in Figure 2.

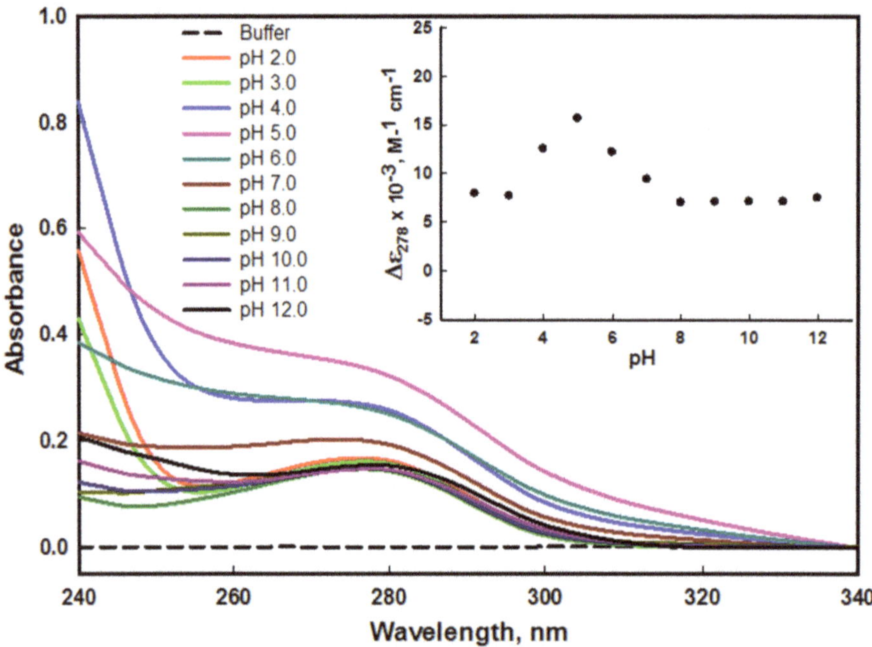

Figure 1. Absorbance spectra of edG at different pH values (2.0−12.0) at 25 °C. The spectra were measured in the range of 340−240 nm. The spectrum at pH 8.0 is considered as a control. The inset shows the denaturation profile of edG from pH 2.0−12.0 followed by observing changes in $\Delta\varepsilon_{278}$ as a function of pH.

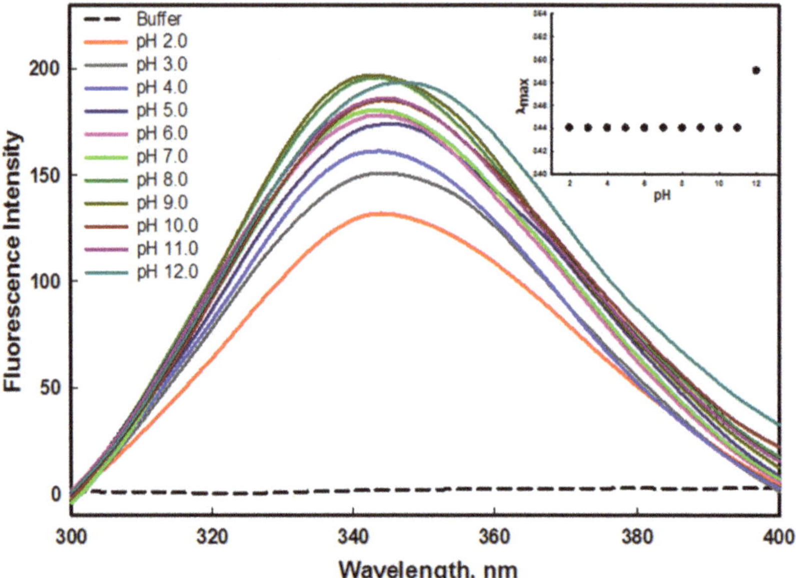

Figure 2. Fluorescence emission spectra of edG in the pH range of 2.0–12.0 at 25 °C. The protein was excited at 280 nm and recorded the 300–400 nm emission spectra. The inset shows the denaturation profile of edG from pH 2.0–12.0, followed by changes in emission maxima (λ_{max}) as a function of pH.

The native conformation of edG at pH 8.0 shows λ_{max} at 344 nm. The emission maxima show no spectral shift, as we are moving from acidic pH to alkaline pH condition (except pH 12.0), attributed that the microenvironment of aromatic residues was significantly disturbed. The decrease in fluorescence intensity was observed as we moved from the physiological pH to acid pH values. The decrease in intensity might be due to the protonation of acidic amino acids or water molecules surrounding the aromatic amino acid residues. Similarly, a substantial decrease in fluorescence intensity was observed as we moved from the physiological pH to alkaline pH values (pH 9.0–12.0). This indicates the deprotonation of essential amino acids present around the intrinsic fluorophore that leads to fluorescence quenching. The deprotonation/protonation of amino acid side chains may lead to disruption of charge in the local environment by interrupting internal salt bridges and electrostatic interactions that are present in the native conformation of the protein [37]. The plot of λ_{max} versus pH shows no significant change in the emission maxima of edG from pH 2.0–11.0. However, a redshift of 5 nm in λ_{max} was observed at pH 12.0 (see inset of Figure 2). The characteristic redshift of 5 nm in emission maxima is indicative of the increased solvent interaction of aromatic amino acid residue upon unfolding of the protein. From our fluorescence measurements, we concluded that the tertiary structure of edG remains similar at pH 8.0–9.0. However, destabilization of charges on protein surface leads to the disruption of the electrostatic interaction at acidic and basic pH values.

The changes in protein structure under various environmental conditions often lead to the exposure of hydrophobic patches normally buried in the native state. The ANS is an extrinsic dye that shows binding with partially unfolded protein in which tertiary structure was distorted and secondary structure of the protein retained. The high binding affinity of ANS to these confirmations confirms the presence of pre molten globule (PMG) or molten globule (MG) state [38,39]. The binding of ANS with hydrophobic patches leads to an increase in fluorescence intensity along with the blue shift in emission maxima [40,41]. Figure 3 shows the change in fluorescence intensity in the presence of ANS at different pH values. We found that the ANS fluorescence intensity decreases as we move towards the basic pH values. The ANS fluorescence intensity of protein at pH 2.0 and 3.0 was much higher than the native state (pH 8.0) with shifting of emission maxima towards the shorter wavelength (blue shift). This might be due to the exposure of buried hydrophobic clusters in non-native states of protein populated at pH 2.0 and 3.0. The above-mentioned features are the well-established characteristics of the molten globule (MG) state (partially folded intermediate state of a protein) that is induced at mild denaturing conditions [42,43]. Hence, the non-native state of edG at pH 2.0 and 3.0 is considered an acid-induced molten globule-like state. With native and alkaline conditions, no significant difference in fluorescence intensity was seen (see inset of Figure 3). This resulted from the solvent inaccessibility of buried hydrophobic clusters, which prevents the ANS binding [44,45].

2.1.3. Urea and GdmCl-Induced Denaturation

The G protein of RSV plays an important role in the host–pathogen interactions. To date, the 3D- structure of G protein is not known, and very limited information is available about its structural properties in solution. For instance, how does the G protein fold and unfold, how a protein behaves in a diverse solvent environment and how much protein is stable? The answer to these questions is still elusive. The behaviour of a protein in different solvent conditions gives information about protein folding and stability. Chemical-induced denaturation is a critical method to determine the structural stability of various proteins [46,47]. The stability and folding mechanism information will improve our knowledge of the behaviour of G protein in different biological conditions. We examined the urea and GdmCl induced denaturation by fluorescence spectroscopy to measure the stability of the edG.

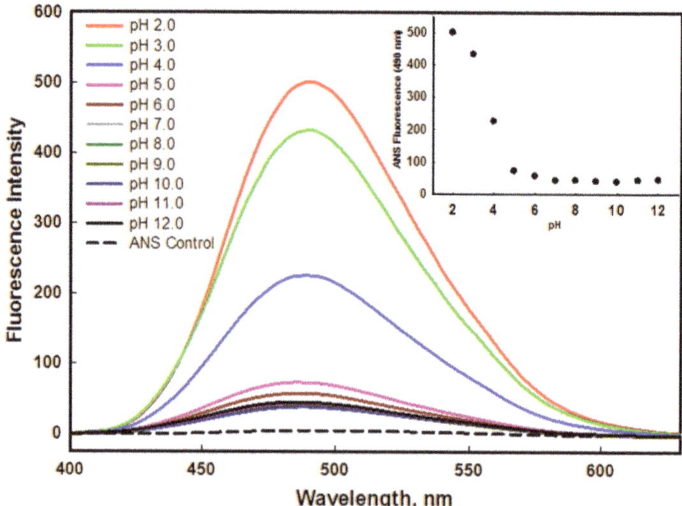

Figure 3. ANS fluorescence spectra of edG in the pH range of 2.0–12.0 at 25 °C. The ANS was excited at 380 nm and recorded the emission spectra from 400–630 nm. The inset shows the ANS profile from pH 2.0–12.0, followed by observing changes in F_{490} as a function of pH.

The protein stability can be measured by equilibrium unfolding studies in the presence of GdmCl or urea [48]. The edG has one tryptophan and five tyrosine residues which are often buried either partially or fully in the hydrophobic core of the folded protein. These aromatic amino acid residues of edG act as markers for the structural integrity of the protein. Figures 4A and 5A shows the fluorescence emission spectra of edG in the presence of increasing concentrations of urea and GdmCl, respectively. The decreased fluorescence spectra were observed with increasing concentrations of urea and GdmCl, with the shifting of λ_{max} of tryptophan residues towards the longer wavelength (redshift). The native edG exhibit an emission maxima peak at 344 nm, and the λ_{max} of the protein shifts to 356 nm at higher denaturant concentrations. From these observations, we confer that the tryptophan residues are shifted from nonpolar to the polar environment, as urea and GdmCl exposes the buried aromatic amino acid residues [49,50].

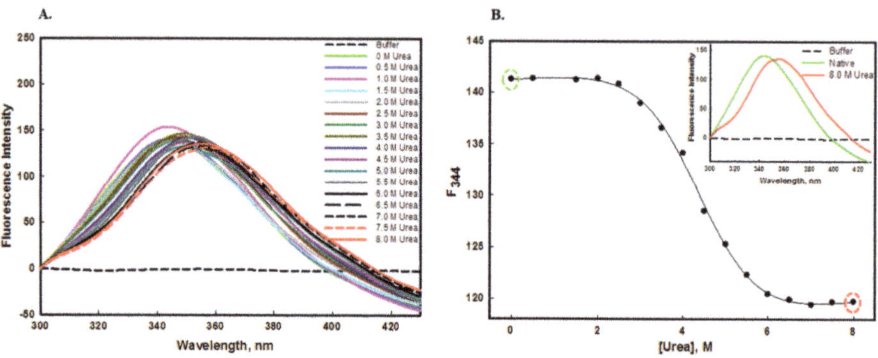

Figure 4. (**A**) Urea-induced denaturation of edG at pH 8.0 at 25 °C measured by intrinsic fluorescence studies. The emission spectra were recorded as a function of increasing urea concentrations (0.0–8.0 M). The protein was excited at 280 nm and recorded the 300–430 nm emission spectra. (**B**) Denaturation curve of edG (plot of F_{344} as a function of [urea]). The inset in figure (**B**) shows the emission spectra of edG in the native and 8 M urea denatured state.

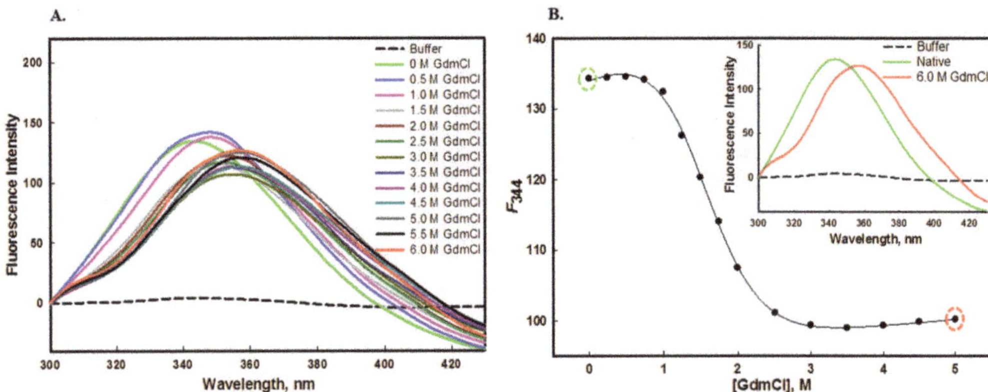

Figure 5. (**A**) GdmCl-induced denaturation of edG at pH 8.0 and 25 °C measured by intrinsic fluorescence studies. The emission spectra were recorded as a function of increasing concentration of GdmCl (0.0–6.0 M). The protein was excited at 280 nm and recorded the emission spectra from 300–430 nm. (**B**) Denaturation curve of edG (plot of F_{344} as a function of [GdmCl]). The inset in figure (**B**) shows the emission spectra of edG in the native and 6 M GdmCl denatured state.

Our observation also suggests that as we increase the concentration of urea and GdmCl, unfolding of edG takes place, which exposes the buried tryptophan to more polar buffer conditions. The changes in the tryptophan microenvironment were monitored by F_{344} (the emission wavelength at 344 nm) as a function of [urea] (Figure 4B) and [GdmCl] (Figure 5B). The plot of F_{344} versus [urea] and [GdmCl] suggested that chemical-induced denaturation of edG occurred in a single step and followed a two-state transition mechanism. The transition curve shown in this figure were analyzed to estimate the stability parameter such as m, ΔG_D^0 and C_m from the denaturation curve using Equation (1). The values of these parameters are mentioned in Table 1. Hence, from these observations, we conclude that the tertiary structure of edG loses cooperatively without the participation of an intermediate state. The equilibrium unfolding transition induced by urea and GdmCl are not always equal; the difference might be due to the ionic character of GdmCl [51].

Table 1. Thermodynamic parameters obtained from urea and GdmCl-induced denaturation of edG at pH 8.0 and 25 ± 0.1 °C.

Probes	Denaturants	Transition	ΔG^0_D, kcal mol^{-1}	m, kcal mol^{-1} M^{-1}	C_m, M
F_{344}	Urea	N↔D	3.76 ± 0.34	0.85 ± 0.08	4.42 ± 0.16
	GdmCl	N↔D	2.53 ± 0.20	1.66 ± 0.09	1.52 ± 0.07

2.2. Binding Interaction Studies of edG with Heparan Sulfate

2.2.1. Fluorescence Quenching Measurements

The intrinsic fluorescence measurements of a protein are susceptible to its microenvironment, making it an important tool to investigate the formation of the complex between the ligand and protein [52,53]. The quenching mechanism of edG with heparan sulfate was studied to know the parameters such as binding constant (K), Stern–Volmer constant (K_{sv}), and the number of binding sites (n). The binding constant of heparan sulfate interacting with protein was determined by exciting the protein at 280 nm, and the change in the fluorescence intensity was recorded in the range of 300–430 nm. The protein excitation at 295 nm was considered as fluorescence of only Tryptophan, while protein exited at 280 nm was considered as the excitation of phenylalanine, tyrosine, and tryptophan [54]. Figure 6A shows the fluorescence spectra of edG in the presence of an increasing concentration of heparan sulfate (0–50 µM). Heparan sulfate did not flourish alone, while protein gave a maxima peak at 344 nm in similar environmental conditions. A progressive decrease in the fluorescence spectra was observed with the addition of HS, indicating the formation of the complex between protein and ligand. The quenching data was analyzed using the Stern–Volmer Equation (2) to calculate the Stern–Volmer constant (K_{sv}). The Stern–Volmer plots of protein quenching in the presence of HS is shown in Figure 6B. The K_{sv} value was determined from the Equation by plotting the fluorescence intensity ratio F_0/F for different concentrations of HS. The value of the bimolecular quenching constant (K_q) was obtained using Equation (3) and further confirmed the mode of quenching. In the presence of HS, the decrease in fluorescence intensity was analyzed by the modified Stern–Volmer Equation (4). Figure 6C shows the fitted experimental data based on the double log relation with the intercept of the plot providing the binding constant. The binding constant (K) value was found to be 3.98×10^6, which confirmed that HS has a high binding affinity to edG. The binding parameters of the edG-HS system calculated from the fluorescence quenching are given in Table 2. Interestingly, our previous binding studies by microscale thermophoresis (MST) and isothermal titration calorimetry (ITC) [30] complement the results of the fluorescence binding study. However, various reports had mentioned the difference in the thermodynamic parameters obtained from ITC and fluorescence quenching. This difference is due to fluorescence quenching, as it measures only the local changes around the fluorophore, whereas the ITC and MST measure the global changes in the thermodynamic properties [55].

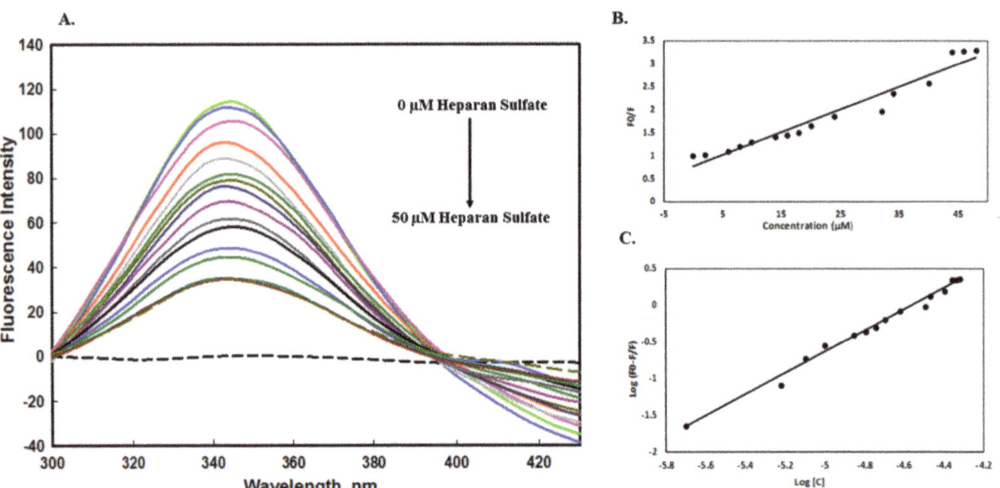

Figure 6. Fluorescence binding studies of the edG with heparan sulfate (HS) at pH 8.0 at 25 °C. The protein was excited at 280 nm and recorded the emission spectra from 300–430 nm. (**A**) Fluorescence emission spectra of protein with increasing concentration (0–50 µM) of HS. (**B**) Stern–Volmer plot for quenching of edG–HS complex. (**C**) Modified Stern–Volmer (double-log relation) plot of the edG–HS complex.

Table 2. Binding parameters of the edG with heparan sulphate were obtained from fluorescence quenching studies at pH 8.0 and 25 °C (298 K).

K_{sv} (10^4 M^{-1})	K_q (10^{13} M^{-1} s^{-1})	K (10^6 M^{-1})	n	R^2
4.96	1.73	3.98	1.44	0.98

2.2.2. Absorbance Binding Measurements

The absorbance spectra of edG give a characteristic peak at 278 nm due to the presence of aromatic residues. The change in the spectra indicates the interaction of the ligand to protein [56]. A gradual decrease in the absorption spectra of edG was observed with an increasing concentration of heparan sulfate (HS). The protein spectra in the absence of ligand (HS) give a peak at 278 nm. As we increase the ligand concentration 0–50 μM, a significant decrease in the spectrum has been observed with some scattering in the range of 340–320 nm (Figure 7). However, the 278 nm peak shifted towards a shorter wavelength (blueshifts) with increasing ligand concentration, confirming that the aromatic amino residues of the protein are exposed to a more polar environment [57].

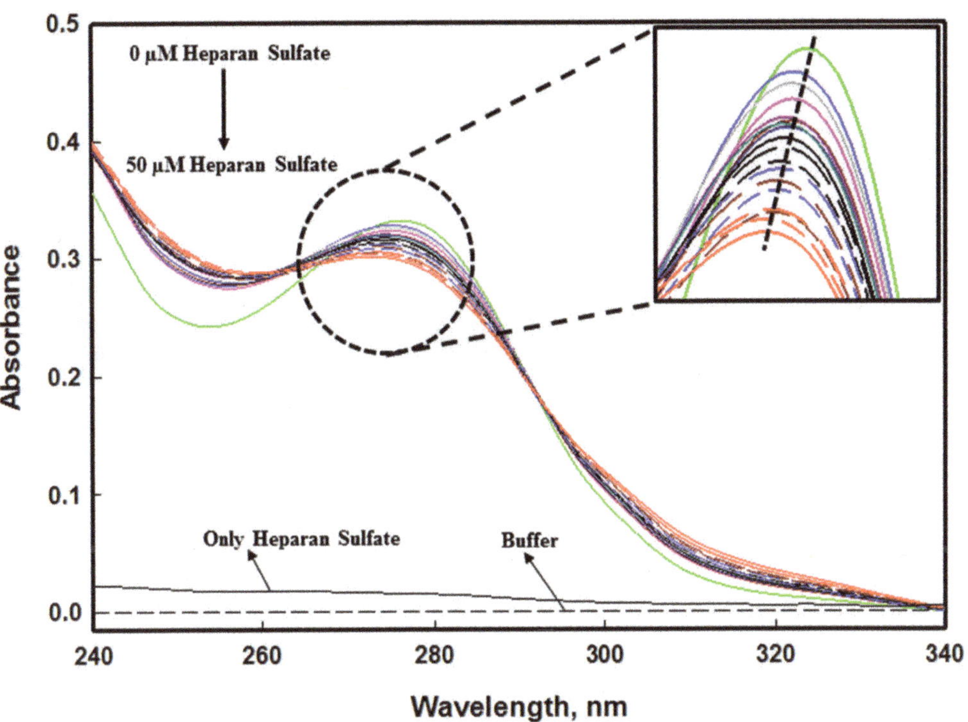

Figure 7. Absorbance binding measurements of the edG with heparan sulfate (HS) at pH 8.0 and 25 °C. The spectra were recorded in the range of 340–240 nm with increasing concentration (0–50 μM) of HS. The inset shows the 278 nm peak and shifting of spectra towards shorter wavelength (blue shift) with increasing concentration of heparan sulfate.

2.2.3. Molecular Docking

To confirm the edG-HS interaction, we performed the molecular docking experiment and estimated the binding energy, intramolecular distance and interacting residues of protein with the heparan sulfate. This analysis evaluated the binding interaction of heparan sulfate with the edG and helped us understand its role as an attractive target. Figure 8A shows the protein (surface view) heparan sulfate (ball and stick) interaction with various amino acid residues. The docking study suggested that heparan sulfate occupies the active binding site of the edG with a strong binding affinity of −6.8 kcal/mol. The protein–ligand complex forms the hydrogen bond interactions with seven key residues: Asn112, Lys117, Thr133, Thr134, Ser191, Arg196, and Glu226 (Figure 8A). Figure 8B shows the bond distance between the ligands (HS) with the interacting amino acid residues of the protein. The HS interact via a single bond with six amino acids viz. Asn112 of distance 2.04 Å, Lys117 of 2.74 Å, Thr133 of 2.92 Å, Thr134 of 2.03 Å, Ser191 of 2.64 Å, Arg196 of 2.27 Å, and via two bonds with Glu226 with an equal distance of 2.13 Å. The distance of the hydrogen bond lies in the range of 2.03–2.92 Å, which indicates the good binding affinity between the protein–

ligand complex. In addition, the protein–ligand complex also forms four carbon–hydrogen bonds with key residues of Leu115, Cys116, Pro132, and Arg196 depicted in the 2D model (Figure 8C). The higher number of hydrogen bonds and lower binding energy suggests the strong binding affinity of the protein–ligand complex.

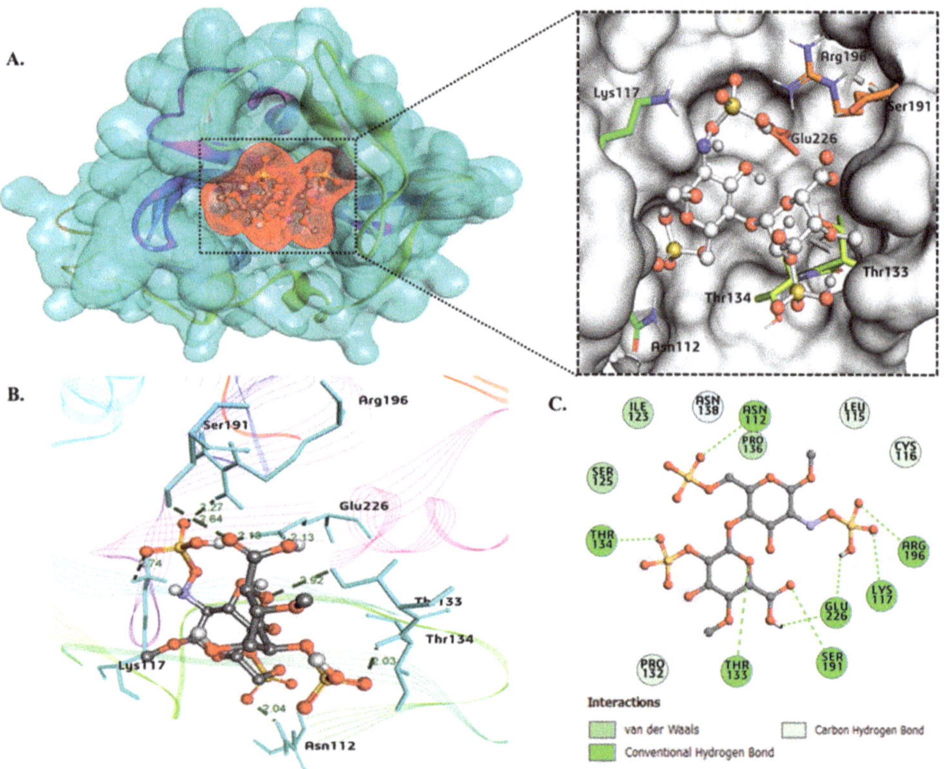

Figure 8. Interaction of edG with heparan-sulfate. (**A**) Surface representation of docked edG-HS complex in the active site pocket and important residues involved in the polar interactions (stick model). (**B**) Representation of edG-HS complex with the distance of hydrogen bonds. (**C**) Detailed 2-dimensional representation showing interactions and types of bonds formed between heparan sulfate and edG.

Several viruses use heparan sulfate proteoglycans (HSPGs) on the cell surface as attachment factors such as vaccinia virus [58], herpes simplex virus [59], hepatitis C virus [60], Sindbis virus [61], human immunodeficiency virus-1 (HIV-1) [62] and HCoV-NL63 [63]. Recently a study reported that the spike protein of SARS COV-2 interacts with heparan sulphate and ACE2 through RBD and promotes the spike–ACE2 interaction [22]. A study by Kalia et al. reported that HS plays a critical role in facilitating HEV infection on target cells because the elimination of heparan sulfate by heparinase hindered pORF2 attachment and blocked infection of HEV to Huh-7 cells [64]. Another study has reported the interaction of ectodomain G protein with quercetin and morin (flavonoids) utilizing fluorescence quenching and suggested it is an antiviral agent against RSV [65]. In our study, we found that the number of hydrogen bonds formed by edG with heparan sulfate is significantly high and hence have an excellent binding affinity. The binding results obtained from fluorescence quenching complement our molecular docking observation that attributed that conventional hydrogen bond and carbon-hydrogen bond was mainly widespread in the protein–ligand complex. The binding parameter of edG–heparan sulfate complex obtained from fluorescence quenching, molecular docking, and previously reported results from ITC and MST [30] are given in Table 3.

Table 3. Binding energy parameters of heparan sulfate with edG obtained from fluorescence and docking studies.

Compound	Binding Constant #(K) M^{-1}	*ΔG (kcal/mol)	Binding Constant @(K) M^{-1} [32]	*K$_d$ (nm) [32]
Heparan sulfate	3.98×10^6	−6.8	10.7×10^4	426

#K calculated from fluorescence quenching, *ΔG calculated from molecular docking, @K calculated from ITC, *K$_d$ calculated from MST.

2.2.4. MD Simulation Studies

We performed extensive MD simulation studies to know the interaction mechanism of edG-HS complex and edG alone for 50 ns. We assumed that HS has a close binding conformation with edG. The system stability was determined by calculating the radius of gyration (R$_g$) and root-mean-square deviations (RMSD) values, which showed that after ~20 ns, the system achieved the equilibrium conformation (Figure 9A,B). A clear difference in the R$_g$ and RMSD values of edG and edG-HS complex suggested that complex form showed higher dynamic value than the unbound form, which can be attributed to the perturbation effect of HS on the structure of edG. The active binding site (CX3C motif and heparin-binding region) of edG lies in the amino acid region 110–130. The root mean square fluctuation (RMSF) plot of the edG and edG-HS complex showed a considerable variation in the structure of these residues (Figure 9C). Compared to the edG-HS complex, the active site residues in the free edG showed a lower degree of mobility in the constituents' residues, indicating lower relative energy. The solvent-accessible surface area (SASA) plot showed no significant changes throughout the 50 ns simulation process, attributing that formation of a stable complex between edG and HS (Figure 9D). The SASA value for edG-HS complex and edG alone was found to be 127 nm^2 and 130 nm^2, respectively.

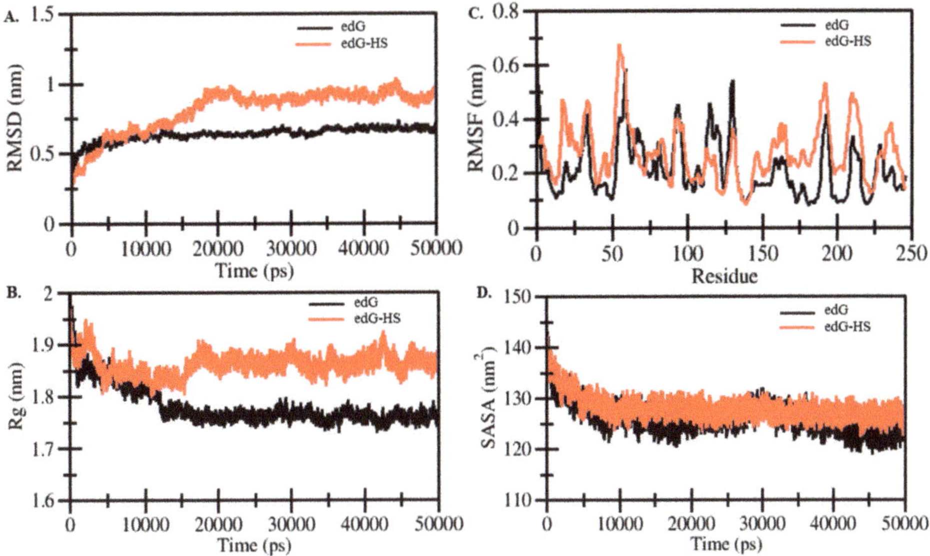

Figure 9. Structural dynamics, compactness and folding of the edG-HS complex as a function of time. (**A**) RMSD plot of edG and edG-HS complex. (**B**) The R$_g$ curves of edG and edG-HS complex showing differences in compactness. (**C**) RMSF plot of edG and edG-HS complex. (**D**) SASA plot of edG before and after binding with HS as a function of time.

The average intermolecular hydrogen bonds were ~3.0 between HS and protein during the MD simulation process (Figure 10A). Furthermore, the total energy presented between edG alone and edG-HS complex was calculated using GROMACS utility (Figure 10B). The total free energy of edG was found −768 kJ/mol, and the edG-HS complex was found −772 kJ/mol, with electrostatic energy accounting for a significant contribution. The structural conformation difference between the edG and edG-HS complex was also studied by analyzing the free energy landscape. No major conformation difference was observed between protein and protein–HS complex (Figure 10C,D). The edG showed an energy-favoured and relatively stable conformation compared to the protein–ligand complex, suggesting that binding of HS not perturbed the structure of the protein.

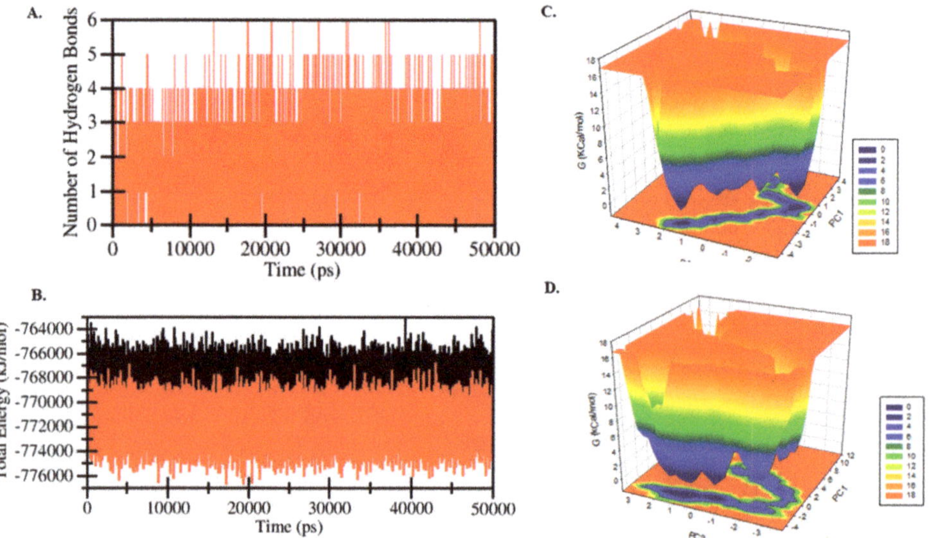

Figure 10. (**A**) The fluctuating curve of hydrogen bonds shows changes in the observed number. (**B**) Generated curves of free energy landscapes showing fluctuations of total energies observed between edG-HS and edG. (**C**) Free energy landscape plot of edG and (**D**) edG-HS complex.

3. Materials and Methods

3.1. Materials

All the consumables used in the experiments were of analytical grade. Luria-Bertani broth, urea, sodium chloride, imidazole, Amicon Ultra10 K device, etc., were purchased from Merck (Darmstadt, Germany). Isopropyl-d-1-thiogalactopyranoside (IPTG) and Kanamycin were obtained from Sigma (Saint Louis, MO, USA). Ni-NTA beads were purchased from Qiagen, Hilden, Germany. The syringe filter (0.22 µm) was purchased from Millipore Corporation (Burlington, MA, USA).

3.2. Expression and Purification of edG

We successfully transformed the edG gene, expressed it into the BL23 (DE3) strain of *E. coli* and purified the protein with some modifications described earlier [30,66,67]. Briefly, the protein was expressed by induction with 0.5 mM IPTG for 12 h at 30 °C. The inclusion bodies (IBs) were prepared, and the protein was purified by Ni-NTA chromatography. The bound protein was eluted with 50 mM Tris buffer (pH 8.0), 100 mM NaCl, 5% glycerol, 0.5% N-lauroylsarcosine, and 150 mM Imidazole. The eluted fractions were analyzed and confirmed by sodium dodecyl sulphate polyacrylamide gel electrophoresis (SDS-PAGE). The eluted fraction with the desired single band was dialyzed against 20 mM Tris buffer (pH 7.5) and 100 mM NaCl. The dialysis buffer was consecutively changed five times at 4 °C with stirring for 24 h to get the refolded protein. The protein concentration was measured using a molar absorbance coefficient of 8730 M^{-1} cm^{-1} by Jasco V-600 UV-visible spectrophotometer at 280 nm [68].

3.3. Sample Preparation

The protein stock solution was filtered with a Millipore filter disc of 0.22 μm. A broad range of different buffers was used to monitor the pH-dependent change in the structure of the edG. The buffer of pH 2.0 and 3.0 was prepared with 50 mM of glycine and adjusted the pH with HCl. For pH 4.0 and 5.0 buffer, we used 50 mM of acetate and adjusted the pH using acetic acid. The buffer of pH 6.0 and 7.0 was made 50 mM phosphate and adjusted the pH with NaH_2PO_4. For pH 8.0 and 9.0, we used 50 mM Tris and 100 mM NaCl and adjusted the pH using HCl. The buffer of pH 10.0, 11.0 and 12.0 was prepared with 50 mM of glycine and adjusted the pH with NaOH. Before taking the spectral measurements, the sample was incubated in a respected buffer for at least three hours to attain equilibrium.

We prepared the 8.7 M GdmCl stock solution and the freshly prepared 10.5 M stock solution of urea to examine the chemical-induced denaturation of the protein. The stock solution of both the chemicals was prepared in 25 mM Tris buffer (pH 8.0). For every measurement, protein concentration was fixed accordingly (0.3–0.6 mg mL^{-1}) with the required volume of buffer and urea/GdmCl. For more accuracy, every time, the concentration of the stock solution of protein was calculated using the molar absorbance coefficient (ε) of protein (8730 M^{-1} cm^{-1}) at 280 nm. The calculated amount of protein and denaturant was taken and mixed correctly, and samples were incubated for at least 3–4 h at $25 \pm 1\,°C$ to ensure that the denaturation process was completed.

3.4. Absorbance Measurements

Absorption spectra of edG were measured to detect alteration in the tertiary structure of the protein at a wide range of pH. The spectral measurement was carried out in Jasco UV/visible spectrophotometer (V-660) with a bandwidth of 0.1 and a scan speed of 100 nm/min at $25 \pm 1\,°C$, equipped with a water bath for temperature control. The spectra were recorded in the range of 240–340 nm using 0.3–0.6 mg mL^{-1} concentration of protein with 1 cm path length cuvettes. All the spectra were recorded in triplicate.

3.5. Fluorescence Measurements

The intrinsic fluorescence spectra of the edG were measured in Jasco spectrofluorometer (FP6200) at $25 \pm 1\,°C$ with a quartz cuvette of 1 cm path. We monitored the changes in the fluorescent emission spectra of the protein in different buffers conditions and in the presence of urea and GdmCl by taking the entrance and exit slits widths at 5 nm and 10 nm, respectively. The protein was excited at 280 nm, and emission spectra were recorded from 300–400 nm in triplicates. The blank values of all the measurements were subtracted from each sample value.

3.6. ANS (8-Anilinonapthalene-1-Sulfonic Acid) Binding Measurements

The ANS fluorescence measurements were carried out using a Jasco spectrofluorometer (FP6200) at $25 \pm 1\,°C$ with a 1 cm path length quartz cuvette. The 1:20 ratio for protein to dye at different pH values was used to know the exposure of hydrophobic surfaces. The excitation and emission slits widths were set at 5 nm with a scanning speed of 125 nm min^{-1}. The sample was incubated in the dark for 30 min before performing the measurements. The ANS emission spectra were recorded in the range of 400–630 nm with an excitation wavelength of 380 nm.

3.7. Analysis of Denaturation Spectral Measurements

The spectral property (F_{344}) plotted against the concentration of urea/GdmCl generated the transition curve. The molar concentration of urea/GdmCl were used to calculate thermodynamic properties such as stability parameters ($\Delta G^0{}_D$), slope (m), and Cm (=$\Delta G^0{}_D/m$) where $\Delta G^0{}_D$ is the Gibbs free energy change without the denaturants, m is

($\partial \Delta G_D / \partial [urea/GdmCl]$), and Cm is the transition midpoint of chemical-induced denaturation curve where $\Delta G_D = 0$. The analysis was done based on the least-square method to fit the denaturation curve using the following Equation (1):

$$y = \frac{yN + yD \times Exp[-(\Delta G°_D - m[urea/GdmCl])/RT]}{(1 + Exp[-(\Delta G°_D - m[urea/GdmCl]/RT])}$$ (1)

where yN and yD are the estimated optical properties of the native protein and the denatured protein, respectively, under the similar experimental condition in which y has been measured, T is the temperature in Kelvin, and R is the universal gas constant.

3.8. Fluorescence Quenching Measurements

The edG binding studies with heparan sulfate were performed by Jasco spectrofluorometer (FP6200) at 25 ± 1 °C in a quartz cuvette with a path length of 1 cm. The stock solution (1 mM) of Heparan sulfate (HS) was prepared in Tris buffer (20 mM) at pH 8.0. The increasing concentration (2 to 50 µM) of heparan sulfate was used to titrate against the fixed protein concentration. The protein excitation was done at 280 nm, and emission spectra were recorded from 300–430 nm with excitation slit at 5 nm and emission slit at 10 nm. The edG showed the emission maxima peak at 344 nm. The final spectra were obtained by subtracting the blank one (heparan sulfate with buffer).

The fluorescence quenching of edG with heparan sulfate was determined to know the different binding parameters such as the Stern–Volmer constant (K_{SV}), the binding constant (K) and the number of binding sites (n).

To determine the Stern–Volmer constant and analyze the quenching data, the Stern–Volmer Equation (2) was used:

$$\frac{F_0}{F} = 1 + K_{SV}[C]$$ (2)

where F_0 denotes the intensity of protein absence of HS, F denotes the intensity of protein at a specific concentration of HS at 344 nm, [C] denotes the different concentrations of HS, and K_{SV} is the obtained Stern–Volmer quenching constant.

The bimolecular quenching constant (Kq) was calculated using Equation (3) to confirm the quenching mode of the protein–HS complex

$$K_q = \frac{K_{SV}}{\tau_0}$$ (3)

where τ_0 is the average integral fluorescence lifetime of tryptophan (2.7×10^{-9} s)

Using Equation (4), the modified Stern–Volmer constant gives a binding constant of the protein–HS complex.

$$log\left(\frac{F_0 - F}{F}\right) = logK + n$$ (4)

where K denotes the binding constant of protein-HS complex and n denotes the number of binding sites.

3.9. Molecular Docking

The molecular docking was performed to identify the interaction of the edG with heparan sulfate. To date, the crystal structure of the RSV G protein is not available. Hence, we have modelled the 3D structure of edG using in silico approach. The detail of the modelled structure of edG has been reported elsewhere in our previous investigation [69]. The chemical structure of heparan sulfate was downloaded from the PubChem database and converted into the pdbqt file using Open Babel software in PyRx. The bioinformatics tool PyRx, Discovery studio, and PyMOL software were used for docking and visualization [70,71] . The docking was structurally blind, where the compound was to be free in motion and search the protein binding sites. Out of nine docked orientations, we selected the one having maximum binding affinity and minimum binding energy in the active site region.

3.10. MD Simulation Studies

MD simulation of edG alone and edG-HS complex was performed using GROMACS version 2018-2 [72]. The topology of protein structure was generated using the GROMOS96 43a2 force-field [73]. The topology of HS was generated using the PRODRG server [74]. After generating the topology of docked complex, salvated through the SPC/E water model and appropriate Na^+ and Cl^- ions were added for neutralization [75]. The system's energy was minimized using the combined steepest descent algorithm, a convergence criterion of 0.005 kcal/mol. The equilibrium condition was developed by NVT (constant volume) and NPT (constant pressure) at a 100 ps time scale. Berendsen weak coupling method was used to maintain the system's temperature at 298 K, and Parrinello–Rahman barostat was used to adjust the pressure of 1 bar in the equilibrium condition. The final conformational production stage of the time scale of 50 ns was generated using the LINCS algorithm, and generated trajectories were analyzed to know the behaviour of the complex in the explicit water milieu. The root-mean-square deviations (RMSD), root-mean square fluctuation (RMSF), the radius of gyration (R_g), solvent-accessible surface area (SASA), Hydrogen bonds, the free energy landscape of complex, and conformational changes were analyzed.

4. Conclusions

The RSV G protein plays a vital role in attaching virion to the host cell membrane. This is the first report that describes the structural and conformational stability of the edG. In this paper, we reported the effect of pH on the structure of edG. We found that the edG was stable at physiological and highly acidic and alkaline conditions. However, a visible aggregation was observed at mild acidic pH values. The urea and GdmCl-induced denaturation studies of edG demonstrated that it follows a two-state transition mechanism. The fluorescence quenching, molecular docking, and MD simulation studies suggested strong binding between the edG and heparan sulfate. Our binding studies elucidate the involvement of heparan sulfate in host–pathogen interaction that is significant for viral infection. Finally, our data suggested that heparan sulfate mimicking compounds can be used to target the effective host–pathogen interaction.

Author Contributions: Experiment design and supervision, S.P., A.I. and A.H.; writing—original draft preparation, S.P., A.I. and A.H.; simulation and data analysis, M.A.I.; writing—review and editing, A.S., M.I.F., F.N.A., T.H., S.P., A.I. and A.A. All authors have read and agreed to the published version of the manuscript.

Funding: This research work was supported by the grant funded by the Council of Scientific and Industrial Research, India [37(1697)17/EMR-II] and Central Council for Research in Unani Medicine, Ministry of Ayurveda, Yoga and Neuropathy, Unani, Siddha and homoeopathy (F.No.3-63/2019-CCRUM/Tech). The authors also extend their appreciation to the Deanship of Scientific Research at King Saud University for funding the work through the Research Group Project No. RG-1439-74.

Institutional Review Board Statement: Not applicable.

Informed Consent Statement: Not applicable.

Data Availability Statement: Not applicable.

Acknowledgments: Abu Hamza acknowledges the Indian Council of Medical Research (VIR/Fellowship/5/2018-ECD-I) and University Grant Commission, Government of India, for providing research fellowship. The authors also acknowledge the FIST program (SR/FST/LSI-541/2012) DST, India and Jamia Millia Islamia for providing equipment for research work.

Conflicts of Interest: The authors declare no conflict of interest.

Sample Availability: Samples of the recombinant plasmids with G protein gene are available from the authors.

References

1. Coultas, J.A.; Smyth, R.; Openshaw, P.J. Respiratory syncytial virus (RSV): A scourge from infancy to old age. *Thorax* **2019**, *74*, 986–993. [CrossRef]
2. Chatzis, O.; Darbre, S.; Pasquier, J.; Meylan, P.; Manuel, O.; Aubert, J.D.; Beck-Popovic, M.; Masouridi-Levrat, S.; Ansari, M.; Kaiser, L.; et al. Burden of severe RSV disease among immunocompromised children and adults: A 10 year retrospective study. *BMC Infect. Dis.* **2018**, *18*, 111. [CrossRef]
3. Schildgen, O. The lack of protective immunity against RSV in the elderly. *Epidemiol. Infect.* **2009**, *137*, 1687–1690. [CrossRef] [PubMed]
4. Stein, R.; Bont, L.J.; Zar, H.; Polack, F.P.; Park, C.; Claxton, A.; Borok, G.; Butylkova, Y.; Wegzyn, C. Respiratory syncytial virus hospitalization and mortality: Systematic review and meta-analysis. *Pediatr. Pulmonol.* **2016**, *52*, 556–569. [CrossRef] [PubMed]
5. Resch, B. Product review on the monoclonal antibody palivizumab for prevention of respiratory syncytial virus infection. *Hum. Vaccin. Immunother.* **2017**, *13*, 2138–2149. [CrossRef]
6. Boyoglu-Barnum, S.; Todd, S.O.; Meng, J.; Barnum, T.R.; Chirkova, T.; Haynes, L.M.; Jadhao, S.J.; Tripp, R.A.; Oomens, A.G.; Moore, M.L.; et al. Mutating the CX3C Motif in the G Protein Should Make a Live Respiratory Syncytial Virus Vaccine Safer and More Effective. *J. Virol.* **2017**, *91*, e02059. [CrossRef]
7. Liang, B. Structures of the Mononegavirales Polymerases. *J. Virol.* **2020**, *94*. [CrossRef]
8. Haynes, L.M.; Jones, L.P.; Barskey, A.; Anderson, L.J.; Tripp, R.A. Enhanced Disease and Pulmonary Eosinophilia Associated with Formalin-Inactivated Respiratory Syncytial Virus Vaccination Are Linked to G Glycoprotein CX3C-CX3CR1 Interaction and Expression of Substance P. *J. Virol.* **2003**, *77*, 9831–9844. [CrossRef]
9. Radu, G.U.; Caidi, H.; Miao, C.; Tripp, R.A.; Anderson, L.J.; Haynes, L.M. Prophylactic Treatment with a G Glycoprotein Monoclonal Antibody Reduces Pulmonary Inflammation in Respiratory Syncytial Virus (RSV)-Challenged Naïve and Formalin-Inactivated RSV-Immunized BALB/c Mice. *J. Virol.* **2010**, *84*, 9632–9636. [CrossRef]
10. Lee, J.; Klenow, L.; Coyle, E.M.; Golding, H.; Khurana, S. Protective antigenic sites in respiratory syncytial virus G attachment protein outside the central conserved and cysteine noose domains. *PLoS Pathog.* **2018**, *14*, e1007262. [CrossRef] [PubMed]
11. Boyoglu-Barnum, S.; Chirkova, T.; Todd, S.O.; Barnum, T.R.; Gaston, K.A.; Jorquera, P.; Haynes, L.M.; Tripp, R.A.; Moore, M.L.; Anderson, L.J. Prophylaxis with a Respiratory Syncytial Virus (RSV) Anti-G Protein Monoclonal Antibody Shifts the Adaptive Immune Response to RSV rA2-line19F Infection from Th2 to Th1 in BALB/c Mice. *J. Virol.* **2014**, *88*, 10569–10583. [CrossRef]
12. Miao, C.; Radu, G.U.; Caidi, H.; Tripp, R.; Anderson, L.J.; Haynes, L.M. Treatment with respiratory syncytial virus G glycoprotein monoclonal antibody or F(ab')2 components mediates reduced pulmonary inflammation in mice. *J. Gen. Virol.* **2009**, *90*, 1119–1123. [CrossRef] [PubMed]
13. Han, J.; Takeda, K.; Wang, M.; Zeng, W.; Jia, Y.; Shiraishi, Y.; Okamoto, M.; Dakhama, A.; Gelfand, E.W. Effects of Anti-G and Anti-F Antibodies on Airway Function after Respiratory Syncytial Virus Infection. *Am. J. Respir. Cell Mol. Biol.* **2014**, *51*, 143–154. [CrossRef] [PubMed]
14. Caidi, H.; Miao, C.; Thornburg, N.J.; Tripp, R.A.; Anderson, L.J.; Haynes, L.M. Anti-respiratory syncytial virus (RSV) G monoclonal antibodies reduce lung inflammation and viral lung titers when delivered therapeutically in a BALB/c mouse model. *Antivir. Res.* **2018**, *154*, 149–157. [CrossRef] [PubMed]
15. Kauvar, L.M.; Harcourt, J.L.; Haynes, L.M.; Tripp, R.A. Therapeutic targeting of respiratory syncytial virus G-protein. *Immunother.* **2010**, *2*, 655–661. [CrossRef]
16. Chirkova, T.; Lin, S.; Oomens, A.G.P.; Gaston, K.A.; Boyoglu-Barnum, S.; Meng, J.; Stobart, C.C.; Cotton, C.U.; Hartert, T.V.; Moore, M.L.; et al. CX3CR1 is an important surface molecule for respiratory syncytial virus infection in human airway epithelial cells. *J. Gen. Virol.* **2015**, *96*, 2543. [CrossRef]
17. Johnson, S.M.; McNally, B.A.; Ioannidis, I.; Flano, E.; Teng, M.N.; Oomens, A.G.; Walsh, E.E.; Peeples, M.E. Respiratory Syncytial Virus Uses CX3CR1 as a Receptor on Primary Human Airway Epithelial Cultures. *PLoS Pathog.* **2015**, *11*, e1005318. [CrossRef]
18. Feldman, S.A.; Hendry, R.M.; Beeler, J.A. Identification of a Linear Heparin Binding Domain for Human Respiratory Syncytial Virus Attachment Glycoprotein G. *J. Virol.* **1999**, *73*, 6610–6617. [CrossRef]
19. Hallak, L.K.; Spillmann, D.; Collins, P.L.; Peeples, M.E. Glycosaminoglycan Sulfation Requirements for Respiratory Syncytial Virus Infection. *J. Virol.* **2000**, *74*, 10508–10513. [CrossRef]
20. Li, J.P.; Kusche-Gullberg, M. Heparan Sulfate: Biosynthesis, Structure, and Function. *Int. Rev. Cell Mol. Biol.* **2016**, *325*, 215–273. [CrossRef]
21. Gomes, P.B.; Dietrich, C.P. Distribution of heparin and other sulfated glycosaminoglycans in vertebrates. *Comp. Biochem. Physiol. Part B: Comp. Biochem.* **1982**, *73*, 857–863. [CrossRef]
22. Clausen, T.M.; Sandoval, D.R.; Spliid, C.B.; Pihl, J.; Perrett, H.R.; Painter, C.D.; Narayanan, A.; Majowicz, S.A.; Kwong, E.M.; McVicar, R.N.; et al. SARS-CoV-2 Infection Depends on Cellular Heparan Sulfate and ACE2. *Cell* **2020**, *183*, 1043–1057.e15. [CrossRef]
23. Fersht, A.R. From the first protein structures to our current knowledge of protein folding: Delights and scepticisms. *Nat. Rev. Mol. Cell Biol.* **2008**, *9*, 650–654. [CrossRef]
24. Hartl, F.U.; Hayer-Hartl, M. Protein folding. Molecular chaperones in the cytosol: From nascent chain to folded protein. *Science* **2002**, *295*, 1852–1858. [CrossRef]
25. Dobson, C.M. Protein folding and misfolding. *Nature* **2003**, *426*, 884–890. [CrossRef] [PubMed]

26. Street, T.O.; Bolen, D.W.; Rose, G.D. A molecular mechanism for osmolyte-induced protein stability. *Proc. Natl. Acad. Sci. USA* **2006**, *103*, 13997–14002. [CrossRef] [PubMed]

27. Beg, I.; Minton, A.P.; Hassan, M.I.; Islam, A.; Ahmad, F. Thermal Stabilization of Proteins by Mono- and Oligosaccharides: Measurement and Analysis in the Context of an Excluded Volume Model. *Biochemistry* **2015**, *54*, 3594–3603. [CrossRef] [PubMed]

28. Do, A.; Ka, D. Solvent denaturation and stabilization of globular proteins. *Biochemistry* **1991**, *30*, 5974–5985. [CrossRef]

29. Lapanje, S. Random Coil Behaviour of Proteins in Concentrated Urea Solutions. *Croat. Chem. Acta* **1969**, *41*, 115–124.

30. Hamza, A.; Shafat, Z.; Parray, Z.A.; Hisamuddin, M.; Khan, W.H.; Ahmed, A.; Almajhdi, F.N.; Farrag, M.A.; Mohammed, A.A.; Islam, A.; et al. Structural Characterization and Binding Studies of the Ectodomain G Protein of Respiratory Syncytial Virus Reveal the Crucial Role of pH with Possible Implications in Host–Pathogen Interactions. *ACS Omega* **2021**, *6*, 10403–10414. [CrossRef]

31. Khan, P.; Parkash, A.; Islam, A.; Ahmad, F.; Hassan, M.I. Molecular basis of the structural stability of hemochromatosis factor E: A combined molecular dynamic simulation and GdmCl-induced denaturation study. *Biopolymers* **2016**, *105*, 133–142. [CrossRef] [PubMed]

32. Battles, M.B.; McLellan, J.S. Respiratory syncytial virus entry and how to block it. *Nat. Rev. Microbiol.* **2019**, *17*, 233–245. [CrossRef]

33. Prasad, S.; Mandal, I.; Singh, S.; Paul, A.; Mandal, B.; Venkatramani, R.; Swaminathan, R. Near UV-Visible electronic absorption originating from charged amino acids in a monomeric protein. *Chem. Sci.* **2017**, *8*, 5416–5433. [CrossRef] [PubMed]

34. Schmid, F.-X. Biological Macromolecules: UV-visible Spectrophotometry. *Encycl. Life Sci.* **2001**, 1–4. [CrossRef]

35. Syed, S.B.; Shahbaaz, M.; Khan, S.H.; Srivastava, S.; Islam, A.; Ahmad, F.; Hassan, I. Estimation of pH effect on the structure and stability of kinase domain of human integrin-linked kinase. *J. Biomol. Struct. Dyn.* **2019**, *37*, 156–165. [CrossRef]

36. Alston, R.W.; Urbanikova, L.; Sevcik, J.; Lasagna, M.; Reinhart, G.D.; Scholtz, J.M.; Pace, C.N. Contribution of Single Tryptophan Residues to the Fluorescence and Stability of Ribonuclease Sa. *Biophys. J.* **2004**, *87*, 4036–4047. [CrossRef]

37. Gasymov, O.K.; Abduragimov, A.R.; Glasgow, B.J. pH-Dependent Conformational Changes in Tear Lipocalin by Site-Directed Tryptophan Fluorescence. *Biochemistry* **2009**, *49*, 582–590. [CrossRef]

38. Ptitsyn, O.; Uversky, V. The molten globule is a third thermodynamical state of protein molecules. *FEBS Lett.* **1994**, *341*, 15–18. [CrossRef]

39. Naiyer, A.; Hassan, M.I.; Islam, A.; Sundd, M.; Ahmad, F. Structural characterization of MG and pre-MG states of proteins by MD simulations, NMR, and other techniques. *J. Biomol. Struct. Dyn.* **2015**, *33*, 2267–2284. [CrossRef]

40. Slavík, J. Anilinonaphthalene sulfonate as a probe of membrane composition and function. *Biochim. Biophys. Acta-Rev. Biomembr.* **1982**, *694*, 1–25. [CrossRef]

41. Gupta, P.; Khan, F.I.; Roy, S.; Anwar, S.; Dahiya, R.; Alajmi, M.F.; Hussain, A.; Rehman, M.T.; Lai, D.; Hassan, M.I. Functional implications of pH-induced conformational changes in the Sphingosine kinase 1. *Spectrochim. Acta Part A Mol. Biomol. Spectrosc.* **2020**, *225*, 117453. [CrossRef] [PubMed]

42. Naseem, F.; Khan, R.H.; Haq, S.K.; Naeem, A. Characterization of molten globule state of fetuin at low pH. *Biochim. Biophys. Acta-Proteins Proteom.* **2003**, *1649*, 164–170. [CrossRef]

43. Bansal, R.; Haque, M.A.; Hassan, M.I.; Ethayathulla, A.S.; Kaur, P. Structural and conformational behavior of MurE ligase from Salmonella enterica serovar Typhi at different temperature and pH conditions. *Int. J. Biol. Macromol.* **2020**, *150*, 389–399. [CrossRef]

44. Haque, M.A.; Ubaid-ullah, S.; Zaidi, S.; Hassan, M.I.; Islam, A.; Batra, J.K.; Ahmad, F. Characterization of pre-molten globule state of yeast iso-1-cytochrome c and its deletants at pH 6.0 and 25 °C. *Int. J. Biol. Macromol.* **2015**, *72*, 1406–1418. [CrossRef]

45. Bychkova, V.E.; Dujsekina, A.E.; Klenin, S.I.; Tiktopulo, E.I.; Uversky, A.V.N.; Ptitsyn, O.B. Molten Globule-Like State of Cytochrome c under Conditions Simulating Those Near the Membrane Surface. *Biochemistry* **1996**, *35*, 6058–6063. [CrossRef]

46. Ahmad, F. Stability of acetylcholinesterase in guanidine hydrochloride solution. *Can. J. Biochem.* **1981**, *59*, 551–555. [CrossRef] [PubMed]

47. Pace, C.N. [14]Determination and analysis of urea and guanidine hydrochloride denaturation curves. *Methods Enzymol.* **1986**, *131*, 266–280. [CrossRef]

48. Idrees, D.; Prakash, A.; Haque, M.A.; Islam, A.; Ahmad, F.; Hassan, M.I. Spectroscopic and MD simulation studies on unfolding processes of mitochondrial carbonic anhydrase VA induced by urea. *J. Biomol. Struct. Dyn.* **2016**, *34*, 1987–1997. [CrossRef]

49. Royer, C.A. Probing Protein Folding and Conformational Transitions with Fluorescence. *Chem. Rev.* **2006**, *106*, 1769–1784. [CrossRef]

50. Vivian, J.T.; Callis, P.R. Mechanisms of Tryptophan Fluorescence Shifts in Proteins. *Biophys. J.* **2001**, *80*, 2093–2109. [CrossRef]

51. Monera, O.D.; Kay, C.M.; Hodges, R.S. Protein denaturation with guanidine hydrochloride or urea provides a different estimate of stability depending on the contributions of electrostatic interactions. *Protein Sci.* **1994**, *3*, 1984–1991. [CrossRef] [PubMed]

52. Roy, S.; Mohammad, T.; Gupta, P.; Dahiya, R.; Parveen, S.; Luqman, S.; Hasan, G.M.; Hassan, M.I. Discovery of Harmaline as a Potent Inhibitor of Sphingosine Kinase-1: A Chemopreventive Role in Lung Cancer. *ACS Omega* **2020**, *5*, 21550–21560. [CrossRef]

53. Susana, S.; Mateus, N.; de Freitas, V. Interaction of Different Polyphenols with Bovine Serum Albumin (BSA) and Human Salivary α-Amylase (HSA) by Fluorescence Quenching. *J. Agric. Food Chem.* **2007**, *55*, 6726–6735. [CrossRef]

54. Anwar, S.; Shamsi, A.; Shahbaaz, M.; Queen, A.; Khan, P.; Hasan, G.M.; Islam, A.; Alajmi, M.F.; Hussain, A.; Ahmad, F.; et al. Rosmarinic Acid Exhibits Anticancer Effects via MARK4 Inhibition. *Sci. Rep.* **2020**, *10*, 10300. [CrossRef]

55. Rehman, M.T.; Shamsi, H.; Khan, A.U. Insight into the Binding Mechanism of Imipenem to Human Serum Albumin by Spectroscopic and Computational Approaches. *Mol. Pharm.* **2014**, *11*, 1785–1797. [CrossRef] [PubMed]

56. Wang, H.-D.; Niu, C.H.; Yang, Q.; Badea, I. Study on protein conformation and adsorption behaviors in nanodiamond particle–proteincomplexes. *Nanotechnology* **2011**, *22*, 145703. [CrossRef]

57. Yu, S.; Perálvarez-Marín, A.; Minelli, C.; Faraudo, J.; Roig, A.; Laromaine, A. Albumin-coated SPIONs: An experimental and theoretical evaluation of protein conformation, binding affinity and competition with serum proteins. *Nanoscale* **2016**, *8*, 14393–14405. [CrossRef]

58. Chung, C.-S.; Hsiao, J.-C.; Chang, Y.-S.; Chang, W. A27L Protein Mediates Vaccinia Virus Interaction with Cell Surface Heparan Sulfate. *J. Virol.* **1998**, *72*, 1577–1585. [CrossRef]

59. Shieh, M.T.; WuDunn, D.; Montgomery, R.I.; Esko, J.D.; Spear, P.G. Cell surface receptors for herpes simplex virus are heparan sulfate proteoglycans. *J. Cell Biol.* **1992**, *116*, 1273–1281. [CrossRef]

60. Barth, H.; Schäfer, C.; Adah, M.I.; Zhang, F.; Linhardt, R.J.; Toyoda, H.; Kinoshita-Toyoda, A.; Toida, T.; Kuppevelt, T.H.; van Depla, E.; et al. Cellular Binding of Hepatitis C Virus Envelope Glycoprotein E2 Requires Cell Surface Heparan Sulfate. *J. Biol. Chem.* **2003**, *278*, 41003–41012. [CrossRef]

61. Byrnes, A.P.; Griffin, D.E. Binding of Sindbis Virus to Cell Surface Heparan Sulfate. *J. Virol.* **1998**, *72*, 7349–7356. [CrossRef]

62. Mondor, I.; Ugolini, S.; Sattentau, Q.J. Human Immunodeficiency Virus Type 1 Attachment to HeLa CD4 Cells Is CD4 Independent and gp120 Dependent and Requires Cell Surface Heparans. *J. Virol.* **1998**, *72*, 3623–3634. [CrossRef]

63. Milewska, A.; Zarebski, M.; Nowak, P.; Stozek, K.; Potempa, J.; Pyrc, K. Human Coronavirus NL63 Utilizes Heparan Sulfate Proteoglycans for Attachment to Target Cells. *J. Virol.* **2014**, *88*, 13221–13230. [CrossRef]

64. Kalia, M.; Chandra, V.; Rahman, S.A.; Sehgal, D.; Jameel, S. Heparan Sulfate Proteoglycans Are Required for Cellular Binding of the Hepatitis E Virus ORF2 Capsid Protein and for Viral Infection. *J. Virol.* **2009**, *83*, 12714–12724. [CrossRef]

65. Machado, V.B.; de Sá, M.J.; Miranda Prado, A.K.; Alves de Toledo, K.; Regasini, L.O.; Pereira de Souza, F.; Caruso, Í.P.; Fossey, M.A. Biophysical and flavonoid-binding studies of the G protein ectodomain of group A human respiratory syncytial virus. *Heliyon* **2019**, *5*, e01394. [CrossRef]

66. Shafat, Z.; Hamza, A.; Deeba, F.; Faizan, M.I.; Khan, N.; Islam, A.; Ahmed, A.; Alamery, S.F.; Parveen, S. Optimization of parameters for expression and purification of G glycoprotein ectodomain of respiratory syncytial virus. *Futur. Virol.* **2020**, *15*, 225–235. [CrossRef]

67. Khan, W.H.; Srungaram, V.L.N.R.; Islam, A.; Beg, I.; Haider, M.S.H.; Ahmad, F.; Broor, S.; Parveen, S. Biophysical characterization of G protein ectodomain of group B human respiratory syncytial virus from *E. coli. Prep. Biochem. Biotechnol.* **2016**, *46*, 483–488. [CrossRef] [PubMed]

68. Pace, C.N.; Vajdos, F.; Fee, L.; Grimsley, G.; Gray, T. How to measure and predict the molar absorption coefficient of a protein. *Protein Sci.* **1995**, *4*, 2411–2423. [CrossRef]

69. Shafat, Z.; Faizan, I.; Tazeen, A.; Farooqui, A.; Deeba, F.; Aftab, S.; Hamza, A.; Parveen, S.; Islam, A.; Broor, S.; et al. In-silico analysis of ectodomain G protein of Respiratory Syncytial Virus. *Indian J. Heal. Sci. Care* **2017**, *4*, 110. [CrossRef]

70. Dassault Systèmes BIOVIA. *Discov. Stud. Model. Environ.* **2017**. *San Diego*. Available online: https://www.3ds.com/products-services/biovia/products/molecular-modeling-simulation/biovia-discovery-studio/ (accessed on 6 December 2021).

71. Schrodinger, L. The PyMOL Molecular Graphics System. Version 1. 2010. Available online: https://sourceforge.net/projects/pymol/files/latest/download (accessed on 6 December 2021).

72. Pronk, S.; Páll, S.; Schulz, R.; Larsson, P.; Bjelkmar, P.; Apostolov, R.; Shirts, M.R.; Smith, J.C.; Kasson, P.M.; van der Spoel, D.; et al. GROMACS 4.5: A high-throughput and highly parallel open source molecular simulation toolkit. *Bioinformatics* **2013**, *29*, 845–854. [CrossRef] [PubMed]

73. Oostenbrink, C.; Villa, A.; Mark, A.E.; Gunsteren, W.F. Van A biomolecular force field based on the free enthalpy of hydration and solvation: The GROMOS force-field parameter sets 53A5 and 53A6. *J. Comput. Chem.* **2004**, *25*, 1656–1676. [CrossRef] [PubMed]

74. Schüttelkopf, A.W.; Van Aalten, D.M.F. PRODRG: A tool for high-throughput crystallography of protein–ligand complexes. *Acta Crystallogr. Sect. D Biol. Crystallogr.* **2004**, *60*, 1355–1363. [CrossRef] [PubMed]

75. Zielkiewicz, J. Structural properties of water: Comparison of the SPC, SPCE, TIP4P, and TIP5P models of water. *J. Chem. Phys.* **2005**, *123*, 104501. [CrossRef] [PubMed]

MDPI
St. Alban-Anlage 66
4052 Basel
Switzerland
Tel. +41 61 683 77 34
Fax +41 61 302 89 18
www.mdpi.com

Molecules Editorial Office
E-mail: molecules@mdpi.com
www.mdpi.com/journal/molecules